D1281735

BIOMECHANICS AND MOTOR CONTROL OF HUMAN MOVEMENT

BIOMECHANICS AND MOTOR CONTROL OF HUMAN MOVEMENT

Third Edition

DAVID A. WINTER
University of Waterloo
Waterloo, Ontario, Canada

JOHN WILEY & SONS, INC.

This book is printed on acid-free paper. ♾

Published by John Wiley & Sons, Inc., Hoboken, New Jersey
Published simultaneously in Canada.

For general information on our other products and services or for technical support, please contact our Customer Care Department within the United States at (800) 762-2974, outside the United States at (317) 572-3993 or fax (317) 572-4002.

Wiley also publishes its books in a variety of electronic formats. Some content that appears in print may not be available in electronic books. For more information about Wiley products, visit our web site at www.wiley.com.

Library of Congress Cataloging-in-Publication Data:

Winter, David A., 1930–
Biomechanics and motor control of human movement / David A. Winter.—3rd ed.
p. cm.
Includes bibliographical references and index.
ISBN 0-471-44989-X (cloth)
1. Human mechanics. 2. Motor ability. 3. Kinesiology. I. Title.

QP303.W59 2004
612′.76—dc22

2004047810

Printed in the United States of America

10 9 8 7 6 5 4 3 2

To my wife and children, and to my colleagues, graduate and undergraduate students, all of whom have encouraged, challenged, and influenced me over the years.

CONTENTS

PREFACE TO THE THIRD EDITION

This text is a revision of the second edition, reflecting the major changes that have taken place in the field of biomechanics over the past decade. The original text, *Biomechanics of Human Movement,* published in 1979, had its title changed, when the new edition was published in 1990, to *Biomechanics and Motor Control of Human Movement* to acknowledge the new directions of the 1980s. In that second edition, five of eight chapters addressed various aspects of muscles and motor systems; this third edition, with its major new addition of three-dimensional (3D) kinematics and kinetics, reflects the continued emphasis on the motor control area.

As in the first two editions, the goal of the text is to fill the gap in the human movement science area where modern science and technology are integrated with anatomy, muscle physiology, and electromyography to assess and understand human movement. The emphasis is on dynamic movements and on live data. A wide spectrum of measurement and analysis techniques are presented and are aimed at those interested in higher level quantitative assessments. The text is intended to appeal to the practitioner as well as the researcher and to those concerned with the physically handicapped, the elite athlete, and the person in the workplace.

This edition has a new Chapter 7 on 3D kinematics and kinetics, and the sample data presented there will allow students to learn or to reinforce their knowledge of matrix transformations. Chapter 2 has been updated to reflect the explosive increase in computer image measurement systems but has not decreased its emphasis on the raw coordinate data. Chapter 3, on anthropometry, has been expanded to reflect new emphasis on 3D center of mass estimates, especially as this measure relates to the posture and balance of the total body in our gravitational environment. The changes in Chapter 4 also reflect models of the kinetics of balance control. Chapter 5, on mechanical energy and power, was drastically updated in the second edition and remains essentially intact. Chapter 6, on synthesis of human movement, was introduced in the second edition and has had some minor additions. Chapter 8 (previously Chapter 7), on muscle mechanics, has undergone minor revisions to reflect more recent research findings on fundamental relationships. The final Chapter 9 (formerly Chapter 8), on kinesiological electromyography (EMG), has had some major additions to reflect new biophysical models of EMG detection, new approaches to the problem of crosstalk, and information on new EMG standards for recording and reporting. The Appendices, which underwent major additions in the second edition, remain intact.

As was stated in the original editions, it is expected that the student has had basic courses in anatomy, mechanics, calculus, and electrical science. The major programs to which the book is directed are: kinesiology, bioengineering (rehabilitation engineering), physical education, ergonomics, physical and occupational therapy; it should also prove valuable to researchers in orthopedics, muscle physiology, and rehabilitation medicine.

The preparation of this third edition has not been a solo effort. I would like to acknowledge the considerable help that I received from Milad Ishac in the preparation of the text and figures for the new 3D Chapter 7. Also, the helpful suggestions and materials from Richard Wells, Stuart McGill, Jack Callaghan, and Sylvia Õunpuu were greatly appreciated.

David A. Winter

Waterloo, Ontario, Canada
January 2004

BIOMECHANICS AND MOTOR CONTROL OF HUMAN MOVEMENT

1

BIOMECHANICS AS AN INTERDISCIPLINE

1.0 INTRODUCTION

Biomechanics of human movement can be defined as the interdiscipline which describes, analyzes, and assesses human movement. A wide variety of physical movements are involved—everything from the gait of the physically handicapped to the lifting of a load by a factory worker to the performance of a superior athlete. The physical and biological principles that apply are the same in all cases. What changes from case to case is the specific movement task and the level of detail that is being asked about the performance of each movement.

The list of professionals and semiprofessionals interested in applied aspects of human movement is quite long: orthopedic surgeons, athletic coaches, rehabilitation engineers, therapists, kinesiologists, prosthetists, psychiatrists, orthotists, sports equipment designers, and so on. At the basic level, the name given to the science dedicated to the broad area of human movement is kinesiology. It is an emerging discipline blending aspects of psychology, motor learning, and exercise physiology as well as biomechanics. Biomechanics, as an outgrowth of both life and physical sciences, is built on the basic body of knowledge of physics, chemistry, mathematics, physiology, and anatomy. It is amazing to note that the first real "biomechanicians" date back to Leonardo DaVinci, Galileo, Lagrange, Bernoulli, Euler, and Young. All these scientists had primary interests in the application of mechanics to biological problems.

1.1 MEASUREMENT, DESCRIPTION, ANALYSIS, AND ASSESSMENT

The scientific approach as applied to biomechanics has been characterized by a fair amount of confusion. Some descriptions of human movement have been passed off as assessments, some studies involving only measurements have been falsely advertised as analyses, and so on. It is therefore important to clarify these terms. Any quantitative assessment of human movement must be preceded by a measurement and description phase, and if more meaningful diagnostics are needed, a biomechanical analysis is usually necessary. Most of the material in this text is aimed at the technology of measurement and description and the modeling process required for analysis. The final interpretation, assessment, or diagnosis is movement specific and is limited to the examples given.

Figure 1.1, which has been prepared for the assessment of the physically handicapped, depicts the relationships between these various phases of assessment. All levels of assessment involve a human being, and are based on his or her visual observation of a patient or subject, recorded data, or some resulting biomechanical analysis. The primary assessment level uses direct observation, which places tremendous "overload" even on the most experienced observer. All measures are subjective and are almost impossible to compare with those obtained previously. Observers are then faced with the tasks of documenting (describing) what they see, monitoring changes, ana-

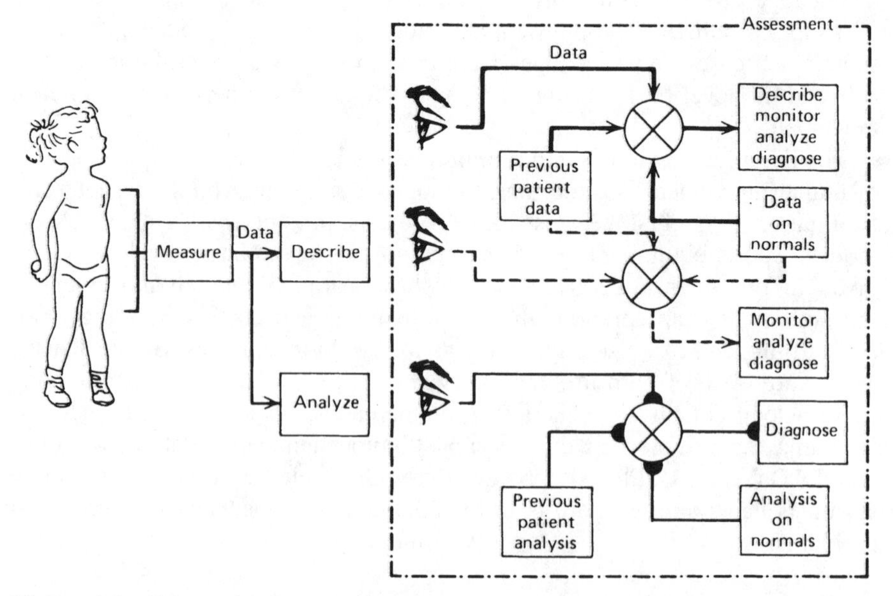

Figure 1.1 Schematic diagram showing the three levels of assessment of human movement.

lyzing the information, and diagnosing the causes. If measurements can be made during the patient's movement, then data can be presented in a convenient manner to describe the movement quantitatively. Here the assessor's task is considerably simplified. He or she can now quantify changes, carry out simple analyses, and try to reach a more objective diagnosis. At the highest level of assessment, the observer can view biomechanical analyses that are extremely powerful in diagnosing the exact cause of the problem, compare these analyses with the normal population, and monitor their detailed changes with time.

The measurement and analysis techniques used in an athletic event could be identical to the techniques used to evaluate an amputee's gait. However, the assessment of the optimization of the energetics of the athlete is quite different from the assessment of the stability of the amputee. Athletes are looking for very detailed but minor changes that will improve their performance by a few percentage points, sufficient to move them from fourth to first place. Their training and exercise programs and reassessment normally continue over an extended period of time. The amputee, on the other hand, is looking for major improvements, probably related to safe walking, but not fine and detailed differences. This person is quite happy to be able to walk at less than maximum capability, although techniques are available to permit training and have the prosthesis readjusted until the amputee reaches some perceived maximum. In ergonomic studies, assessors are likely looking for maximum stresses in specific tissues during a given task, to thereby ascertain whether the tissue is working within safe limits. If not, they will analyze possible changes in the workplace or task in order to reduce the stress or fatigue.

1.1.1 Measurement, Description, and Monitoring

It is difficult to separate the two functions of measurement and description. However, for clarity the student should be aware that a given measurement device can have its data presented in a number of different ways. Conversely, a given description could have come from several different measurement devices.

Earlier biomechanical studies had the sole purpose of describing a given movement, and any assessments that were made resulted from visual inspection of the data. The description of the data can take many forms: pen recorder curves, plots of body coordinates, stick diagrams, or simple outcome measures such as gait velocity, load lifted, or height of a jump. A movie camera, by itself, is a measurement device, and the resulting plots from the description of the event in time and space. Figure 1.2 illustrates a system incorporating a cine camera and two different descriptive plots. The coordinates of key anatomical landmarks can be extracted and plotted at regular intervals in time. Time history plots of one or more coordinates are useful in describing detailed changes of a particular landmark. They also can reveal to the trained eye

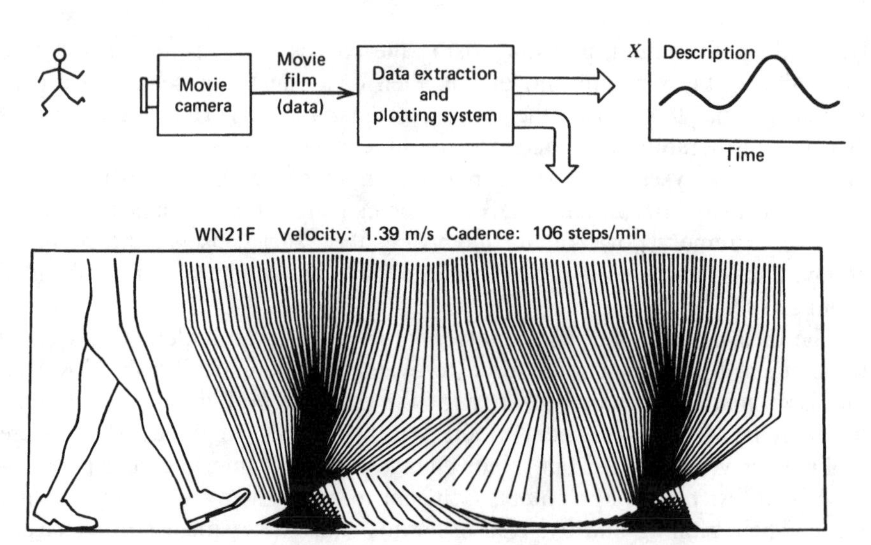

Figure 1.2 Flow of data from a camera system and plotting of data in two different forms, each yielding a different description of the same event.

changes in velocity and acceleration. A total description in the plane of the movement is the stick diagram, in which each body segment is represented by a straight line or stick. Joining the sticks together gives the spatial orientation of all segments at any point in time. Repetition of this plot at equal intervals of time gives a pictorial and anatomical description of the dynamics of the movement. Here trajectories, velocities, and accelerations can by visualized. To get some idea of the volume of the data present in a stick diagram, the student should note that one full page of coordinate data is required to make this complete plot for the description of the event. The coordinate data can be used directly for any desired analysis: reaction forces, muscle moments, energy changes, efficiency, and so on. Conversely, an assessment can occasionally be made directly from the description. A trained observer, for example, can scan a stick diagram and extract useful information that will give some directions for training or therapy, or give the researcher some insight into basic mechanisms of movement.

The term *monitor* needs to be introduced in conjunction with the term *describe*. To monitor means to note changes over time. Thus a physical therapist will monitor the progress (or the lack of it) for each physically disabled person undergoing therapy. Only through accurate and reliable measurements will the therapist be able to monitor any improvement and thereby make inferences to the validity of the current therapy. What monitoring does not tell us is why an improvement is or is not taking place; it merely documents the change. All too many coaches or therapists document the changes with the inferred assumption that their intervention has been the cause. However, the scientific rationale behind such inferences is missing. Unless a detailed analysis is done, we cannot document the detailed motor level changes that will reflect the results of therapy or training.

1.1.2 Analysis

The measurement system yields data that are suitable for analysis. This means that data have been calibrated and are as free as possible from noise and artifacts. *Analysis* can be defined as any mathematical operation that is performed on a set of data to present them in another form or to combine the data from several sources to produce a variable that is not directly measurable. From the analyzed data, information may be extracted to assist in the assessment stage. In some cases the mathematical operation can be very simple, such as the processing of an electromyographic signal to yield an *envelope* signal (Figure 1.3). The mathematical operation performed here can be described in two stages. The first is a full-wave *rectifier* (the electronic term for a circuit that gives the absolute value). The second stage is a low-pass filter (which mathematically has the same transfer function as that between a neural pulse and its resultant muscle twitch). A more complex biomechanical analysis could involve a link-segment model, and with appropriate kinematic, anthropometric, and kinetic output data, we can carry out analyses that could yield a multitude of significant time-course curves. Figure 1.4 depicts the relationships between some of these variables. The output of the movement is what we see. It can be described by a large number of kinematic variables: displacements, joint angles, velocities, and accelerations. If we have an accurate model of the human body in terms of anthropometric variables, we can develop a reliable link-segment model. With this model and accurate kinematic data, we can predict the net forces and muscle moments that caused the movement we just observed. Such an analysis technique is called an *inverse solution.* It is extremely valuable, as it allows us to estimate variables such as joint reaction forces and moments of force. Such variables are not measurable directly. In a similar manner, individual muscle forces might be predicted through the development of a mathematical model of a muscle which could have neural drive, length, velocity, and cross-sectional area as inputs.

1.1.3 Assessment and Interpretation

The entire purpose of any assessment is to make a positive decision about a physical movement. An athletic coach might ask, "Is the mechanical energy of the movement better or worse than before the new training program was instigated, and why?" Or the orthopedic surgeon may wish to see the improvement in the knee muscle moments of a patient a month after surgery. Or a basic researcher may wish to interpret the motor changes resulting from certain perturbations and thereby verify or negate different theories of neural control. In all cases, if the questions asked yield no answers, it can be said

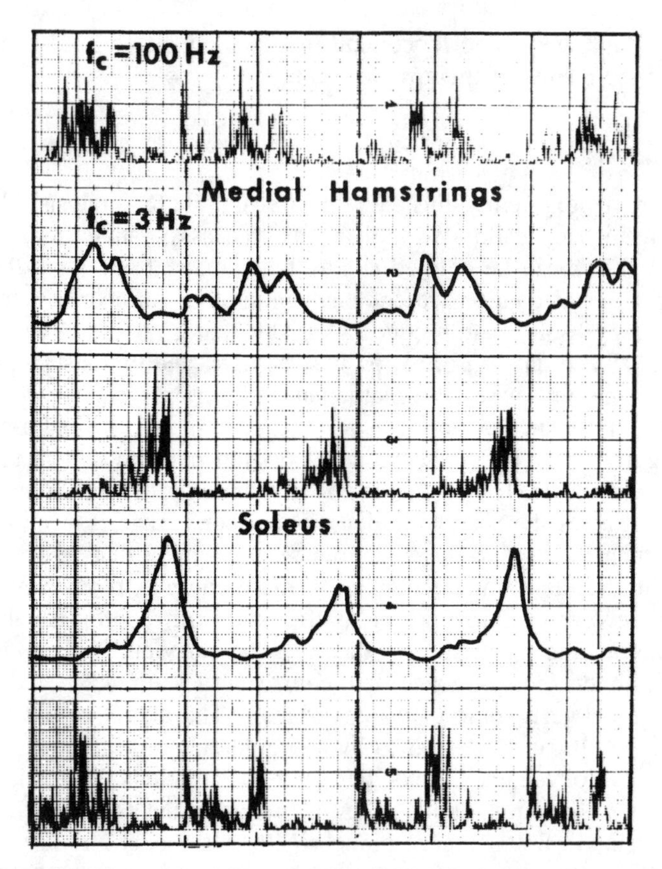

Figure 1.3 Processing of raw electromyogram (EMG) signals to present the variable in a different form. Traces 1 and 3 show the full-wave rectified EMG of the medial hamstrings and soleus muscles during walking. A cutoff frequency (f_c = 100 Hz) is indicated for the rectified signal because this is the bandwidth of the pen recorder. In traces 2 and 4, the linear envelope signal (low-pass filter with f_c = 3 Hz) is presented.

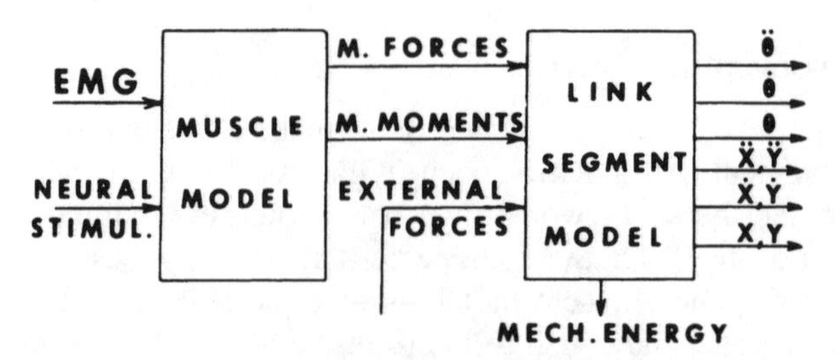

Figure 1.4 Schematic diagram to show the relationship between the neural, kinetic, and kinematic variables required to describe and analyze human movement.

that there was no information present in the analysis. The decision may be positive in that it may confirm that the coaching, surgery, or therapy has been correct and should continue exactly as before. Or, if this is an initial assessment, the decision may be to proceed with a definite plan based on new information from the analysis. The information can also cause a negative decision, for example, to cancel a planned surgical procedure and to prescribe therapy instead.

Some biomechanical assessments involve a look at the description itself rather than some analyzed version of it. Commonly, ground reaction force curves from a force plate are examined. This electromechanical device gives an electrical signal that is proportional to the weight (force) of the body acting downward on it. Such patterns appear in Figure 1.5. A trained observed can detect pattern changes as a result of pathological gait and may come to some conclusions as to whether the patient is improving, but he or she will not be able to assess why. At best, this approach is speculative and yields little information regarding the underlying cause of the observed patterns.

1.2 BIOMECHANICS AND ITS RELATIONSHIP WITH PHYSIOLOGY AND ANATOMY

Because biomechanics has been a recent entry on the research scene, it is important to identify its interaction with other areas of movement science: neurophysiology, exercise physiology and anatomy. The neuromuscular system acts to control the release of metabolic energy for the purpose of generating controlled patterns of tension at the tendon. That tension waveform is a function of the physiological characteristics of the muscle (i.e., fiber type) and of its metabolic state (rested vs. fatigued). The tendon tension is generated in the presence of passive anatomical structures (ligaments, articulating surfaces, and skeletal structures). Figure 1.6 depicts the relationship between the sensory system, the neurological pathways, the muscles, the skeletal system,

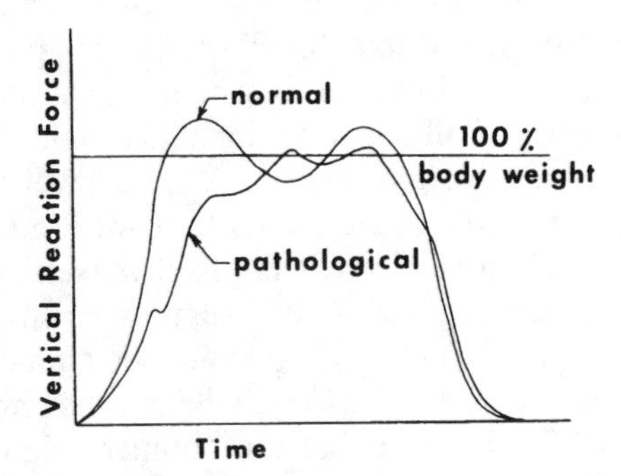

Figure 1.5 Example of a ground reaction force curve that has sometimes been used in the diagnostic assessment of pathological gait.

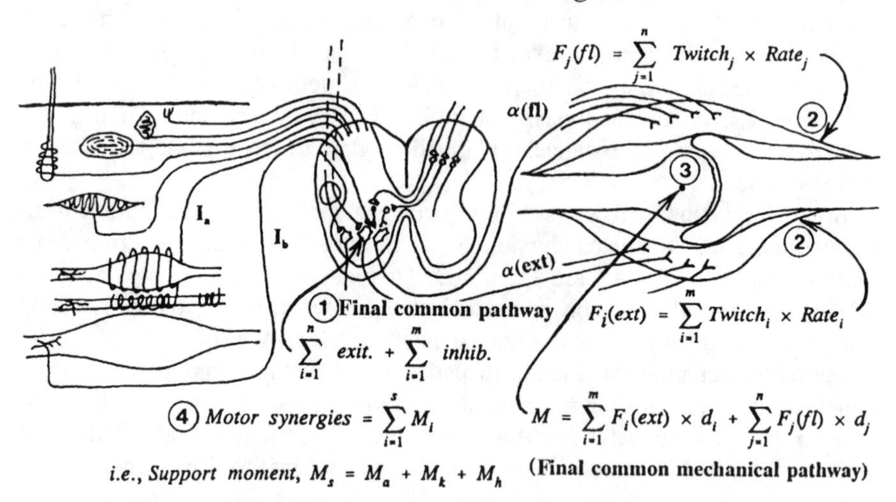

Figure 1.6 Four levels of integration in the neuromusculoskeletal system provide control of human movement. The first is the neural summation of all excitatory/inhibitory inputs to the α motoneuron ①. The second is the summation of all motor twitches from the recruitment of all active motor units within the muscle and is seen as a tendon force ②. The third is the algebraic summation of all agonist and antagonist muscle moments at the joint axis ③. Finally, integrations are evident in combined moments acting synergistically toward a common goal ④.

and the link–segment model that we analyze. The essential characteristic of this total system is that it is converging in nature. The structure of the neural system has many excitatory and inhibitory synaptic junctions, all summing their control on a final synaptic junction in the spinal cord to control individual motor units. The α motoneuron ①, which is often described as the final common pathway, has its synapse on the motor end point of the muscle motor unit. A second level of convergence is the summation of all twitches from all active motor units at the level of the tendon ②. This summation results from the neural recruitment of motor units based on the size principle (cf. DeLuca et al., 1982; Henneman and Olson, 1965). The resultant tension is a temporal superposition of twitches of all active motor units, modulated by the length and velocity characteristics of the muscle. A third level of musculoskeletal integration at each joint center where the moment-of-force ③ is the algebraic summation of the force/moment products of all muscles crossing that joint plus the moments generated by the passive anatomical structures at the joint. The moments we routinely calculate include the net summation of all agonist and antagonist muscles crossing that joint, whether they are single or double joint muscles. In spite of the fact that this moment signal has mechanical units ($N \cdot m$), we must consider the moment signal as a neurological signal because it represents the final desired central nervous system (CNS) control.

Finally, an intersegment integration may be present when the moments at two or more joints collaborate toward a common goal. This collaboration is called a synergy. One such synergy ④, referred to as the support moment, quantifies the integrated activity of all muscles of the lower limb in their defense against a gravity-induced collapse during walking (Winter, 1980, 1984).

Bernstein (1967) predicted that the CNS exerts control at the level of the joints or at the synergy level when he postulated the "principle of equal simplicity" because "it would be incredibly complex to control each and every muscle." One of the by-products of these many levels of integration and convergence is that there is considerably more variability at the neural (EMG) level than at the motor level and more variability at the motor level than at the kinematic level. The resultant variability can frustrate researchers at the neural (EMG) level, but the positive aspect of this redundancy is that the neuromuscular system is therefore very adaptable (Winter, 1984). This adaptability is very meaningful in pathological gait as a compensation for motor or skeletal deficits. For example, a major adaptation took place in a patient who underwent a knee replacement because of osteoarthritic degeneration (Winter, 1989). For years prior to the surgery, this patient had refrained from using her quadriceps to support her during walking; the resultant increase in bone-on-bone forces induced pain in her arthritic knee joint. She compensated by using her hip extensors instead of her knee extensors and maintained a near-normal walking pattern; these altered patterns were retained by her CNS long after the painful arthritic knee was replaced. Therefore, this moment-of-force must be considered the final desired pattern of CNS control, or in the case of pathological movement, it must be interpreted either as a disturbed pattern or as a CNS adaptation to the disturbed patterns. This adaptability is discussed further in Chapter 4, on kinetics.

1.3 SCOPE OF THE TEXTBOOK

The best way to outline the scope of any scientific text is to describe the topics covered. In this text, the biomechanics of human movement has been defined as the mechanics and biophysics of the musculoskeletal system as it pertains to the performance of any movement skill. The neural system is also involved, but it is limited to electromyography and its relationship to the mechanics of the muscle. The variables that are used in the description and analysis of any movement can be categorized as follows: kinematics, kinetics, anthropometry, muscle mechanics, and electromyography. A summary of these variables and how they interrelate now follows.

1.3.1 Kinematics

Kinematic variables are involved in the description of the movement, independent of forces that cause that movement. They include linear and angular displacements, velocities, and accelerations. The displacement data are taken from any anatomical landmark: center of gravity of body segments, centers of rotation of joints, extremes of limb segments, or key anatomical prominances. The spatial reference system can be either relative or absolute. The former requires that all coordinates by reported relative to an anatomical coordinate system which changes from segment to segment. An absolute system means that the coordinates are referred to an external spatial reference system. The same applies to angular data. Relative angles mean joint angles; absolute angles are referred to the external spatial reference. For example, in a two-dimensional (2D) system, horizontal to the right is 0°, and counterclockwise is a positive angular displacement.

The basic kinematic concepts are taught on a 2D basis in one plane. All kinematic displacement and rotational variables are vectors. However, in any given direction or rotation, they are considered as scalar signals and can be processed and analyzed as such. In three-dimensional (3D) analysis, we add an additional vector direction but we now have three planes to analyze. Each segment in 3D analyses has its own axis system; thus, the 3D orientation of the planes for one segment are not necessarily the same as those for the adjacent segments.

1.3.2 Kinetics

The general term given to the forces that cause the movement is *kinetics.* Both internal and external forces are included. Internal forces come from muscle activity, ligaments, or from friction in the muscles and joints. External forces come from the ground or from external loads, from active bodies (e.g., those forces exerted by a tackler in football), or from passive sources (e.g., wind resistance). A wide variety of kinetic analyses can be done. The moments of force produced by muscles crossing a joint, the mechanical power flowing to or from those same muscles, and the energy changes of the body that result from this power flow are all considered part of kinetics. It is here that a major focus of the book is made, because it is in the kinetics that we can really get at the cause of the movement and therefore get some insight into the mechanisms involved and into movement strategies and compensations of the neural system. A large part of the future of biomechanics lies in kinetic analyses, because the information present permits us to make very definitive assessments and interpretations.

As with the kinematics, all basic kinetic concepts will be covered in detail in 2D analyses. Three-dimensional analysis adds an additional force vector in the global reference system (GRS), but because of the two additional planes there are two additional moment vectors. However, within any of these three planes the analysis and interpretation is the same as in 2D analyses.

1.3.3 Anthropometry

Many of the earlier anatomical studies involving body and limb measurements were not considered to be of interest to biomechanics. However, it is impossible to evolve a biomechanical model without data regarding masses of limb segments, location of mass centers, segment lengths, centers of rotation, angles of pull of muscles, mass and cross-sectional area of muscles, moments of inertia, and so on. The accuracy of any analysis depends as much on the quality and completeness of the anthropometric measures as on the kinematics and kinetics.

1.3.4 Muscle and Joint Biomechanics

One body of knowledge that is not included in any of the preceding categories is the mechanical characteristics of the muscle itself. How does its tension vary with length and with velocity? What are the passive characteristics of the muscle—mass, elasticity, and viscosity? What are the various characteristics of the joints? What are the advantages of double joint muscles? What are the differences in muscle activity during lengthening versus shortening? How does the neural recruitment affect the muscle tension? What kind of mathematical models best fit a muscle? How can we calculate the center of rotation of a joint? The final assessment of the many movements cannot ignore the influence of active and passive characteristics of the muscle, nor can it disregard the passive role of the articulating surfaces in stabilizing joints and limiting ranges of movement.

1.3.5 Electromyography

The neural control of movement cannot be separated from the movement itself, and in the electromyogram (EMG) we have information regarding the final control signal of each muscle. The EMG is the primary signal to describe the input to the muscular system. It gives information regarding which muscle or muscles are responsible for a muscle moment or whether antagonistic activity is taking place. Because of the relationship between a muscle's EMG and its tension, a number of biomechanical models have evolved. The EMG also has information regarding the recruitment of different types of muscle fibers and the fatigue state of the muscle.

1.3.6 Synthesis of Human Movement

Most biomechanical modeling involves inverse solutions to predict variables such as reaction forces, moments of force, mechanical energy, and power, none of which are directly measurable in humans. The reverse of this analysis is called *synthesis,* which assumes a similar biomechanical model, and using assumed moments of force (or muscle forces) as forcing functions, the kinematics are predicted. The ultimate goal, once a valid model has been developed, is to ask the question, "What would happen if?" Only through such modeling are we able to make predictions that are impossible to create in vivo in a human experiment. The influence of abnormal motor patterns can be predicted, and the door is now open to determine optimal motor patterns.

Although synthesis has great potential payoff, the usefulness of such models to date has been very poor and has been limited to very simplistic movements. The major problem is that the models that have been proposed are not very valid; they lack the correct anthropometrics and degrees of freedom to make their predictions very useful. However, because of its potential payoff, it is important that students have an introduction to the process, in the hope that useful models will evolve as a result of what we learn from our minor successes and major mistakes.

1.4 REFERENCES

Bernstein, N. A. *The Coordination and Regulation of Movements.* (Pergaman Press. Oxford, UK, 1967).

DeLuca, C. J., R. A. LeFever, M. P. McCue, and A. P. Xenakis. "Control Scheme Governing Concurrently Active Motor Units During Voluntary Contractions," *J. Physiol.* **329**:129–142, 1982.

Henneman, E. and C. B. Olson. "Relations Between Structure and Function in the Design of Skeletal Muscle," *J. Neurophysiol.* **28**:581–598, 1965.

Winter, D. A. "Overall Principle of Lower Limb Support During Stance Phase of Gait," *J. Biomech.* **13**:923–927, 1980.

Winter, D. A. "Kinematic and Kinetic Patterns in Human Gait: Variability and Compensating Effects," *Human Movement Sci.* **3**:51–76, 1984.

Winter, D. A. "Biomechanics of Normal and Pathological Gait: Implications for Understanding Human Locomotor Control," *J. Motor Behav.* **21**:337–355, 1989.

2

KINEMATICS

2.0 HISTORICAL DEVELOPMENT AND COMPLEXITY OF PROBLEM

Interest in the actual patterns of movement of humans and animals goes back to prehistoric times and was depicted in cave drawings, statues, and paintings. Such replications were subjective impressions of the artist. It was not until a century ago that the first motion picture cameras recorded locomotion patterns of both humans and animals. Marey, the French physiologist, used a photographic "gun" in 1885 to record displacements in human gait and chronophotographic equipment to get a stick diagram of a runner. About the same time, Muybridge in the United States triggered 24 cameras sequentially to record the patterns of a running man. Progress has been rapid during this century, and we now can record and analyze everything from the gait of a child with cerebral palsy to the performance of an elite athlete.

The term used for these descriptions of human movement is *kinematics.* Kinematics is not concerned with the forces, either internal or external, that cause the movement, but rather with the details of the movement itself. A complete and accurate quantitative description of the simplest movement requires a huge volume of data and a large number of calculations, resulting in an enormous number of graphic plots. For example, to describe the movement of the lower limb in the sagittal plane during one stride can require up to 50 variables. These include linear and angular displacements, velocities, and accelerations. It should be understood that any given analysis may use only a small fraction of the available kinematic variables. An assessment of a running broad jump, for example, may require only the velocity and height of the body's center of mass. On the other hand, a mechanical power analysis of an amputee's gait may require almost all the kinematic variables that are available.

2.1 KINEMATIC CONVENTIONS

In order to keep track of all the kinematic variables, it is important to establish a convention system. In the anatomical literature, a definite convention has been established, and we can completely describe a movement using terms such as *proximal, flexion,* and *anterior*. It should be noted that these terms are all relative, that is, they describe the position of one limb relative to another. They do not give us any idea as to where we are in space. Thus, if we wish to analyze movement relative to the ground or the direction of gravity, we must establish an absolute spatial reference system. Such conventions are mandatory when imaging devices are used to record the movement. However, when instruments are attached to the body, the data become relative, and we lose information about gravity and the direction of movement.

2.1.1 Absolute Spatial Reference System

Several spatial reference systems have been proposed. The one utilized throughout the text is the one often used for human gait. The vertical direction is Y, the direction of progression (anterior–posterior) is X, and the sideways direction (medial–lateral) is Z. Figure 2.1 depicts this convention. The positive direction is as shown. Angles must also have a zero reference and a positive direction. Angles in the XY plane are measured from 0° in the X direction, with positive angles being counterclockwise. Similarly, in the YZ plane, angles start at 0° in the Y direction and increase positively counterclockwise. The convention for velocities and accelerations follows correctly if we maintain the spatial coordinate convention:

$\dot{x}$ = velocity in the X direction, positive when X is increasing

$\dot{y}$ = velocity in the Y direction, positive when Y is increasing

$\dot{z}$ = velocity in the Z direction, positive when Z is increasing

$\ddot{x}$ = acceleration in the X direction, positive when $\dot{x}$ is increasing

$\ddot{y}$ = acceleration in the Y direction, positive when $\dot{y}$ is increasing

$\ddot{z}$ = acceleration in the Z direction, positive when $\dot{z}$ is increasing

The same applies for angular velocities and angular accelerations. A counterclockwise angular increase is a positive angular velocity ω. When ω is increasing, we calculate a positive angular acceleration, α.

An example taken from the data on a human subject during walking will illustrate the convention. The kinematics of the right leg segment (as viewed from the right side) and its center of mass were analyzed as follows:

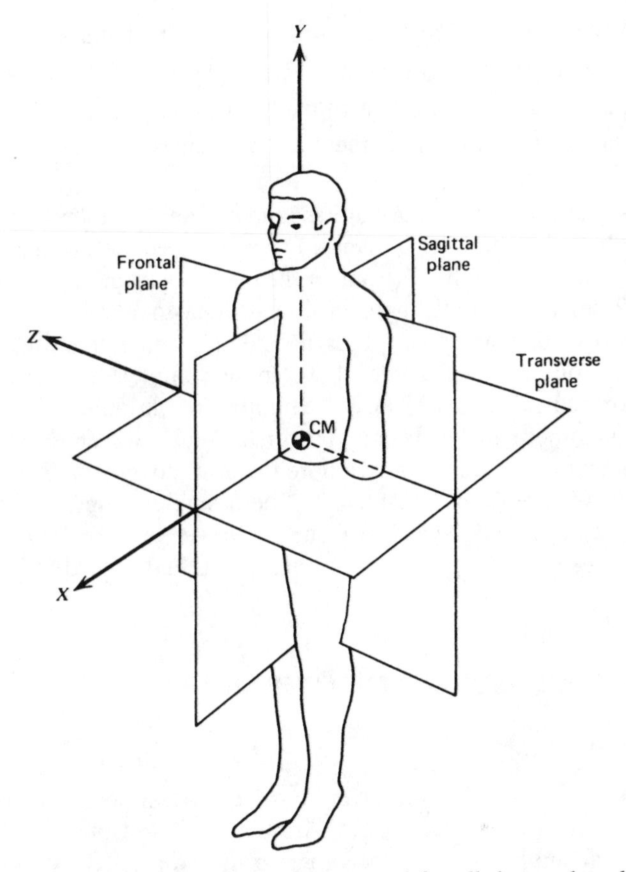

Figure 2.1 Spatial coordinate system used for all data and analysis.

$$\omega = -2.34 \text{ rad/s}, \ \alpha = 14.29 \text{ rad/s}^2, \ v_x = 0.783 \text{ m/s}$$

$$a_x = -9.27 \text{ m/s}^2, \ v_y = 0.021 \text{ m/s}, \ a_y = -0.31 \text{ m/s}^2$$

This means that the leg segment is rotating clockwise but is decelerating (accelerating in a counterclockwise direction). The velocity of the leg's center of mass is forward and very slightly upward, but is decelerating in the forward direction and accelerating downward.

2.1.2 Total Description of a Body Segment in Space

The complete kinematics of any body segment requires 15 data variables, all of which are changing with time:

1. Position (x, y, z) of segment center of mass
2. Linear velocity $(\dot{x}, \dot{y}, \dot{z})$ of segment center of mass

3. Linear acceleration ($\ddot{x}$, $\ddot{y}$, $\ddot{z}$) of segment center of mass
4. Angle of segment in two planes, θ_{xy}, θ_{yz}
5. Angular velocity of segment in two planes, ω_{xy}, ω_{yz}
6. Angular acceleration of segment in two planes, α_{xy}, α_{yz}

Note that the third angle data are redundant; any segment's direction can be completely described in two planes. For a complete description of the total body (feet + legs + thighs + trunk + head + upper arms + forearms and hands = 12 segments) movement in three-dimensional (3D) space required $15 \times 12 = 180$ data variables. It is no small wonder that we have yet to describe, let alone analyze, some of the more complex movements. Certain simplifications can certainly reduce the number of variables to a manageable number. In symmetrical level walking, for example, we can assume sagittal plane movement and can normally ignore the arm movement. The head, arms, and trunk (HAT) are often considered to be a single segment, and assuming symmetry, we need to collect data from one lower limb only. The data variables in this case (four segments, one plane) can be reduced to a more manageable 36.

2.2 DIRECT MEASUREMENT TECHNIQUES

2.2.1 Goniometers*

A goniometer is a special name given to the electrical potentiometer that can be attached to measure a joint angle. As such, one arm of the goniometer is attached to one limb segment, the other to the adjacent limb segment, and the axis of the goniometer is aligned to the joint axis. In Figure 2.2 we see the fitting of the goniometer to a knee joint along with the equivalent electrical circuit. A constant voltage E is applied across the outside terminals, and the wiper arm moves to pick off a fraction of the total voltage. The fraction of the voltage depends on the joint angle θ. Thus the voltage on the wiper arm is $v = kE\theta = k_1\theta$ volts. Note that a voltage proportional to θ requires a potentiometer whose resistance varies linearly with θ. A goniometer designed for clinical studies is shown fitted on a patient in Figure 2.3.

Advantages

1. A goniometer is generally inexpensive.
2. Output signal is available immediately for recording or conversion into a computer.
3. Planar rotation is recorded independent of the plane of movement of the joint.

*Representative paper: Finley and Karpovich, 1964.

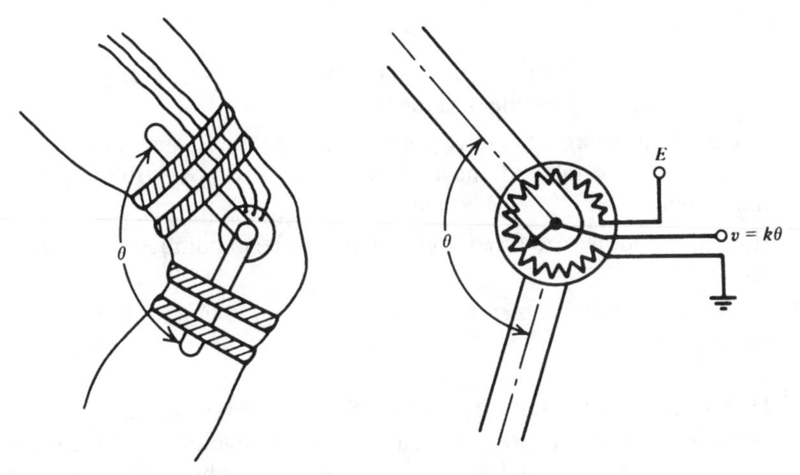

Figure 2.2 Mechanical and electrical arrangement of a goniometer located at the knee joint. Voltage output is proportional to the joint angle.

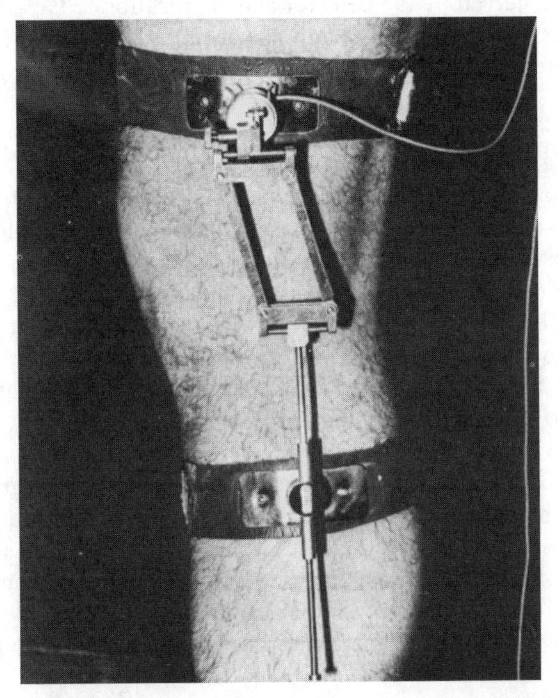

Figure 2.3 Electrogoniometer designed to accommodate changes in center of rotation of the knee joint, shown here fitted on a patient. (Reproduced by permission of Chedoke-McMaster Medical Center, Hamilton, Ont., Canada.)

Disadvantages

1. Relative angular data are given, not absolute angles, thus severely limiting the data's assessment value.
2. It may require an excessive length of time to fit and align, and the alignment over fat and muscle tissue can vary over the time of the movement.
3. If a large number are fitted, movement can be encumbered by the straps and cables.
4. More complex goniometers are required for joints that do not move as hinge joints.

2.2.1.1 Special Joint Angle Measuring Systems. More recently in the area of ergonomics, a special glove system has been developed to measure the kinematics of the fingers and the thumb. Figure 2.4 shows the construction of the glove transducer, which comprises a lightweight elastic glove with sensors on the proximal two joints of each finger and thumb plus a thumb abductor sensor. Each transducer is a loop of fiber optic cable with a constant infrared source and is etched in the region of the joint of interest. As the joint flexes, the fiber bends and light escapes; the greater the bend, the more light escapes. Thus, the flexion angular displacement is detected as a reduction in light intensity received by the detector and is precalibrated against the bending angle. A major use for such a system has been in the study of repetitive strain injuries (cf. Moore et al., 1991).

2.2.2 Accelerometers*

As indicated by its name, an accelerometer is a device that measures acceleration. Most accelerometers are nothing more than force transducers designed to measure the reaction forces associated with a given acceleration. If the acceleration of a limb segment is a and the mass inside is m, then the force exerted by the mass is $F = ma$. This force is measured by a force transducer, usually a strain gauge or piezoresistive type. The mass is accelerated against a force transducer that produces a signal voltage V, which is proportional to the force, and since m is known and constant, then V is also proportional to the acceleration. The acceleration can be toward or away from the face of the transducer; this is indicated by a reversal in sign of the signal. In most movements, there is no guarantee that the acceleration vector will act at right angles to the face of the force transducer. The more likely situation is depicted in Figure 2.5, with the acceleration vector having a component normal to the transducer and another component tangent to the transducer face. Thus the accelerometer measures the a_n component. Nothing is known

*Representative paper: Morris, 1973.

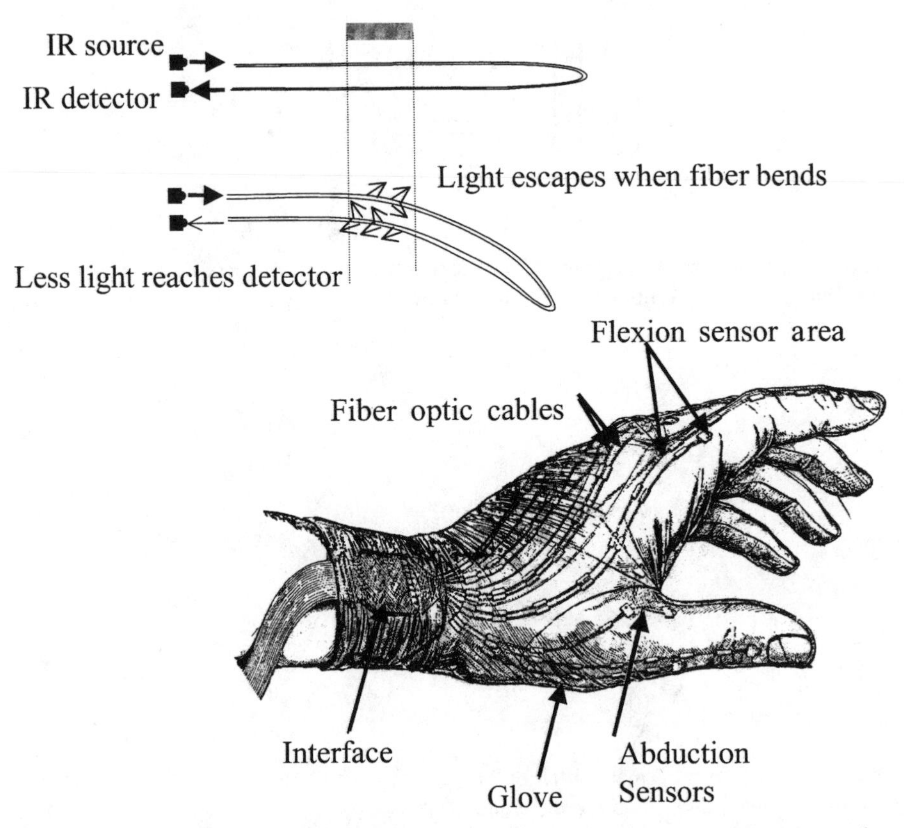

Figure 2.4 Construction and operation of a glove transducer to measure angular displacement of the fingers. Transducer is a loop of fiber optic cable; the amount of light returning to the detector decreases with increased finger flexion. Each cable is calibrated for angular displacement versus detected light intensity. (Courtesy of the Ergonomics Laboratory, Department of Kinesiology, University of Waterloo, Ont., Canada.)

about a_t or a unless a triaxial accelerometer is used. Such a 3D transducer is nothing more than three individual accelerometers mounted at right angles to each other, each one then reacting to the orthogonal component acting along its axis. Even with a triaxial accelerometer mounted on a limb, there can be problems because of limb rotation, as indicated in Figure 2.6. In both cases, the leg is accelerating in the same *absolute* direction as indicated by vector a. The measured acceleration component a_n is quite different in each case. Thus, the accelerometer is limited to those movements whose direction in space does not change drastically or to special contrived movements, such as horizontal flexion of the forearm about a fixed elbow joint.

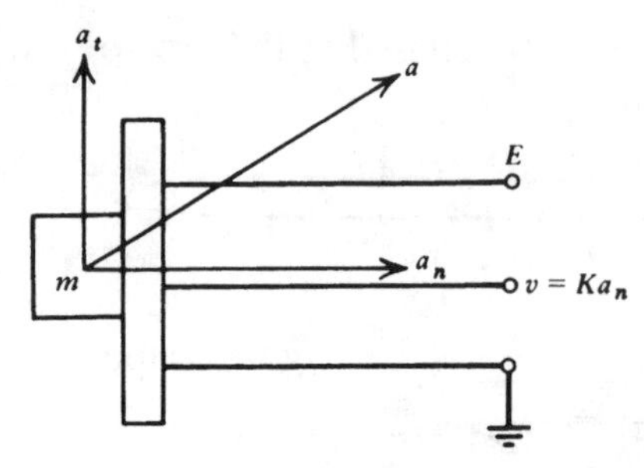

Figure 2.5 Schematic diagram of an accelerometer, showing acceleration with normal and tangential components. Voltage output is proportional to the normal component of acceleration a_n.

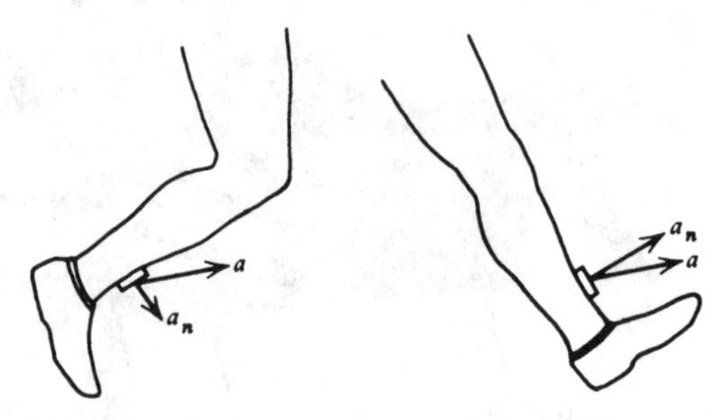

Figure 2.6 Two movement situations where the acceleration in space is identical but the normal component is quite different.

A typical electric circuit of a piezoresistive accelerometer is shown in Figure 2.7. It comprises a half-bridge consisting of two equal resistors R_1. Within the transducer, resistors R_a and R_b change their resistances proportional to the acceleration acting against them. With no acceleration, $R_a = R_b = R_1$, and with the balance potentiometer properly adjusted, the voltage at terminal 1 is the same as that at terminal 2. Thus, the output voltage is $V = 0$. With the acceleration in the direction shown, R_b increases and R_a decreases; thus, the voltage at terminal 1 increases. The resultant unbalance in the bridge circuit results in voltage V proportional to the acceleration. Conversely, if the acceleration is upward, R_b decreases and R_a increases; the bridge unbalances in the reverse direction, giving a signal of the opposite polarity. Thus, over the dynamic range of the accelerometer, the signal is proportional to both the magnitude and the direction of acceleration acting along the axis of the accelerometer. However, if the balance potentiometer is not properly set, we have an unbalanced bridge and we could get a voltage–acceleration relationship like that indicated by the dashed lines.

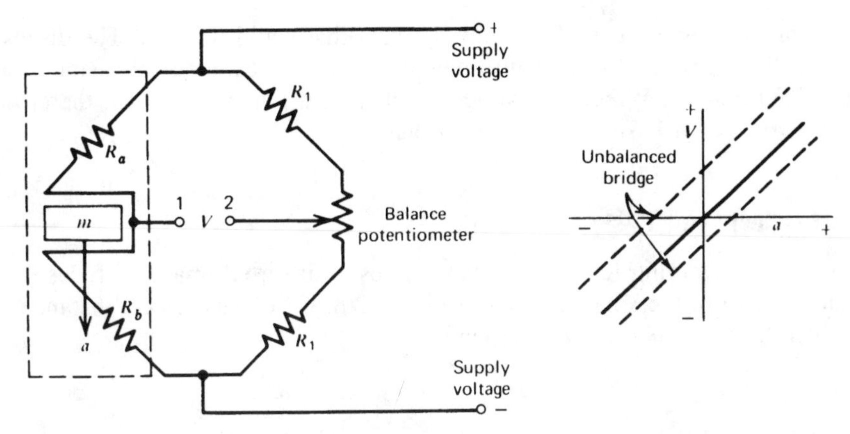

Figure 2.7 Electrical bridge circuit used in most force transducers and accelerometers. See text for detailed operation.

Advantages

1. Output signal is available immediately for recording or conversion into a computer.

Disadvantages

1. Acceleration is relative to its position on the limb segment.
2. Cost of accelerometers can be excessive if a large number are used; also the cost of the recorder or analog-to-digital converter may be high.
3. If a large number are used, they can encumber movement.
4. Many types of accelerometers are quite sensitive to shock and are easily broken.
5. The mass of the accelerometer may result in a movement artifact, especially in rapid movements or movements involving impacts.

2.3 IMAGING MEASUREMENT TECHNIQUES

The Chinese proverb "A picture is worth more than ten thousand words" holds an important message for any human observer, including the biomechanics researcher interested in human movement. Because of the complexity of most movements, the only system that can possibly capture all the data is an imaging system. Given the additional task of describing a dynamic activity, we are further challenged by having to capture data over an extended period of time. This necessitates taking many images at regular intervals during the event.

There are many types of imaging systems that could be used. The discussion will be limited to three different types: movie camera, television, and optoelectric types. Whichever system is chosen, a lens is involved; therefore a short review of basic optics is given here.

2.3.1 Review of Basic Lens Optics

A simple converging lens is one that creates an inverted image in focus at a distance v from the lens. As seen in Figure 2.8, if the lens–object distance is u, then the focal length f of the lens is

$$\frac{1}{f} = \frac{1}{v} + \frac{1}{u} \tag{2.1}$$

The imaging systems used for movement studies are such that the object–lens distance is quite large compared with the lens–image distance. Therefore,

$$\frac{1}{u} \approx 0, \frac{1}{f} = \frac{1}{v}, \text{ or } f = v \tag{2.2}$$

Thus, if we know the focal length of the lens system, we can see that the image size is related to the object size by a simple triangulation. A typical focal length is 25 mm, a wide-angle lens is 13 mm, and a telephoto lens is 150 mm. A zoom lens is just one in which the focal length is infinitely variable over a given range. Thus, as L increases, the focal length must increase proportionately to produce the same image size. Figure 2.9 illustrates this principle. For maximum accuracy, it is highly desirable that the image be as large as possible. Thus, it is advantageous to have a zoom lens rather than a series of fixed lenses; individual adjustments can be readily made for each movement to be studied, or even during the course of the event.

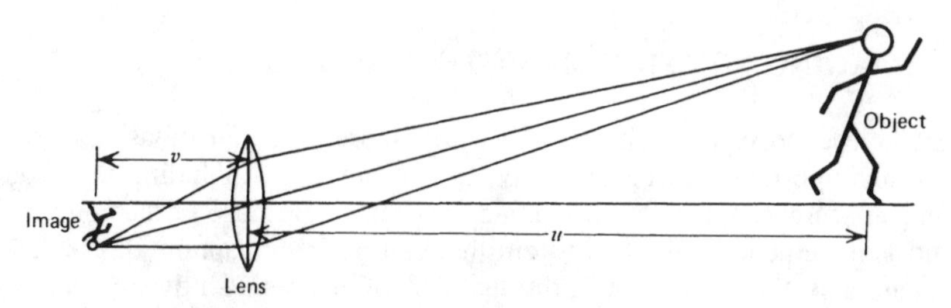

Figure 2.8 Simple focusing lens system showing relationship between object and image.

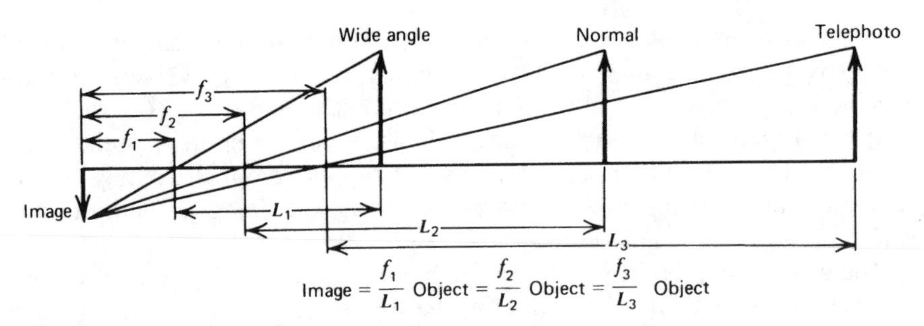

Figure 2.9 Differences in focal length of wide-angle, normal, and telephoto lenses result in an image that is the same size.

2.3.2 *f*-Stop Setting and Field of Focus

The amount of light entering the lens is controlled by the lens opening, which is measured by its *f*-stop (*f* means fraction of lens aperture opening). The larger the opening, the lower the *f*-stop setting. Each *f*-stop setting corresponds to a proportional change in the amount of light allowed in. A lens may have the following settings: 22, 16, 11, 8, 5.6, 4, 2.8, and 2. $f/22$ is 1/22 of the lens diameter and $f/11$ is 1/11 of the lens diameter. Thus $f/11$ lets in four times the light that $f/22$ does. The fractions are arranged so that each one lets in twice the light of the adjacent higher setting (e.g., $f/2.8$ is twice the light of $f/4$).

To keep the lighting requirements to a minimum, it is obvious that the lens should be opened as wide as possible with a low *f* setting. However, problems occur with the field of focus. This is defined as the maximum and minimum range of the object that will produce a focused image. The lower the *f* setting, the narrower the range over which an object will be in focus. For example, if we wish to photograph a movement that is to move over a range from 10 to 30 ft, we cannot reduce the *f*-stop below 5.6. The range set on the lens would be about 15 ft, and everything between 10 and 30 ft would remain in focus. The final decision regarding *f*-stop depends on the shutter speed of the movie camera and the film speed.

2.3.3 Cinematography*

Many different sizes of movie cameras are available; 8-mm cameras are the smallest. (They actually use 16-mm film, which is run through the camera twice, then split into two 8-mm strips after it is developed.) Then there are 16 mm, 35 mm, and 70 mm. The image size of 8 mm is somewhat small for

*Representative paper: Eberhart and Inman, 1951.

accurate measurements, while 35-mm and 70-mm movie cameras are too expensive to buy and operate. Thus, 16-mm cameras have evolved as a reasonable compromise, and most high-speed movie cameras are 16 mm. There are several types of 16-mm cameras available. Some are spring driven; others are motor driven by either batteries or power supplies from alternating current sources. Battery-driven types have the advantage of being portable to sites where power is not available.

The type of film required depends on the lighting available. The ASA rating is a measure of the *speed* of the film; the higher the rating, the less light is required to get the same exposure. 4-X reversal film with an ASA rating of 400 is a common type. Higher ASA ratings are also available and are good for a qualitative assessment of movement, especially faster moving sporting events. However, the coarse grain of these higher ASA films introduces inaccuracies in quantitative analyses.

The final factor that influences the lighting required is the shutter speed of the camera. The higher the frame rate, the less time is available to expose film. Most high-speed cameras have rotating shutters that open once per revolution for a period of time to expose a new frame of unexposed film. The arc of the opening, as depicted in Figure 2.10, and the speed of rotation of the shutter decide the exposure time. For example, at 60 frames per second, using a 3 factor shutter, the exposure time is 1/180 s. The amount of light entering will be the same as a normal (still) camera set to a speed of 1/180 s.

To make the final settings, we use an exposure meter to measure the light intensity on the human subject. For a given filming, the variables that are

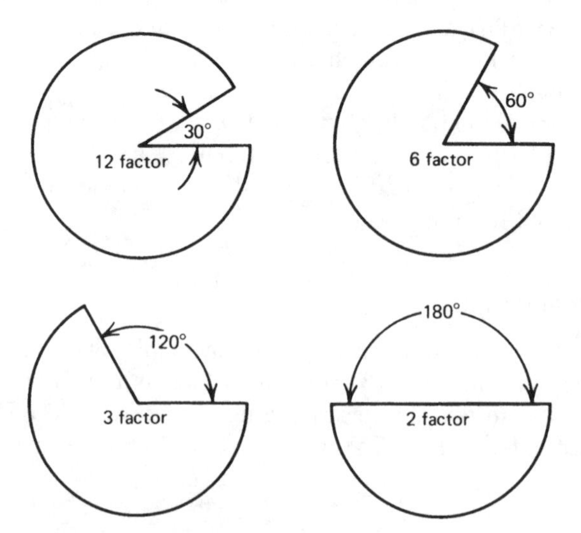

Figure 2.10 Various-factor shutters used in movie cameras. Film is exposed during the opening arc and is advanced while the shutter is closed.

preset are film ASA, shutter factor, and frame rate. The frame rate is set low enough to capture the desired event, but not too high to require extra lighting or result in film wastage. To understand the problem associated with the selection of an optimal rate, the student is referred to Section 2.5.3. The final variable to decide is the f-stop. The light meter gives an electrical meter reading proportional to the light intensity, such that when the film ASA and exposure time are set, the correct f-stop can be determined. Thus, with the movie camera set at the right frame rate, f-stop, and range, the filming is ready to commence.

2.3.4 Television*

The major difference between television and cinematography is the fact that television has a fixed frame rate. The name given to each television image is a field. In North America there are 60 fields per second, in Europe the standard is 50 fields per second. Thus, television has a high enough field rate for most movements, but probably too low for a quantitative analysis of rapid athletic events. The f-stop, focus, and lighting for television can be adjusted by watching the television monitor as the controls are varied. Many television cameras have electronic as well as optical controls that influence brightness and contrast, and some have built-in strobe lighting. Also, focus can be adjusted electronically as well as optically. The major advantage of television is the capability for instant replay, which serves both as a quality control check and as an initial qualitative assessment. Secondly, the television signal can be digitally converted by a "frame grabber" for immediate analysis.

2.3.4.1 Television Imaging Cameras. Some technical problems can result from the use of standard vidicon television cameras. The strong signal from a reflective marker produces a distinct circular image when the marker is not moving. However, when there is a rapid marker movement, the circular image blurs and produces a trailing edge. Thus, the triggering threshold for conversion into two levels (black and white) must be carefully set to ensure a circular digitized image. A more reliable way to get rid of the blur is to use a strobe system, which results in the exposure of the TV imaging tube for a millisecond or less. The strobe in effect acts as an electronic shutter. Strobe systems also eliminate a second problem associated with a continuously exposed imaging tube: skewing of the marker coordinates because of the time delay in the scanning from the top of the image to the bottom of the image. It takes about 15 ms to scan one TV field; thus, a head marker could be scanned 10 ms before a foot marker. The strobe system freezes all marker images at the same point in time in the same way that a movie camera does. Newer charge coupled diodes (CCD) cameras have mechanical or electronic shutter controls

*Representative paper: Winter et al., 1972.

that eliminate both blurring and skewing. A further development is the infrared camera, which does not use visible light and is not influenced by reflections from light sources other than those sources required to get the desired circular reflection from the markers. Figure 2.11 shows a typical infrared camera mounted permanently from the ceiling in a clinical gait laboratory. The active infrared lights form a "donut" shape about the camera lens and are pulsed at 120 Hz for a period of less than a millisecond. This camera is

Figure 2.11 Typical infrared television camera mounted permanently in a clinical gait laboratory. The infrared lights form a "donut" shape around the lens and are pulsed for a short time each frame to freeze the image prior to scanning. (Courtesy of the Gait Analysis Laboratory, Connecticut Children's Medical Center, Hartford, CT.)

one of six to twelve cameras that could be mounted around the gait laboratory. Thus, the reflected infrared light from the markers is the only light that is picked up by the camera, and since it is a pulsed source, the marker images are "frozen" in time. Figure 2.12 depicts such an arrangement in a clinical gait laboratory along with the spherical reflective markers mounted on a young patient.

2.3.4.2 Historical Development of Television Digitizing Systems. Almost all of the movement analysis television systems were developed in university research laboratories. In the late 1960s, the first reports of television-based

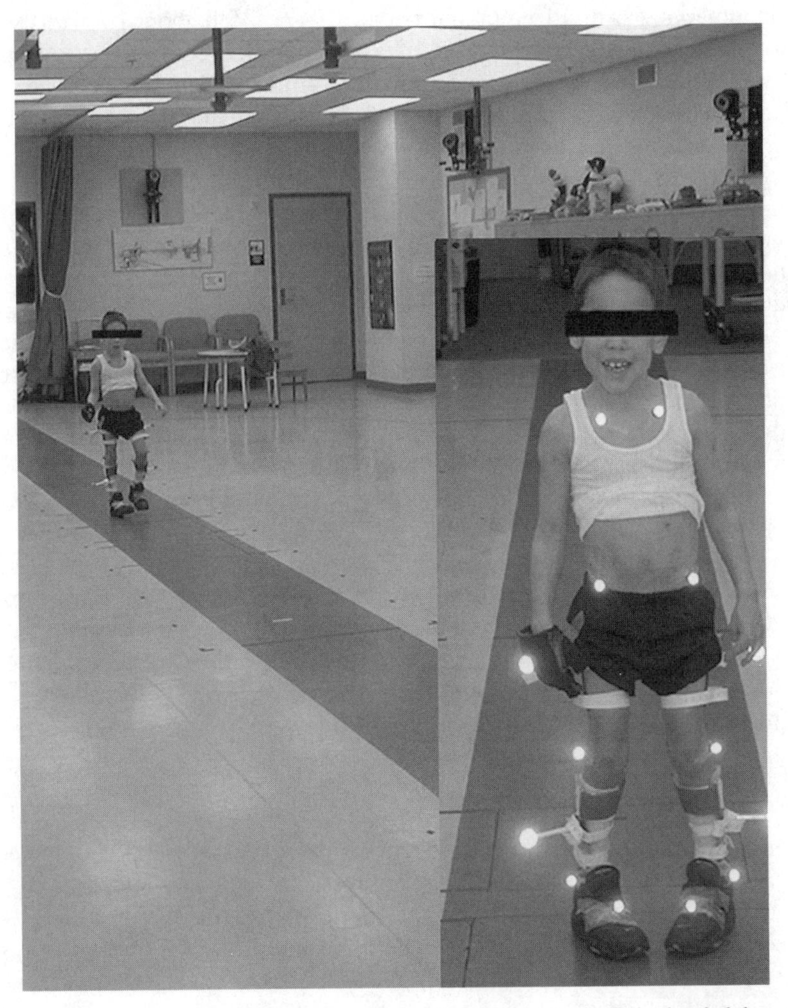

Figure 2.12 Gait assessment of a cerebral palsy patient in a clinical gait laboratory. Infrared cameras are mounted on the ceiling and walls to capture the reflected light from the spherical reflective markers mounted on both sides of the body. (Courtesy of the Gait Analysis Laboratory, Connecticut Children's Medical Center, Hartford, CT.)

systems started to appear: at Delft University of Technology in The Netherlands (Furnée, 1967, according to Woltring, 1987) and at the Twenty-First Conference EMB in Houston, TX (Winter et al., 1968). The first published paper on an operational system was by Dinn et al. (1970) from the Technical University of Nova Scotia in Halifax, N.S., Canada. It was called CINTEL (Computer INterface for TELivision), and was developed for digitizing angiographic images at 4 bits (16 grey levels) to determine the time course of left ventricular volume (Trenholm et al., 1972). It was also used for gait studies at the University of Manitoba in Winnipeg, Man., Canada, where, with higher spatial resolution and a one-bit (black/white) conversion, the circular image of a reflective hemispheric ping pong ball attached on anatomical landmarks was digitized (Winter et al., 1972). With about 10 pixels within each marker image, it was possible by averaging their coordinates to improve the spatial precision of each marker from 1 cm (distance between scan lines of each field) to about 1 mm. The 3M Scotch® material that was used as reflective material has been used by most subsequent experimental and commercial systems.

Jarett et al. (1976) reported a system that detected the left edge of the image of a small reflective marker that occupied one or two scan lines. Unfortunately, the spatial precision was equal to the scan line distance, which is about 1 cm. This system was adopted and improved by VICON (VIdeo CONvertor) in their commercial system. Both left and right edges of the marker image were detected, and subsequently the detected points were curve fitted via software AMASS to a circle (Macleod et al., 1990). Based on the circle fit, the centroid was calculated. Other commercial systems, such as that developed by the Motion Analysis Corporation, use patented edge-detection techniques (Expert Vision). Shape recognition of the entire marker image, rather than edge detection, was used by the ELITE (Elaboratore di Immagini Televisive) system developed in Milan, Italy. A dedicated computer algorithm operating in real time used a cross-correlation pattern recognition technique based on size and shape (Ferrigno and Pedotti, 1985). This system uses all grey levels in the shape detection, thus improving the spatial resolution to 1/2800 of the viewing field. Considering the field height to be about 2.5 m, this represents a precision of about 0.9 mm.

2.3.5 Optoelectric Techniques*

In the past few years, there have been several developments in optoelectric imaging systems that have some advantages over cinematography and television. The first commercial system was developed by Northern Digital in Waterloo, Ont., Canada, and was called Watsmart. It was an active system that required the subject to wear tiny infrared lights on each desired anatom-

*Representative paper: Winter et al., 2003.

ical landmark. The lights were flashed sequentially, and the light flash was detected on a special camera. The camera consisted of a standard lens focusing the light flash onto a special semiconductor diode surface. More recent development of this active system has evolved into a 3D camera system called OPTOTRAK. Such a system consists of three cameras mounted in line on a rigid frame, as shown in Figure 2.13. The left and right lenses are mounted to face slightly inward and their linear diode arrays are mounted horizontally. Thus, their scan of the pulsed light will define the location of a marker in a vertical plane. The middle lens is mounted with its diode array mounted vertically, and its scan will define a horizontal plane. Figure 2.14 depicts this arrangement. The left and right detectors each define the location of all markers in a vertical plane; the intersection of these two vertical planes is a vertical line. Thus, any markers on this vertical line will record the same signal on the left and right cameras. The middle camera has its lens facing directly ahead with its diode array mounted vertically. Thus, this camera will define all markers in a horizontal plane. The intersection of this horizontal plane with the vertical line defined by the other two cameras is a unique 3D point in space. Thus, as each infrared diode (IRED) pulses, its x, y, z coordinates in the global reference system (GRS) are recorded. The pulsed light from a

Figure 2.13 An OPTOTRAK system with three lenses, each with a linear diode array. The outside two lenses face slightly inward and each defines a vertical plane, while the middle lens defines a horizontal plane. See Figure 2.14 to see how these three diode arrays define a marker in 3D space. (Courtesy of Gait and Posture Laboratory, Department of Kinesiology, University of Waterloo, Waterloo, Ont., Canada.)

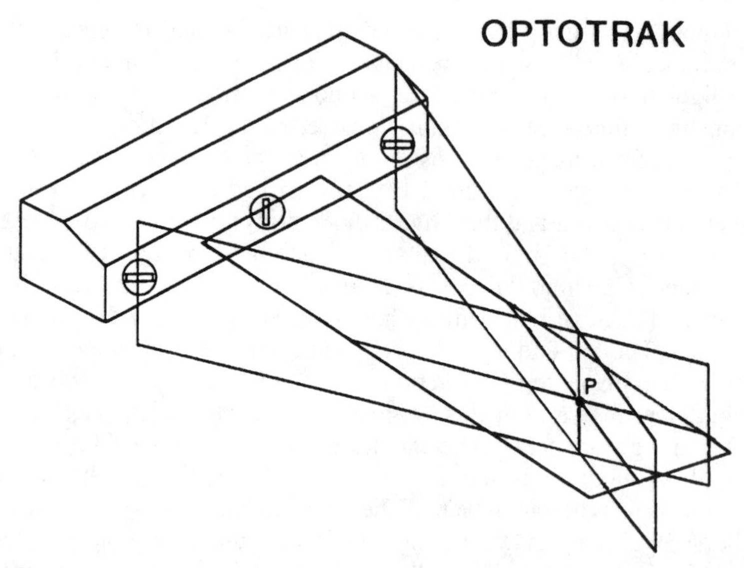

Figure 2.14 OPTOTRAK system with two outside lenses facing slightly inward with their diode arrays each defining markers in a vertical plane. Any marker on the intersection of these two planes will define all markers on this vertical line. The middle lens array defines all markers in a horizontal plane. Thus, the intersection of a marker on this horizontal plane with the vertical line will define the unique coordinates of a marker in 3D space.

second light source yields a different vertical line and horizontal plane, and thus a different set of x, y, z coordinates. There are some unique advantages of such an active system. There is no specialized software required (as in television) to identify which marker is which. Thus, in laboratories where the number and location of markers is changing day-to-day, there are no problems with marker labeling, which makes this system flexible for the changing research requirements. Also, because of the precision of the IRED array, the precision of the x, y, z coordinates is better than that of TV systems, which are constrained by the distance between the scan lines. The precision for an OPTOTRAK camera mounted as shown in Figure 2.13 at a distance of 4 m from the subject is 0.03 mm with noise = 0.015 mm (Gage et al., 2004). Possible disadvantages are the number of IREDs that can be mounted and the potential encumbrance of the cables connecting the power source to the active IREDs.

2.3.6 Advantages and Disadvantages of Optical Systems

Advantages

1. All data are presented in an absolute spatial reference system, in a plane normal to the optical axis of the camera.

2. Most systems (cine, TV) are not limited as to the number of markers used.
3. Encumbrance to movement is minimal for most systems that use lightweight reflective markers (cine, TV), and the time to apply the markers is minimal.
4. TV cameras and VCRs are reasonably inexpensive.
5. Cine and TV systems can be replayed for teaching purposes or for qualitative analysis of the total body movement.

Disadvantages

1. Most multiple-camera systems are expensive (cine, TV, optoelectric), as are the digitizing and conversion systems for all imaging sources.
2. For film, the turnaround time for development may be a problem, and the labor to digitize film coordinates may also be a constraint. The digitizing errors, however, are less than those from many commercial imaging systems.
3. Encumbrance and time to fit wired light sources (e.g., IREDs) can be prohibitive in certain movements, and the number of light sources is limited.
4. Some imaging systems (e.g., IREDs) cannot be used outside in daylight.

2.3.7 Summary of Various Kinematic Systems

Each laboratory must define its special requirements before choosing a particular system. A clinical gait lab may settle on TV because of the encumbrance of optoelectric systems and because of the need for a qualitative assessment, rapid turnaround, and for teaching. Ergonomic and athletic environments may require instant or near-instant feedback to the subject or athlete, thus dictating the need for an automated system. Basic researchers do not require a rapid turnaround and may need a large number of coordinates; thus, they may opt for movie cameras or an optoelectric system. And, finally, the cost of hardware and software may be the single limiting factor that may force a compromise as to the final decision.

2.4 DATA CONVERSION TECHNIQUES

2.4.1 Analog-to-Digital Converters

To students not familiar with electronics, the process that takes place during conversion of a physiological signal into a digital computer can be somewhat mystifying. A short schematic description of that process is now given. An electrical signal representing a force, an acceleration, an electromyographic (EMG) potential, or the like is fed into the input terminals of the analog-to-digital converter. The computer controls the rate at which the signal is sampled; the optimal rate is governed by the sampling theorem (see Section 2.5.3).

Figure 2.15 depicts the various stages in the conversion process. The first is a sample/hold circuit in which the analog input signal is changed into a series of short-duration pulses, each one equal in amplitude to the original analog signal at the time of sampling. (These times are specified by the computer operator.) The final stage of conversion is to translate the amplitude and polarity of the sampled pulse into digital format. This is usually a binary code in which the signal is represented by a number of bits. For example, a 12-bit code represents $2^{12} = 4096$ levels. This means that the original sampled analog signal can be broken into 4096 discrete amplitude levels with a unique code representing each of these levels. Each coded sample (consisting of 0s and 1s) forms a 12-bit "word," which is rapidly stored in computer memory for recall at a later time. If a 5-s signal were converted at a sampling rate of 100 times per second, there would be 500 data words stored in memory to represent the original 5-s signal.

2.4.2 Movie Conversion Techniques

As 16-mm movie cameras are the most common form of data collection, it is important to be aware of various coordinate extraction techniques. Each system that has evolved requires the projection of each movie frame on some form of screen. The most common type requires the operator to move a mechanical xy coordinate system until a point, light, or cross hair lies over the desired anatomical landmark. Then the x and y coordinates can be read

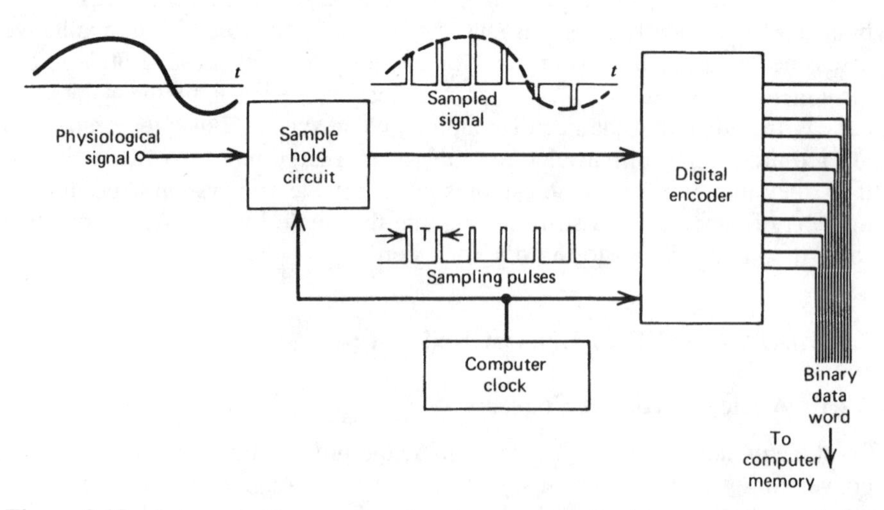

Figure 2.15 Schematic diagram showing the steps involved in an analog-to-digital conversion of a physiological signal.

off or transferred to a computer at the push of a button. Figure 2.16 shows the component parts of such a conversion system.

A second type of system involves the projection of the film image onto a special grid system. When the operator touches the grid with a special pen, the coordinates are automatically transferred into a computer. Both systems are limited to the speed and accuracy of the human operator. Our experience indicates that an experienced operator can convert an average of 15 coordinate pairs per minute. Thus a 3-s film record filmed at 50 frames per second could have five markers converted in 30 min.

The human error involved in this digitizing has been found to be random and quite small. For a camera 4 m from a subject, the root-mean-square (rms) "noise" present in the converted data has been measured at 1–1.5 mm.

2.4.3 Television Conversion

Each of the commercial television systems has it own unique technique for identifying the presence of a marker and determining its centroid, and for labeling markers from a multi-camera system. Students are referred back to Section 2.3.4.2 for a summary of and references to the more common systems.

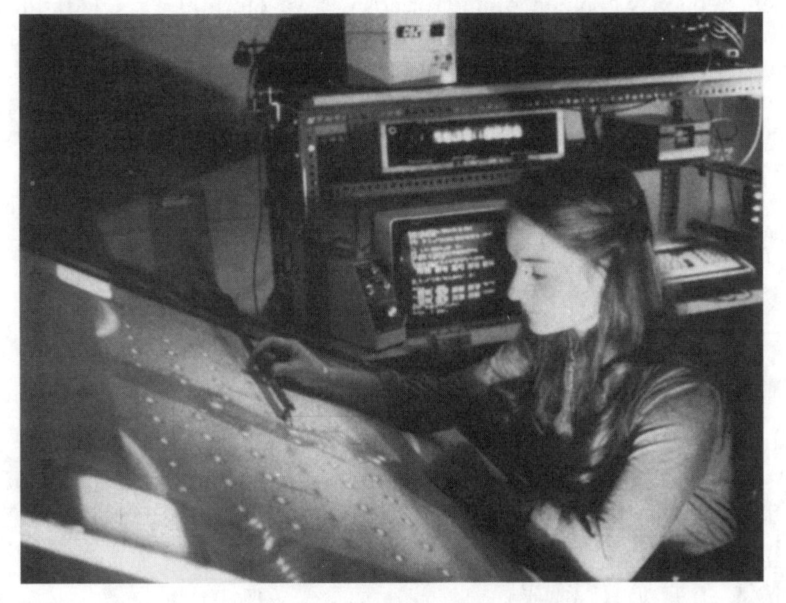

Figure 2.16 Typical arrangement for the microcomputer digitization of data coordinated from movie film. Foot pedal (not shown) allows operator to transfer coordinate data into a digital computer at the rate of about 10 coordinate pairs per minute. Digitizing error is about 1mm rms with the camera located 4 m from the subject.

2.5 PROCESSING OF RAW KINEMATIC DATA

2.5.1 Nature of Unprocessed Data

The converted coordinate data from film or television are called *raw* data. This means that they contain additive noise from many sources: electronic noise in optoelectric devices, spatial precision of the TV scan or film digitizing system, or human error in film digitizing. All of these will result in random errors in the converted data. It is therefore essential that the raw data be smoothed, and in order to understand the techniques used to smooth the data, an appreciation of harmonic (or frequency) analysis is necessary.

2.5.2 Harmonic (Fourier) Analysis

1. *Alternating Signals.* An alternating signal (often called ac, for alternating current) is one that is continuously changing with time. It may be periodic or completely random, or a combination of both. Also, any signal may have a dc (direct current) component, which may be defined as the bias value about which the ac component fluctuates. Figure 2.17 shows example signals.

2. *Frequency Content.* Any of these signals can also be discussed in terms of their frequency content. A sine (or cosine) waveform is a single frequency; any other waveform can be the sum of a number of sine and cosine wave.

Note that the Fourier transformation (Figure 2.18) of periodic signals has discrete frequencies, while nonperiodic signals have a continuous spectrum defined by its lowest frequency f_1 and its highest frequency f_2. To analyze a

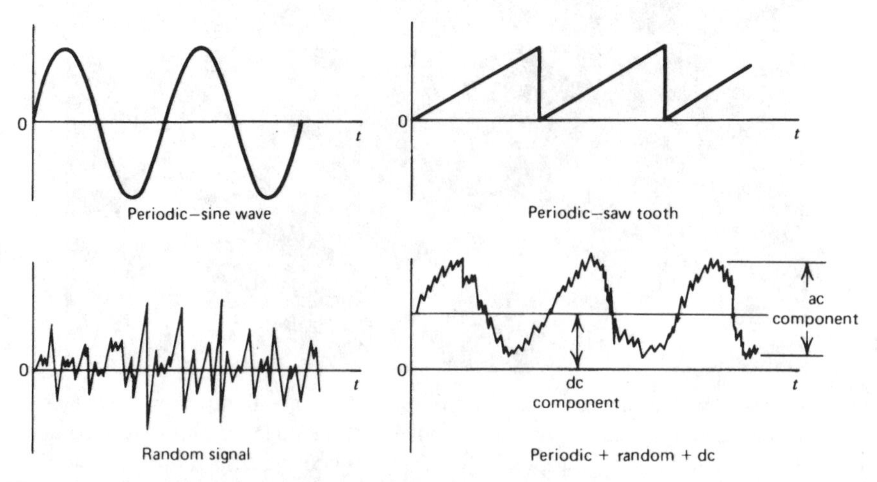

Figure 2.17 Time-related waveforms to demonstrate the different types of signals that may be processed.

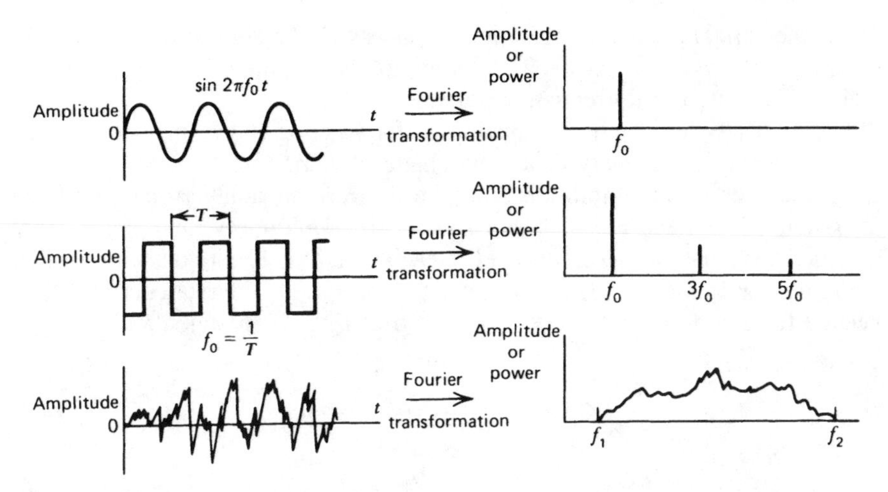

Figure 2.18 Relationship between a signal as seen in the time domain and its equivalent in the frequency domain.

periodic signal, we must express the frequency content in multiples of the fundamental frequency f_0. These higher frequencies are called *harmonics.* The third harmonic is $3f_0$ and the tenth harmonic is $10f_0$. Any perfectly periodic signal can be broken down into its harmonic components. The sum of the proper amplitudes of these harmonics is called a *Fourier series.*

Thus, a given signal $V(t)$ can be expressed as

$$V(t) = V_{dc} + V_1 \sin(\omega_0 t + \theta_1) + V_2 \sin(2\omega_0 t + \theta_2) + \cdots + V_n \sin(n\omega_0 t + \theta_n) \tag{2.3}$$

where $\omega_0 = 2\pi f_0$, and θ_n is the phase angle of the nth harmonic.

For example, a square wave of amplitude V can be described by the Fourier series

$$V(t) = \frac{4V}{\pi}\left(\sin \omega_0 t + \frac{1}{3}\sin 3\omega_0 t + \frac{1}{5}\sin 5\omega_0 t + \cdots\right) \tag{2.4}$$

A triangular wave of duration $2t$ and repeating itself every T seconds is

$$V(t) = \frac{2Vt}{T}\left[\frac{1}{2} + \left(\frac{2}{\pi}\right)^2 \cos \omega_0 t + \left(\frac{2}{3\pi}\right)^2 \cos 3\omega_0 t + \cdots\right] \tag{2.5}$$

Several names are given to the graph showing these frequency components: *spectral plots, harmonic plots,* or *spectral density functions.* Each shows the amplitude or power of each frequency component plotted against frequency;

the mathematical process to accomplish this is called a *Fourier transformation* or *harmonic analysis.* Figure 2.18 shows plots of time-domain signals and their equivalents in the frequency domain.

Care must be used when analyzing or interpreting the results of any harmonic analysis. Such analyses assume that each harmonic component is present with a constant amplitude and phase over the analysis period. Such consistency is evident in Equation (2.3), where amplitude V_n and phase θ_n are assumed constant. However, in real life each harmonic is not constant in either amplitude or phase. A look at the calculation of the Fourier coefficients is needed for any signal x(t). Over the period of time T, we calculate:

$$a_n = \frac{2}{T}\int_0^T x(t) \cos n\omega_0 t \, dt \tag{2.6}$$

$$b_n = \frac{2}{T}\int_0^T x(t) \sin n\omega_0 t \, dt \tag{2.7}$$

$$c_n = \sqrt{a_n^2 + b_n^2}$$

$$\theta_n = \tan^{-1}\left(\frac{a_n}{b_n}\right) \tag{2.8}$$

It should be noted that a_n and b_n are calculated *average* values over the period of time T. Thus, the amplitude c_n and the phase θ_n of the nth harmonic are average values as well. A certain harmonic may be present only for part of the time T, but the computer analysis will return an average value assuming it is present over the entire time.

Figure 2.19 is presented to illustrate this assumption. It represents the Fourier reconstitution of the vertical trajectory of the heel of an adult walking his or her natural cadence. A total of nine harmonics are represented here because the addition of higher harmonics did not improve the curve of the original data. As can be seen, the harmonic reconstitution is visibly different from the original, sufficiently so as to cause reasonable errors in subsequent biomechanical analyses.

In spite of the negative comments about Fourier's reconstructions, there is considerable information in harmonic or Fourier analyses as to the bandwidth of the signal and the processing of that signal in both analog and digital systems. The first value of such analyses relates to the sampling theorem.

2.5.3 Sampling Theorem

Film and television are sampling processes. They capture the movement event for a short period of time, after which no further changes are recorded until the next field or frame. Playing a movie film back slowly demonstrates this

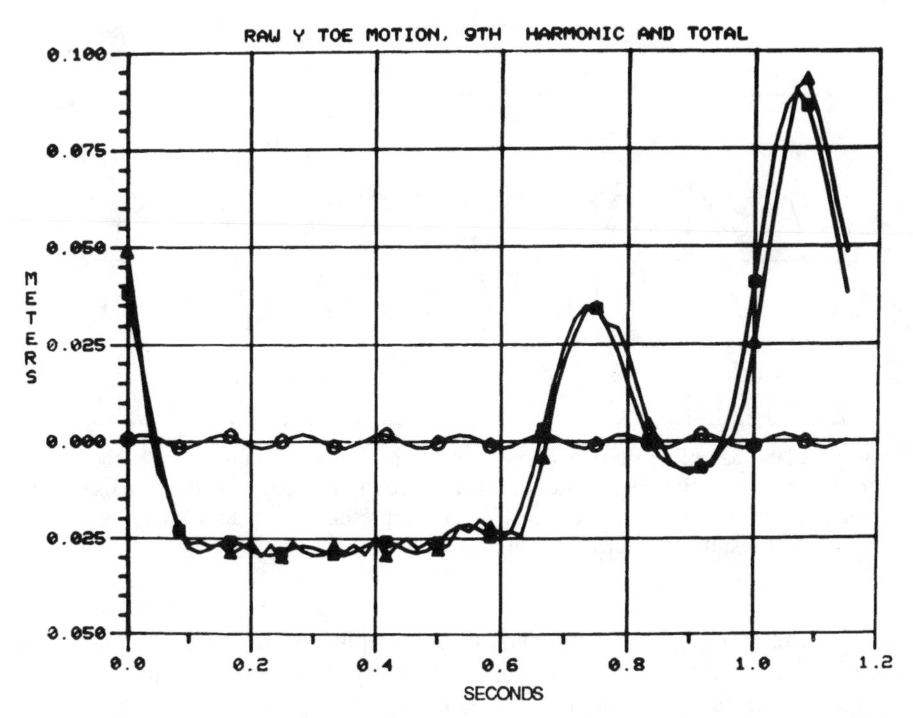

Figure 2.19 Fourier reconstruction of the vertical trajectory of a toe marker during one walking stride. The actual trajectory is shown by the open square, the reconstruction from the first 9 harmonics is plotted with open triangles, and the contribution of the 9th harmonic is plotted with open circles. The difference between the actual and the reconstructed waveforms is due to the lack of stationarity in the original signal.

phenomenon: the image jumps from one position to the next in a distinct step rather than a continuous process. The only reason film or television does not appear to jump at normal projection speeds (24 per second for film, 60 per second for television) is because the eye can retain an image for a period of about 1/15 s. The eye's short-term "memory" enables the human observer to average or smooth out the jumping movement.

In the processing of any time-varying data, no matter what their source, the sampling theorem must not be violated. Without going into the mathematics of the sampling process, the theorem states that "the process signal must be sampled at a frequency at least twice as high as the highest frequency present in the signal itself." If we sample a signal at too low a frequency, we get aliasing errors. This results in false frequencies, frequencies that were not present in the original signal, being generated in the sample data. Figure 2.20 illustrates this effect. Both signals are being sampled at the same interval T. Signal 1 is being sampled about 10 times per cycle, while signal 2 is being sampled less than twice per cycle. Note that the amplitudes of the samples

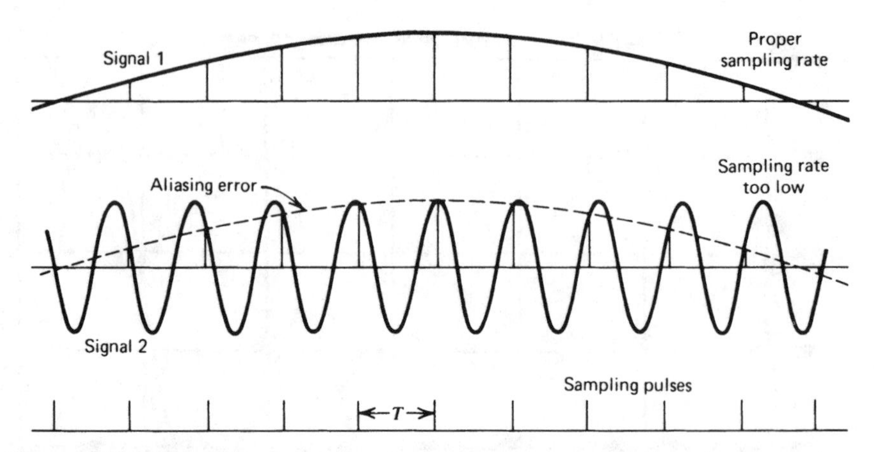

Figure 2.20 Sampling of two signals, one at a proper rate, the other at too low a rate. Signal 2 is sampled at a rate less than twice its frequency, such that its sampled amplitudes are the same as for signal 1. This represents a violation of the sampling theorem and results in an error called *aliasing.*

taken from signal 2 are identical to those sampled from signal 1. A false set of sampled data has been generated from signal 2 because the sample rate is too low—the sampling theorem has been violated.

The tendency of those using film is to play it safe and film at too high a rate. Usually there is a cost associated with such a decision. The initial cost is probably in the equipment required. A high-speed movie camera can cost four or five times as much as a standard model (24 frames per second). Or a special optoelectric system complete with the necessary computer can be a $70,000 decision. In addition to these capital costs, there are the higher operational costs of converting the data and running the necessary kinematic and kinetic computer programs. Except for higher speed running and athletic movements, it is quite adequate to use a standard movie or television camera. For normal and pathological gait studies, it has been shown that kinetic and energy analyses can be done with negligible error using a standard 24-frame per second movie camera (Winter, 1982). Figure 2.21 compares the results of kinematic analysis of the foot during normal walking, where a 50-Hz film rate was compared with 25 Hz. The data were collected at 50 Hz and the acceleration of the foot was calculated using every frame of data, then reanalyzed again, using every second frame of converted data. It can be seen that the difference between the curves is minimal; only at the peak negative acceleration was there a noticeable difference. The final decision as to whether this error is acceptable should not rest in this curve, but in your final goal. If, for example, the final anlaysis was a hip and knee torque analysis, the acceleration of the foot segment may not be too important, as is evident from another walking trial, shown in Figure 2.22. The minor differences in no way interfere with the general pattern of joint torques over the stride period, and

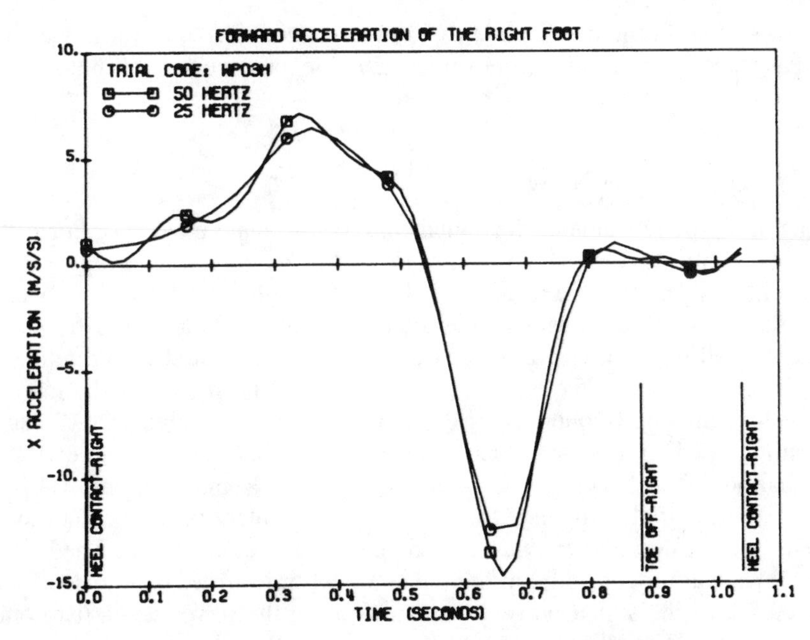

Figure 2.21 Comparison of the forward acceleration of the right foot during walking using the same data sampled at 50 Hz and at 25 Hz (using data from every second frame). The major pattern is maintained with minor errors at the peaks.

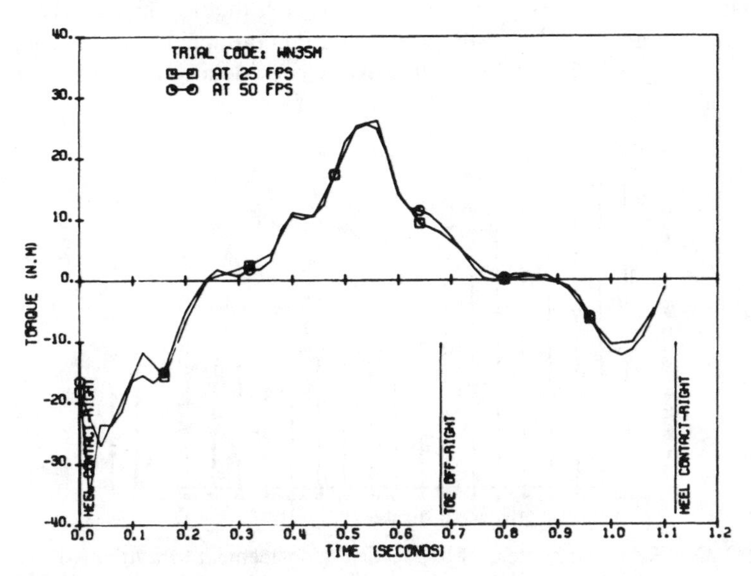

Figure 2.22 Comparison of the hip moment of force during level walking using the same data sampled at 50 Hz and at 25 Hz. The residual error is quite small because the joint reaction forces dominate the inertial contributions to the net moment of force.

the assessment of the motor patterns would be identical. Thus, for movements such as walking or for slower movements, an inexpensive camera at 24 frames per second appears to be quite adequate.

2.5.4 Signal Versus Noise

In the study of movement, the signal may be an anatomical coordinate that changes with time. For example, in running, the *Y* (vertical) coordinate of the heel will have certain frequencies that will be higher than those associated with the vertical coordinate of the knee or trunk. Similarly, the frequency content of all trajectories will decrease in walking compared with running. In repetitive movements, the frequencies present will be multiples (harmonics) of the fundamental frequency (stride frequency). When walking at 120 steps per minute (2 Hz), the stride frequency is 1 Hz. Therefore we can expect to find harmonics at 2 Hz, 3 Hz, 4 Hz, and so on. Normal walking has been analyzed by digital computer, and the harmonic content of the trajectories of seven leg and foot markers was determined (Winter et al., 1974). The highest harmonics were found to be in the toe and heel trajectories, and it was found that 99.7% of the signal power was contained in the lower seven harmonics (below 6 Hz). The harmonic analysis for the heel marker for 20 subjects is shown in Figure 2.23. Above the seventh harmonic, there was still some signal power, but it had the characteristics of "noise." Noise is the term used to describe components of the final signal that are not due to the process itself (in this case, walking). Sources of noise were noted in Section 2.5.1, and if the total effect of all these errors is random, then the true signal will have an added random component. Usually the random component is high frequency, as is borne out in Figure 2.23. Here we see evidence of higher frequency

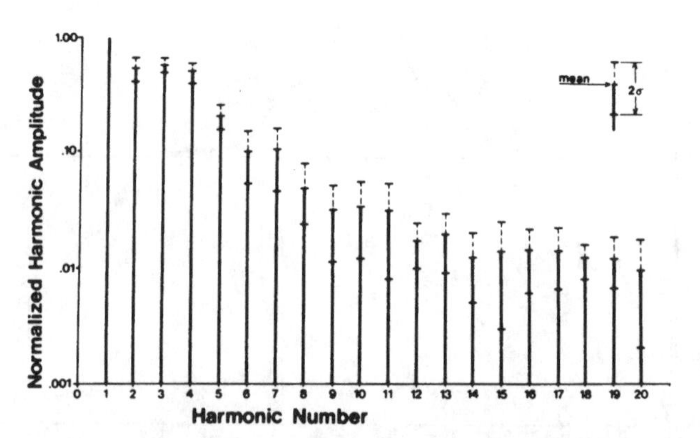

Figure 2.23 Harmonic content of the vertical displacement of a toe marker from 20 subjects during normal walking. Fundamental frequency (harmonic number = 1) is normalized at 1.00. Over 99% of power is contained below the 7th harmonic. (Reproduced by permission from the *Journal of Biomechanics*.)

components extending up to the twentieth harmonic, which was the highest frequency analyzed. The presence of the higher frequency noise is of considerable importance when we consider the problem of trying to calculate velocities and accelerations. Consider the process of time differentiation of a signal containing additive higher frequency noise. Suppose the signal can be represented by a summation of N harmonics:

$$x = \sum_{n=1}^{N} X_n \sin (n\omega_0 t + \theta_n) \tag{2.9}$$

where ω_0 = fundamental frequency
n = harmonic number
X_n = amplitude of nth harmonic
θ_n = phase of nth harmonic

To get the velocity in the x direction V_x, we differentiate with respect to time:

$$V_x = \frac{dx}{dt} = \sum_{n=1}^{N} n\omega_0 X_n \cos (n\omega_0 t + \theta_n) \tag{2.10}$$

Similarly, the acceleration A_x is

$$A_x \frac{dV_x}{dt} = -\sum_{n=1}^{N} (n\omega_0)^2 X_n \sin (n\omega_0 t + \theta_n) \tag{2.11}$$

Thus, the amplitude of each of the harmonics increases with its harmonic number; for velocities they increase linearly, and for accelerations the increase is proportional to the square of the harmonic number. This phenomenon is demonstrated in Figure 2.24 where the fundamental, second, and third harmonics are shown along with their first and second time derivatives. Assuming that the amplitude x of all three components is the same, we can see that the first derivative (velocity) of harmonics increases linearly with increasing frequency. The first derivative of the third harmonic is now three times that of the fundamental. For the second time derivative, the increase repeats itself, and the third harmonic acceleration is now nine times that of the fundamental.

In the trajectory data for gait, x_1 might be 5 cm and x_{20} = 0.5 mm. The twentieth harmonic noise is hardly perceptible in the displacement plot. In the velocity calculation, the twentieth harmonic increases 20-fold so that it is now one-fifth that of the fundamental. In the acceleration calculation, the twentieth harmonic increases another factor of 20 and now is four times the magnitude of the fundamental. This effect is shown in Figure 2.25 which plots the acceleration of the toe during walking. The random-looking signal is the raw data differentiated twice. The smooth signal is the acceleration

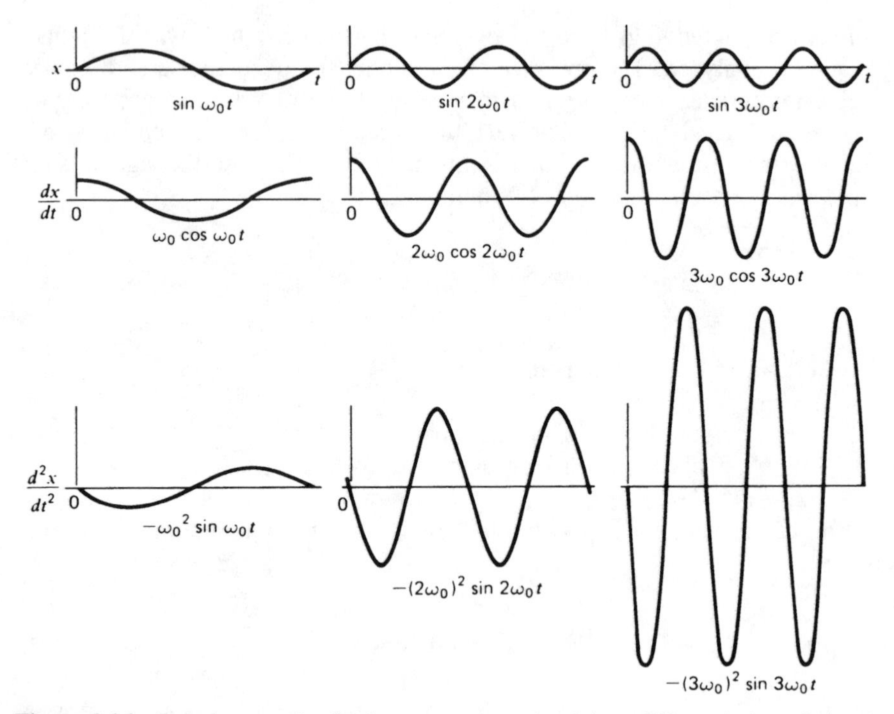

Figure 2.24 Relative amplitude changes as a result of time differentiation of signals of increasing frequency. First derivative increases amplitude proportional to frequency; second derivative increases amplitude proportional to frequency squared. Such a rapid increase has severe implications in calculating accelerations when the original displacement signal has high-frequency noise present.

calculated after most of the higher frequency noise has been removed. Techniques to accomplish this are now discussed.

2.5.5 Smoothing and Fitting of Data

The removal of noise can be accomplished in several ways. The aims of each technique are basically the same. However, the results differ somewhat.

2.5.5.1 Curve-Fitting Techniques. The basic assumption here is that the trajectory signal has a predetermined shape and that by fitting the assumed shape to a "best fit" with the raw data, a smooth signal will result. For example, it may be assumed that the data are a certain order polynomial:

$$x(t) = a_0 + a_1 t + a_2 t^2 + a_3 t^3 + \cdots + a_n t^n \tag{2.12}$$

By computer techniques, the coefficients $a_0, \cdots, a_n$ can be selected to give a best fit using such criteria as minimum mean square error. The complexity

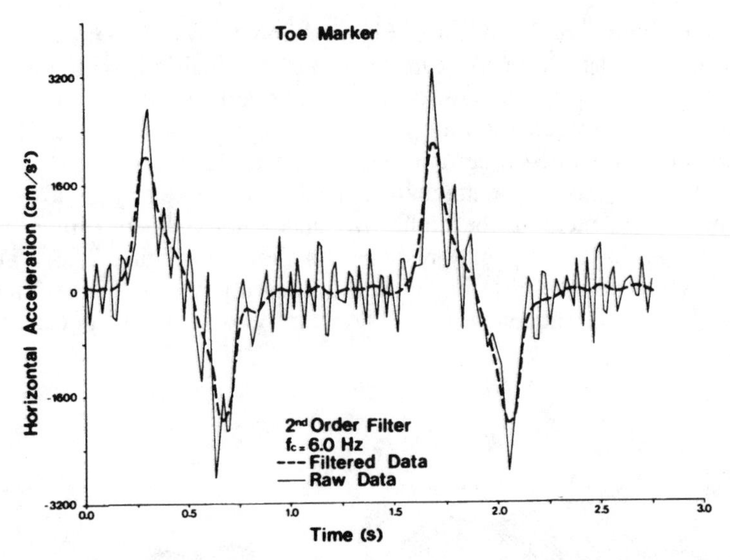

Figure 2.25 Horizontal acceleration of the toe marker during normal walking as calculated from displacement data from television. Solid line is acceleration based on unprocessed "raw" data; dotted line is that calculated after data have been filtered twice with a second-order low-pass filter. (Reproduced by permission from the *Journal of Biomechanics.*)

of the curve fitting can be quite restrictive in terms of computer time. A similar fit can be made assuming a certain number of harmonics. Reconstituting the final signal as a sum of N lowest harmonics,

$$x(t) = a_0 + \sum_{n=1}^{N} a_n \sin(n\omega_0 t + \theta_n) \tag{2.13}$$

This model has a better basis, especially in repetitive movement, while the polynomial may be better in certain nonrepetitive movement such as broad jumping. However, there are severe assumptions regarding the consistency (stationarity) of a_n and θ_n, as was demonstrated in Figure 2.19. A third technique, spline curve fitting, is a modification of the polynomial technique. The curve to be fitted is broken into sections, each section starting and ending with an inflection point, with special fitting being done between adjacent sections. The major problem with this technique, other than computer time, is the error introduced by improper selection of the inflection points. These inflection points must be determined from the noisy data and, as such, are strongly influenced by the very noise that we are trying to eliminate.

2.5.5.2 Digital Filtering. Technological advances in digital filtering have opened up a much more promising and less restrictive solution to the noise reduction. The basic approach can be described by analyzing the frequency spectrum of both signal and noise. Figure 2.26*a* shows a schematic plot of a signal and noise spectrum. As can be seen, the signal is assumed to occupy the lower end of the frequency spectrum and overlaps with the noise, which is usually higher frequency. Filtering of any signal is aimed at the selective rejection, or attenuation, of certain frequencies. In the above case, the obvious filter is one that passes, unattenuated, the lower frequency signals while at the same time attenuating the higher frequency noise. Such a filter, called a *low-pass filter,* has a frequency response as shown in Figure 2.26*b*. The frequency response of the filter is the ratio of the output $X_o(f)$ of the filter to its input $X_i(f)$ at each frequency present. As can be seen, the response at lower

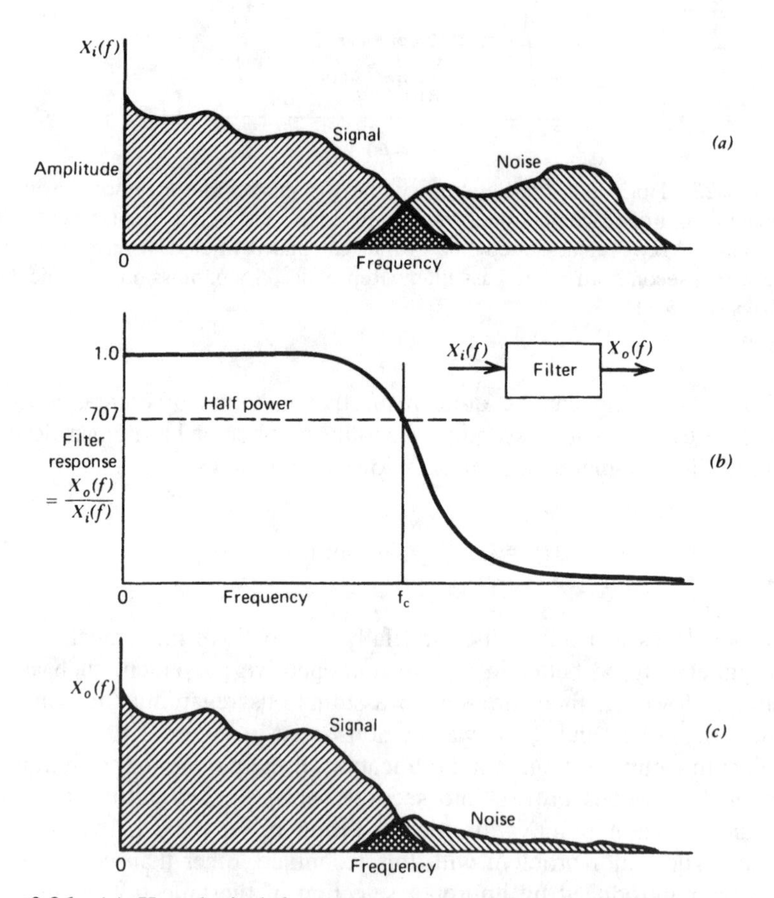

Figure 2.26 (*a*) Hypothetical frequency spectrum of a waveform consisting of a desired signal and unwanted higher frequency noise. (*b*) Response of low-pass filter $X_o(f)/X_i(f)$, introduced to attenuate the noise. (*c*) Spectrum of the output waveform, obtained by multiplying the amplitude of the input by the filter response at each frequency. Higher frequency noise is severely attenuated, while the signal is passed with only minor distortion in transition region around f_c.

frequencies is 1.0. This means that the input signal passes through the filter unattenuated. However, there is a sharp transition at the cutoff frequency f_c so that the signals above f_c are severely attenuated. The net result of the filtering process can be seen by plotting the spectrum of the output signal $X_o(f)$ as seen in Figure 2.26*c*. Two things should be noted. First, the higher frequency noise has been severely reduced but not completely rejected. Second, the signal, especially in the region where the signal and noise overlap (usually around f_c) is also slightly attenuated. This results in a slight distortion of the signal. Thus, a compromise has to be made in the selection of the cutoff frequency. If f_c is set too high, less signal distortion occurs, but too much noise is allowed to pass. Conversely, if f_c is too low, the noise is reduced drastically, but at the expense of increased signal distortion. A sharper cutoff filter will improve matters, but at an additional expense. In digital filtering, this means a more complex digital filter and, thus, more computer time.

The theory behind digital filtering (Radar and Gold, 1967) will not be covered, but the application of low-pass digital filtering to kinematic data processing will be described in detail. First, it must be assessed what the signal spectrum is as opposed to the noise spectrum. This can readily be done, as is seen in the harmonic analysis presented previously in Figure 2.23. As a result of the previous discussion for these data on walking, the cutoff frequency of a digital filter should be set at about 6 Hz. The format of a recursive digital filter that processes the raw data in time domain is as follows:

$$\begin{aligned} X^1(nT) &= a_0X(nT) + a_1X(nT - T) + a_2X(nT - 2T) \\ &\quad + b_1X^1(nT - T) + b_2X^1(nT - 2T) \end{aligned} \tag{2.14}$$

where X^1 = filtered output coordinates
X = unfiltered coordinate data
nT = nth sample
$(nT - T)$ = $(n - 1)$th sample
$(nT - 2T)$ = $(n - 2)$th sample
$a_0, \cdots, b_2, \cdots$ = filter coefficients

These filter coefficients are constants that depend on the type and order of the filter, the sampling frequency, and the cutoff frequency. As can be seen, the filter output $X^1(nT)$ is a weighted version of the immediate and past raw data plus a weighted contribution of past filtered output.

The order of the filter decides the sharpness of the cutoff. The higher the order, the sharper the cutoff, but the larger the number of coefficients. For example, a Butterworth-type low-pass filter of second order is to be designed to cut off at 6 Hz using film data taken at 60 Hz (60 frames per second). All that is required to determine these coefficients is the ratio of sampling frequency to cutoff frequency. In this case it is 10. The design of such a filter would yield the following coefficients:

$$a_0 = 0.06746, \quad a_1 = 0.13491, \quad a_2 = 0.06746,$$

$$b_1 = 1.14298, \quad b_2 = -0.41280$$

Note that the algebraic sum of all the coefficients equals 1.0000. This gives a response of unity over the passband. Tables 2.1 and 2.2 list the coefficients for use in a critically damped filter and a second-order Butterworth filter for various values of f_s/f_c. Note that the same filter coefficients could be used in many different applications, as long as the ratio f_s/f_c is the same. For example, an EMG signal sampled at 2000 Hz with cutoff desired at 400 Hz would have the same coefficients as one employed for movie film coordinates where the film rate was 30 Hz and cutoff was 6 Hz.

If more exact cutoff frequencies are required, the exact equations to calculate the coefficients for a Butterworth or a critically damped filter are as follows:

TABLE 2.1 Coefficients for Critically Damped Low-Pass Filter

f_s/f_c	a_0	a_1	a_2	b_1	b_2
4	0.25000	0.50000	0.25000	0.00000	0.00000
5	0.17708	0.35416	0.17708	0.31677	−0.02509
6	0.13397	0.26795	0.13397	0.53590	−0.07180
7	0.10565	0.21130	0.10565	0.69983	−0.12244
8	0.08579	0.17157	0.08579	0.82843	−0.17157
9	0.07121	0.14241	0.07121	0.93262	−0.21744
10	0.06014	0.12028	0.06014	1.01905	−0.25962
11	0.05152	0.10304	0.05152	1.09208	−0.29816
12	0.04466	0.08932	0.04466	1.15470	−0.33333
13	0.03910	0.07820	0.03910	1.20904	−0.36545
14	0.03453	0.06906	0.03453	1.25668	−0.39481
15	0.03073	0.06146	0.03073	1.29882	−0.42173
16	0.02753	0.05505	0.02753	1.33636	−0.44646
17	0.02480	0.04961	0.02480	1.37003	−0.46925
18	0.02247	0.04494	0.02247	1.40042	−0.49029
19	0.02045	0.04090	0.02045	1.42797	−0.50978
20	0.01869	0.03739	0.01869	1.45309	−0.52786
21	0.01716	0.03431	0.01716	1.47607	−0.54469
22	0.01580	0.03160	0.01580	1.49718	−0.56039
23	0.01460	0.02920	0.01460	1.51665	−0.57506
24	0.01353	0.02707	0.01353	1.53465	−0.58879
25	0.01258	0.02516	0.01258	1.55136	−0.60168

TABLE 2.2 Coefficients for Butterworth Low-Pass Filter

f_s/f_c	a_0	a_1	a_2	b_1	b_2
4	0.29289	0.58579	0.29289	0.00000	−0.17157
5	0.20657	0.41314	0.20657	0.36953	−0.19582
6	0.15505	0.31010	0.15505	0.62021	−0.24041
7	0.12123	0.24247	0.12123	0.80303	−0.28796
8	0.09763	0.19526	0.09763	0.94281	−0.33333
9	0.08042	0.16085	0.08042	1.05333	−0.37502
10	0.6746	0.13491	0.06746	1.14298	−0.41280
11	0.05742	0.11484	0.05742	1.21719	−0.44687
12	0.04949	0.09898	0.04949	1.27963	−0.47759
13	0.04311	0.08621	0.04311	1.33291	−0.50533
14	0.03789	0.07578	0.03789	1.37889	−0.53045
15	0.03357	0.06714	0.03357	1.41898	−0.55327
16	0.02995	0.05991	0.02995	1.45424	−0.57406
17	0.02689	0.05379	0.02689	1.48550	−0.59307
18	0.02428	0.04856	0.02428	1.51338	−0.61051
19	0.02203	0.04407	0.02203	1.53842	−0.62655
20	0.02008	0.04017	0.02008	1.56102	−0.64135
21	0.01838	0.03676	0.01838	1.58152	−0.65505
22	0.01689	0.03378	0.01689	1.60020	−0.66776
23	0.01557	0.03114	0.01557	1.61730	−0.67958
24	0.01440	0.02880	0.01440	1.63299	−0.69060
25	0.01336	0.02672	0.01336	1.64746	−0.70090

$$\omega_c = \frac{(\tan(\pi f_c/f_s))}{C} \tag{2.15}$$

where C is the correction factor for number of passes required, to be explained shortly. For a single-pass filter with the coefficients shown in Tables 2.1 and 2.2, $C = 1$.

$$K = \sqrt{2}\omega_c \text{ for a Butterworth filter,}$$
$$\text{or, } 2\omega_c \text{ for a critically damped filter}$$

$$K_2 = \omega_c^2, \qquad a_0 = \frac{K_2}{(1 + K_1 + K_2)}, \qquad a_1 = 2a_0, \qquad a_2 = a_0$$

$$K_3 = \frac{2a_0}{K_2}, \qquad b_1 = -2a_0 + K_3$$

$$b_2 = 1 - 2a_0 - K_3, \qquad \text{or} \qquad b_2 = 1 - a_0 - a_1 - a_2 - b_1$$

As well as attenuating the signal, there is a phase shift of the output signal relative to the input. For this second-order filter there is a 90° phase lag at the cutoff frequency. This will cause a second form of distortion, called *phase distortion,* to the higher harmonics within the bandpass region. Even more phase distortion will occur to those harmonics above f_c, but these components are mainly noise, and they are being severely attenuated. This phase distortion may be more serious than the amplitude distortion that occurs to the signal in the transition region. To cancel out this phase lag, the once-filtered data can be filtered again, but this time in the reverse direction of time (Winter et al., 1974). This introduces an equal and opposite phase lead so that the net phase shift is zero. Also, the cutoff of the filter will be twice as sharp as that for single filtering. In effect, by this second filtering in the reverse direction we have created a fourth-order zero-phase-shift filter, which yields a filtered signal that is back in phase with the raw data, but with most of the noise removed.

In Figure 2.27 we see the frequency response of a second-order Butterworth filter normalized with respect to the cutoff frequency. Superimposed on this curve is the response of the fourth-order zero-phase-shift filter. Thus, the new cutoff frequency is lower than that of the original single-pass filter; in this case, it is about 80% of the original. The correction factor for each additional pass of a Butterworth filter is $C = (2^{1/n} - 1)^{0.25}$, where n is the number of passes. Thus, for a dual pass, $C = 0.802$. For a critically damped filter, $C = (2^{1/2n} - 1)^{0.5}$; thus, for a dual pass, $C = 0.435$. This correction factor is applied to Equation (2.15) and results in the cutoff frequency for the original single-pass filter being set higher, so that after the second pass the

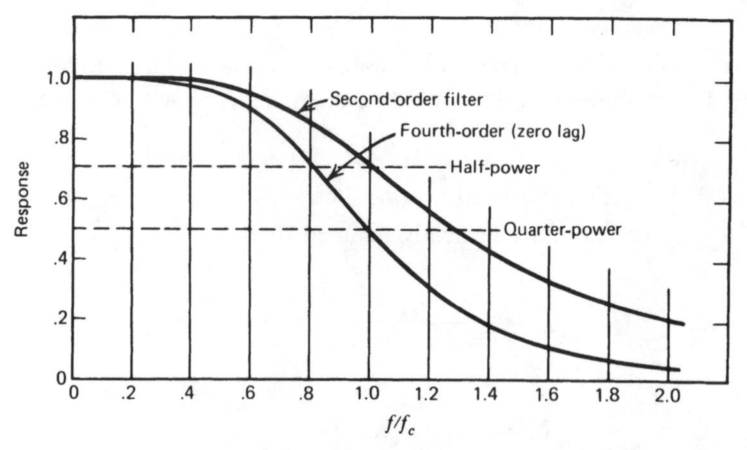

Figure 2.27 Response of a second-order low-pass digital filter. Curve is normalized at 1.0 at cutoff frequency f_c. Because of phase-lag characteristics of the filter, a second filtering is done in the reverse direction of time, which results in a fourth-order zero-lag filter.

desired cutoff frequency is achieved. The major difference between these two filters is a compromise in the response in the time domain. Butterworth filters have a slight overshoot in response to step or impulse type inputs, but they have a much shorter rise time. Critically damped filters have no overshoot, but suffer from a slower rise time. Because impulsive type inputs are rarely seen in human movement data, the Butterworth filter is preferred.

The application of one of these filters in smoothing raw coordinate data can now be seen by examining the data that yielded the harmonic plot of Figure 2.23. The horizontal acceleration of this toe marker, as calculated by finite differences from the filtered data, is plotted in Figure 2.25. Note how repetitive the filtered acceleration is and how it passes through the "middle" of the noisy curve, as calculated using the unfiltered data. Also, note that there is no phase lag in these filtered data because of the dual forward and reverse filtering processes.

2.5.5.3 Choice of Cutoff Frequency—Residual Analysis. There are several ways to choose the best cutoff frequency. The first is to carry out a harmonic analysis as depicted in Figure 2.23. By analyzing the power in each of the components, a decision can be made as to how much power to accept and how much to reject. However, such a decision assumes that the filter is ideal and has an infinitely sharp cutoff. A better method is to do a residual analysis of the difference between filtered and unfiltered signals over a wide range of cutoff frequencies. In this way, the characteristics of the filter in the transition region are reflected in the decision process. Figure 2.28 shows a theoretical

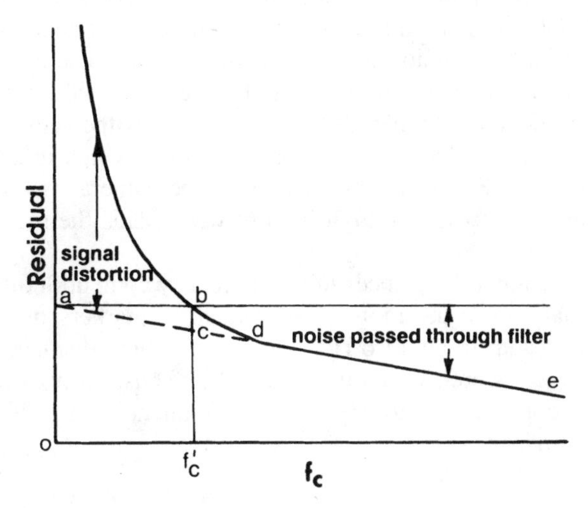

Figure 2.28 Plot of the residual between a filtered and an unfiltered signal as a function of the filter cutoff frequency. See text for the interpretation as to where to set the cutoff frequency of the filter.

plot of residual versus frequency. The residual at any cutoff frequency is calculated as follows for a signal of N sample points in time:

$$R(f_c) = \sqrt{\frac{1}{N}\sum_{i=1}^{N}(X_i - \hat{X}_i)^2} \tag{2.16}$$

where X_i = raw data at ith sample
$\hat{X}_i$ = filtered data at the ith sample

If our data contained no signal, just random noise, the residual plot would be a straight line decreasing from an intercept at 0 Hz to an intercept on the abscissa at the Nyquist frequency ($0.5\ f_s$). The line *de* represents our best estimate of that noise residual. The intercept a on the ordinate (at 0 Hz) is nothing more than the rms value of the noise, because $\hat{X}_i$ for a 0-Hz filter is nothing more than the mean of the noise over the N samples. When the data consist of true signal plus noise, the residual will be seen to rise above the straight (dashed) line as the cutoff frequency is reduced. This rise above the dashed line represents the signal distortion that is taking place as the cutoff is reduced more and more.

The final decision is where f_c should be chosen. The compromise is always a balance between the amount of signal distortion versus the amount of noise allowed through. If we decide that both should be equal, then we simply project a line horizontally from a to intersect the residual line at b. The frequency chosen is f_c^1, and at this frequency the signal distortion is represented by *bc*. This is also an estimate of the noise that is passed through the filter. Figure 2.29 is a plot of the residual of four markers from one stride of gait data, and both vertical and horizontal coordinates were analyzed. As can be seen, the straight regression line that represents the noise is essentially the same for both coordinates on all markers. This tells us that the noise content, mainly introduced by the human digitizing process, is the same for all markers. This regression line has an intercept of 1.8 mm, which indicates that the rms of the noise is 1.8 mm. In this case, the cine camera was 5 m from the subject and the image was 2 m high by 3 m wide. Thus the rms noise is less than one part in 1000.

Also, we see distinct differences in the frequency content of different markers. The residual shows the more rapidly moving markers on the heel and ball to have power up to about 6 Hz, while the vertical displacements of the rib and hip markers were limited to about 3 Hz. Thus, through this selection technique, we could have different cutoff frequencies specified for each marker displacement.

2.5.6 Comparison of Some Smoothing Techniques

It is valuable to see the effect of several different curve-fitting techniques on the same set of noisy data. The following summary of a validation experiment,

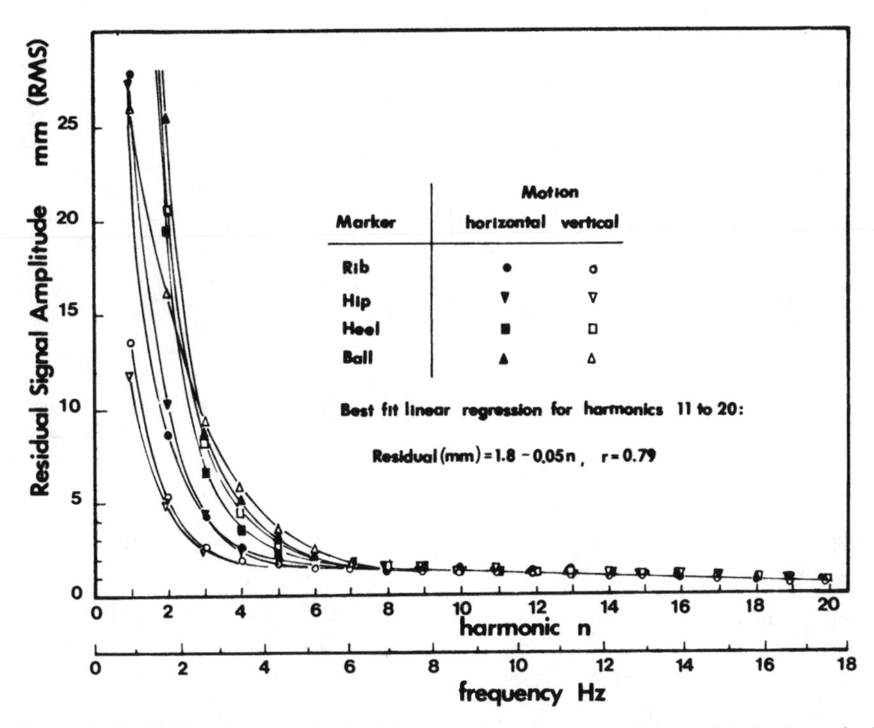

Figure 2.29 Plot of the residual of four markers from a walking trial; both vertical and horizontal displacement data. Data were digitized from movie film with the camera 5 m from the subject.

which was conducted to compare (Pezzack et al., 1977) three commonly used techniques, illustrates the wide differences in the calculated accelerations.

Data obtained from the horizontal movement of a lever arm about a vertical axis were recorded three different ways. A goniometer on the axis recorded angular position, an accelerometer mounted at the end of the arm gave tangential acceleration and thus angular acceleration, and cinefilm data gave image information that could be compared with the angular and acceleration records. The comparisons are given in Figure 2.30. Figure 2.30*a* compares the angular position of the lever arm as it was manually moved from rest through about 130° and back to the original position. The goniometer signal and the lever angle as analyzed from the film data are plotted and compare closely. The only difference is that the goniometer record is somewhat noisy compared with the film data.

Figure 2.30*b* compares the directly recorded angular acceleration, which can be calculated by dividing the tangential acceleration by the radius of the accelerometer from the center of rotation, with the angular acceleration as calculated via the second derivative of the digitally filtered coordinate data (Winter et al., 1974). The two curves match extremely well, and the finite-difference acceleration exhibits less noise than the directly recorded acceler-

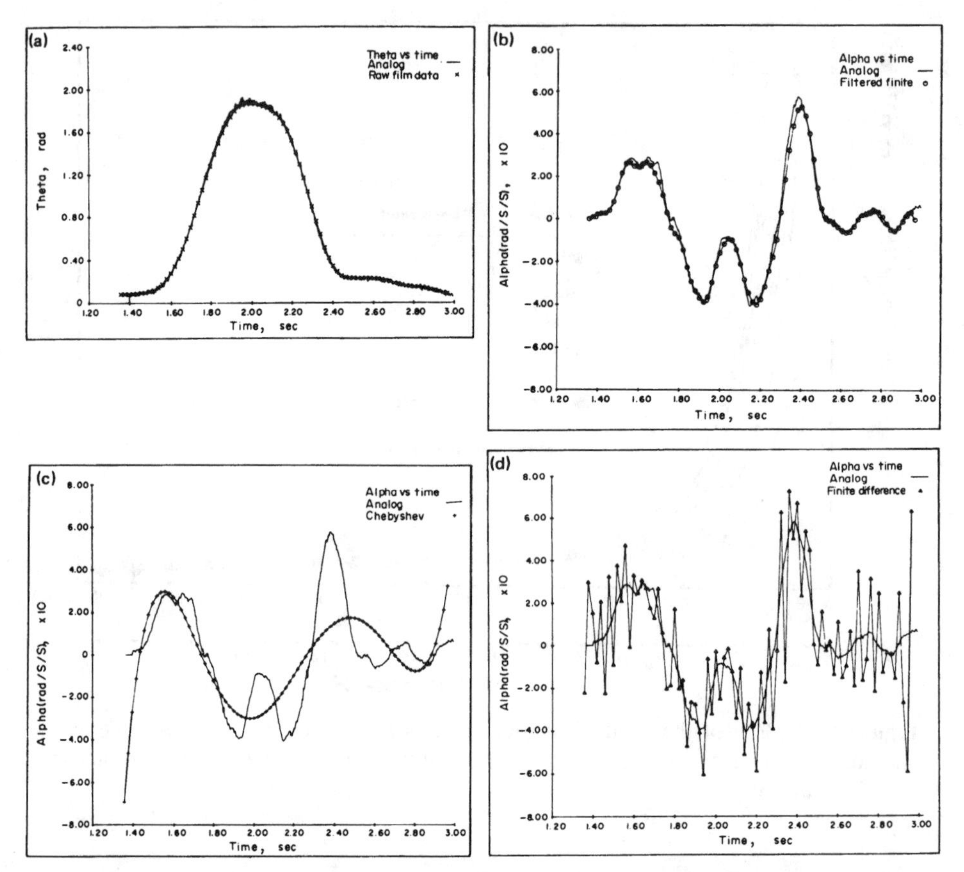

Figure 2.30 Comparison of several techniques used to determine the acceleration of a movement based on film displacement data. (*a*) Displacement angle of a simple extension/flexion as plotted from film and goniometer data. (*b*) Acceleration of movement in (*a*) as measured by accelerometer, calculated from film coordinates after digital filtering. (*c*) Acceleration as determined from a 9th-order polynomial fit of the displacement data compared with directly recorded acceleration. (*d*) Acceleration as determined by finite-difference techniques compared with measured curve. (Reproduced by permission from the *Journal of Biomechanics.*)

ation. Figure 2.30*c* compares the directly recorded acceleration with the calculated angular acceleration using a polynomial fit on the raw angular data. A ninth-order polynomial was fitted to the angular displacement curve to yield the following fit:

$$\theta(t) = 0.064 + 2.0t - 35t^2 + 210t^3 - 430t^4 + 400t^5 - 170t^6 + 25t^7 + 2.2t^8 - 0.41t^9 \quad \text{rad} \tag{2.17}$$

Note that θ is in radians and t in seconds. To get the curve for angular acceleration, all we need to do is take the second time derivative to yield

$$\alpha(t) = -70 + 1260t - 5160t^2 + 8000t^3 - 5100t^4 + 1050t^5 + 123t^6 - 29.5t^7 \quad \text{rad/s}^2 \tag{2.18}$$

This acceleration curve, compared with the accelerometer signal, shows considerable discrepancy, enough to cast doubt on the value of the polynomial fit technique. The polynomial is fitted to the displacement data in order to get an analytic curve, which can be differentiated to yield another smooth curve. Unfortunately, it appears that a considerably higher order polynomial would be required to achieve even a crude fit, and the computer time might become too prohibitive.

Finally, in Figure 2.30*d* we see the accelerometer signal plotted against angular acceleration as calculated by second-order finite-difference techniques. The plot speaks for itself—the accelerations are too noisy to mean anything.

2.6 CALCULATION OF ANGLES FROM SMOOTHED DATA

2.6.1 Limb-Segment Angles

Given the coordinate data from anatomical markers at either end of a limb segment, it is an easy step to calculate the absolute angle of that segment in space. It is not necessary that the two markers be at the extreme ends of the limb segment, so long as they are in line with the long-bone axis. Figure 2.31 shows the outline of a leg with seven anatomical markers in a four-segment three-joint system. Markers 1 and 2 define the thigh in the sagittal plane. Note that by convention all angles are measured in a counter clockwise direction starting with the horizontal equal to 0°. Thus θ_{43} is the angle of the leg in space and can be calculated from

$$\theta_{43} = \arctan \frac{y_3 - y_4}{x_3 - x_4} \tag{2.19}$$

or, in more general notation,

$$\theta_{ij} = \arctan \frac{y_j - y_i}{x_j - x_i} \tag{2.20}$$

As has already been noted, these segment angles are absolute in the defined spatial reference system. It is therefore quite easy to calculate the joint angles from the angles of the two adjacent segments.

2.6.2 Joint Angles

Each joint has a convention for describing its magnitude and polarity. For example, when the knee is fully extended, it is described as 0° flexion, and

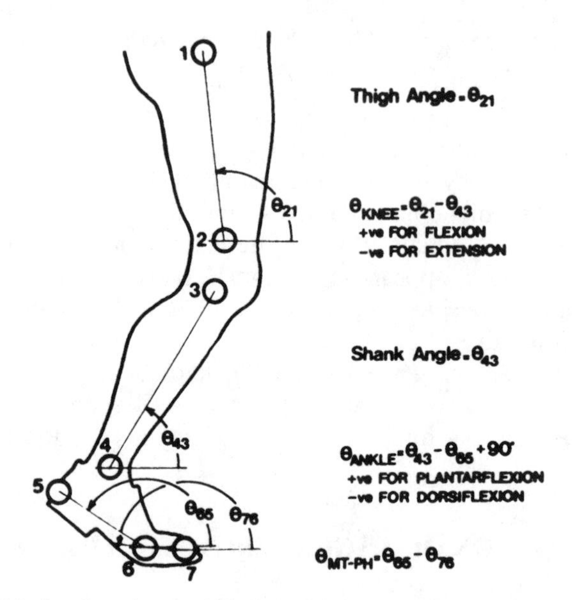

Figure 2.31 Marker location and limb and joint angles as defined using an established convention. Limb angles in the spatial reference system are defined using counterclockwise from the horizontal as positive. Thus, angular velocities and accelerations are also positive in a counterclockwise direction in the plane of movement, which is essential for consistent use in subsequent kinetic analysis. Convention for joint angles (which are relative) is subject to wide variations among researchers, but the convention must be clarified.

when the leg moves in a posterior direction relative to the thigh, the knee is said to be in flexion. In terms of the absolute angles described previously,

$$\text{knee angle} = \theta_k = \theta_{21} - \theta_{43}$$

If $\theta_{21} > \theta_{43}$, the knee is flexed; if $\theta_{21} < \theta_{43}$, the knee is extended.

The convention for the ankle is slightly different in that 90° between the leg and the foot is boundary between plantarflexion and dorsiflexion. Therefore,

$$\text{ankle angle} = \theta_a = \theta_{43} - \theta_{65} + 90°$$

If θ_a is positive, the foot is plantarflexed; if θ_a is negative, the foot is dorsiflexed.

2.7 CALCULATION OF VELOCITY AND ACCELERATION

2.7.1 Velocity Calculation

As was seen in Section 2.5.4, there can be severe problems associated with the determination of velocity and acceleration information. For the reasons outlined, we will assume that the raw displacement data have been suitably smoothed by digital filtering and we have a set of smoothed coordinates and angles to operate upon. To calculate the velocity from displacement data, all that is needed is to take the finite difference. For example, to determine the velocity in the x direction, we calculate $\Delta x/\Delta t$, where $\Delta x = x_{i+1} - x_i$, and Δt is the time between adjacent samples x_{i+1} and x_i.

The velocity calculated this way does not represent the velocity at either of the sample times. Rather, it represents the velocity of a point in time halfway between the two samples. This can result in errors later on when we try to relate the velocity-derived information to displacement data, and both results do not occur at the same point in time. A way around this problem is to calculate the velocity and accelerations on the basis of $2\Delta t$ rather than Δt. Thus, the velocity at the ith sample is

$$Vx_i = \frac{x_{i+1} - x_{i-1}}{2\Delta t} \tag{2.21}$$

Note that the velocity is at a point halfway between the two samples, as depicted in Figure 2.32. The assumption is that the line joining x_{i-1} to x_{i+1} has the same slope as the line drawn tangent to the curve at x_i.

2.7.2 Acceleration Calculation

Similarly, the acceleration is

$$Ax_i = \frac{Vx_{i+1} - Vx_{i-1}}{2\Delta t} \tag{2.22}$$

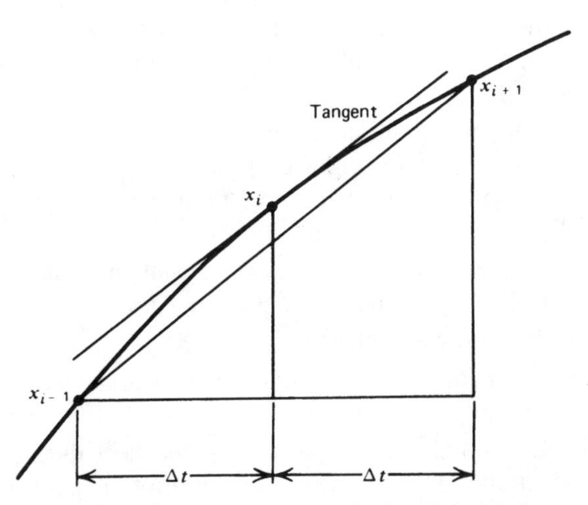

Figure 2.32 Finite-difference techniques for calculated slope of curve in the ith sample point.

Note that Equation (2.22) requires displacement data from samples $i + 2$ and $i - 2$; thus, a total of five successive data points go into the acceleration. An alternative and slightly better calculation of acceleration uses only three successive data coordinates and utilizes the calculated velocities halfway between sample times:

$$Vx_{i+1/2} = \frac{x_{i+1} - x_i}{\Delta t} \tag{2.23a}$$

$$Vx_{i-1/2} = \frac{x_i - x_{i-1}}{\Delta t} \tag{2.23b}$$

Therefore,

$$Ax_i = \frac{x_{i+1} - 2x_i + x_{i-1}}{\Delta t^2} \tag{2.23c}$$

2.8 PROBLEMS BASED ON KINEMATIC DATA

1. From Tables A.1 and A.2 in Appendix A, plot the vertical displacement of the raw and filtered data (in centimeters) for the greater trochanter (hip) marker for frames 1 to 30. Use a vertical scale as large as possible so as to identify the noise content of the raw data. In a few lines, describe the results of the smoothing by the digital filter.
2. Using filtered coordinate data (Table A.2), plot the vertical displacement of the heel marker from TOR (frame 1) to the next TOR (frame 70).
   (a) Estimate the instant of heel-off during midstance. (*Hint:* Consider the elastic compression and release of the shoe material when arriving at your answer.)
   (b) Determine the maximum height of the heel above ground level during swing. When does this occur during the swing phase? (*Hint:* Consider the lowest displacement of the heel marker during stance as an indication of ground level.)
   (c) Describe the vertical heel trajectory during the latter half of swing (frames 14–27), especially the four frames immediately prior to HRC.
   (d) Calculate the vertical heel velocity at HRC.
   (e) Calculate from the horizontal displacement data the horizontal heel velocity at HCR.
   (f) From the horizontal coordinate data of the heel during the first foot flat period (frames 35–40) and the second foot flat period (frames 102–106), estimate the stride length.

(g) If one stride period is 69 frames, estimate the forward velocity of this subject.

3. Plot the trajectory of the trunk marker (rib cage) over one stride (frames 28–97).
   (a) Is the shape of this trajectory what you would expect in walking?
   (b) Is there any evidence of conservation of mechanical energy over the stride period? (That is, is potential energy being converted to kinetic energy and vice versa?)

4. Determine the vertical displacement of the toe marker when it reaches its lowest point in late stance and compare that with the lowest point during swing, and thereby determine how much toe clearance took place.
   Answer: y_{toe} (fr.13) = 0.0485 m, y_{toe} (fr.66) = 0.0333 m, clearance = 0.152 m = 1.52 cm.

5. From the filtered coordinate data (Table A.2), calculate the following and check your answer with that listed in the appropriate listings (Tables A.2, A.3, and A.4).
   (a) The velocity of the knee in the X direction for frame 10.
   (b) The acceleration of the knee in the X direction for frame 10.
   (c) The angle of the thigh and leg in the spatial reference system for frame 30.
   (d) From (c) calculate the knee angle for frame 30.
   (e) The absolute angular velocity of the leg for frame 30 (use angular data, Table A.3).
   (f) Using the tabulated vertical velocities of the toe, calculate its vertical acceleration for frames 25 and 33.

6. From the filtered coordinate data in Table A.2, calculate the following and check your answer from the results tabulated in Table A.3.
   (a) The center of mass of the foot segment for frame 80.
   (b) The velocity of the center of mass of the leg for frame 70. Give the answer in both coordinate and polar form.

2.9 REFERENCES

Dinn, D. F., D. A. Winter, and B. G. Trenholm. "CINTEL-Computer Interface for Television," *IEEE Trans. Computers* **C-19**:1091–1095, 1970.

Eberhart, H. D. and V. T. Inman. "An Evaluation of Experimental Procedures Used in a Fundamental Study of Human Locomotion," *Ann. NY Acad. Sci.* **5**:1213–1228, 1951.

Ferrigno, G. and A. Pedotti. "ELITE: A Digital Dedicated Hardware System for Movement Analysis Via Real-Time TV Signal Processing," *IEEE Trans. Biomed. Eng.* **32**:943–950, 1985.

Finley, F. R. and P. V. Karpovich. "Electrogoniometric Analysis of Normal and Pathological Gaits," *Res. Quart.* **35**:379–384, 1964.

Furnée, E. H. 1967. *See* Woltring, 1987.

Gage, W. G., D. A. Winter, J. S. Frank, and A. L. Adkin. "Kinematic and Kinetic Validation of Inverted Pendulum Model in Quiet Standing," *Gait and Posture,* **19**: 124–132, 2004.

Jarett, M. O., B. J. Andrews, and J. P. Paul. "A Television/Computer System for the Analysis of Human Locomotion," *Proc. IERE Conf. on Applications of Electronics in Medicine,* Southhampton, England, 1976.

Macleod, A., J. R. W. Morris, and M. Lyster. "Close-Range Photogrammetry Meets Machine Vision," *SPIE Vol. 1395,* A. Gruen and E. Baltsavias, Eds. 1990, pp. 12–17. Bellingham, WA.

Moore, A., R. Wells, and D. Ranney. "Quantifying Exposure in Occupational Manual Tasks with Cumulative Trauma Disorder Potential," *Ergonomics* **34**:1433–1453, 1991.

Morris, J. R. W. "Accelerometry—A Technique for the Measurement of Human Body Movements," *J. Biomech.* **6**:729–736, 1973.

Pezzack, J. C., R. W. Norman, and D. A. Winter. "An Assessment of Derivative Determining Techniques Used for Motion Analysis," *J. Biomech.* **10**:377–382, 1977.

Radar, C. M. and B. Gold. "Digital Filtering Design Techniques in the Frequency Domain," *Proc. IEEE* **55**:149–171, 1967.

Trenholm, B. G., D. A. Winter, D. Mymin, and E. L. Lansdown, "Computer Determination of Left Ventricular Volume Using Videodensitometry," *Med. & Biol. Eng.* **10**:163–173, 1972.

Winter, D. A. "Camera Speeds for Normal and Pathological Gait Analyses," *Med. Biol. Eng. Comput.* **20**:408–412, 1982.

Winter, D. A., S. A. Malcolm, and B. G. Trenholm, "System for Real-Time Conversion of Video Images of Physiological Events," *Proc. 21st Conf. Engineering in Medicine & Biol.,* Houston, Texas, 1968.

Winter, D. A., R. K. Greenlaw, and D. A. Hobson. "Television-Computer Analysis of Kinematics of Human Gait," *Computers Biomed. Res.* **5**:498–504, 1972.

Winter, D. A., H. G. Sidwall, and D. A. Hobson. "Measurement and Reduction of Noise in Kinematics of Locomotion," *J. Biomech.* **7**:157–159, 1974.

Winter, D. A., A. E. Patla, M. G. Ishac, and W. H. Gage. "Motor Mechanisms of Balance During Quiet Standing," *J. Electromyogr. Kinesiol.* **13**:49–56, 2003.

Woltring, H. J. "Data Acquisition and Processing Systems in Functional Movement Analysis." *Minerva Orthop. Traumatol.* **38**:703–716, 1987.

3

ANTHROPOMETRY

3.0 SCOPE OF ANTHROPOMETRY IN MOVEMENT BIOMECHANICS

Anthropometry is the major branch of anthropology that studies the physical measurements of the human body to determine differences in individuals and groups. A wide variety of physical measurements are required to describe and differentiate the characteristics of race, sex, age, and body type. In the past the major emphasis of these studies has been evolutionary and historical. However, more recently a major impetus has come from the needs of technological developments, especially man–machine interfaces: workspace design, cockpits, pressure suits, armor, and so on. Most of these needs are satisfied by basic linear, area, and volume measures. However, human movement analysis requires kinetic measures as well: masses, moments of inertia, and their locations. There exists also a moderate body of knowledge regarding the joint centers of rotation, the origin and insertion of muscles, the angles of pull of tendons, and the length and cross-sectional area of muscles.

3.0.1 Segment Dimensions

The most basic body dimension is the length of the segments between each joint. These vary with body build, sex, and racial origin. Dempster and coworkers (1955, 1959) have summarized estimates of segment lengths and joint center locations relative to anatomical landmarks. An average set of segment lengths expressed as a percentage of body height was prepared by Drillis and Contini (1966) and is shown in Figure 3.1. These segment proportions serve

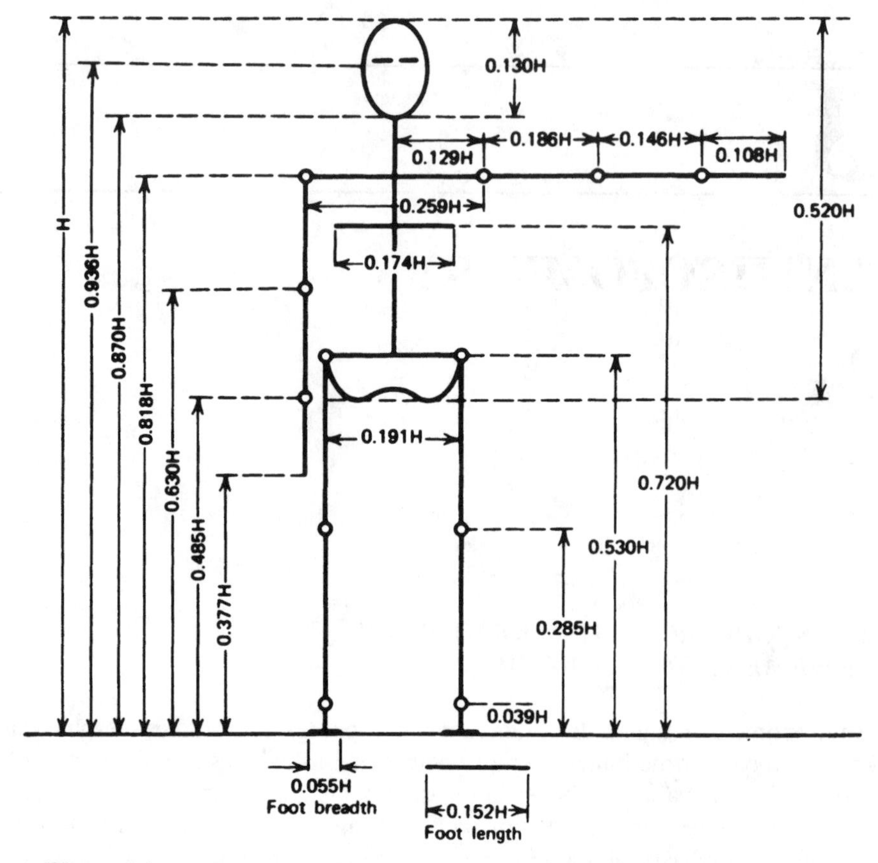

Figure 3.1 Body segment lengths expressed as a fraction of body height *H*.

as a good approximation in the absence of better data, preferably measured directly from the individual.

3.1 DENSITY, MASS, AND INERTIAL PROPERTIES

Kinematic and kinetic analyses require data regarding mass distributions, mass centers, moments of inertia, and the like. Some of these measures have been determined directly from cadavers; others have utilized measured segment volumes in conjunction with density tables, and more modern techniques use scanning systems that produce the cross-sectional image at many intervals across the segment.

3.1.1 Whole-Body Density

The human body consists of many types of tissue, each with a different density. Cortical bone has a specific gravity greater than 1.8, muscle tissue is just over 1.0, fat is less than 1.0, and the lungs contain light respiratory gases. The average density is a function of body build, called *somatotype.* Drillis and Contini (1966) developed an expression for body density d as a function of ponderal index $c = h/w^{1/3}$, where w is body weight (pounds) and h is body height (inches):

$$d = 0.69 + 0.0297c \quad \text{kg/l} \tag{3.1}$$

The equivalent expression in metric units, where body mass is expressed in kilograms and height in meters, is

$$d = 0.69 + 0.9c \quad \text{kg/l} \tag{3.2}$$

It can be seen that a short fat person has a lower ponderal index than a tall skinny person and therefore has a lower body density.

Example 3.1. Using Equations (3.1) and (3.2), calculate the whole-body density of an adult whose height is 5′10″ and who weighs 170 lb,

$$c = h/w^{1/3} = 70/170^{1/3} = 12.64$$

Using Equation (3.1),

$$d = 0.69 + 0.0297c = 0.69 + 0.0297 \times 12.64 = 1.065 \text{ kg/l}$$

In metric units,

$$h = 70/39.4 = 1.78 \text{ m}, \quad w = 170/2.2 = 77.3 \text{ kg}, \quad \text{and}$$
$$c = 1.78/77.3^{1/3} = 0.418$$

Using Equation (3.2),

$$d = 0.69 + 0.9c = 0.69 + 0.9 \times 0.418 = 1.066 \text{ kg/l}$$

3.1.2 Segment Densities

Each body segment has a unique combination of bone, muscle, fat, and other tissue, and the density within a given segment is not uniform. Generally, because of the higher proportion of bone, the density of distal segments is greater than that of proximal segments, and individual segments increase their densities as the average body density increases. Figure 3.2 shows these trends for six limb segments as a function of whole-body density, as calculated by Equations (3.1) or (3.2) or as measured directly (Drillis and Contini, 1966; Contini, 1972).

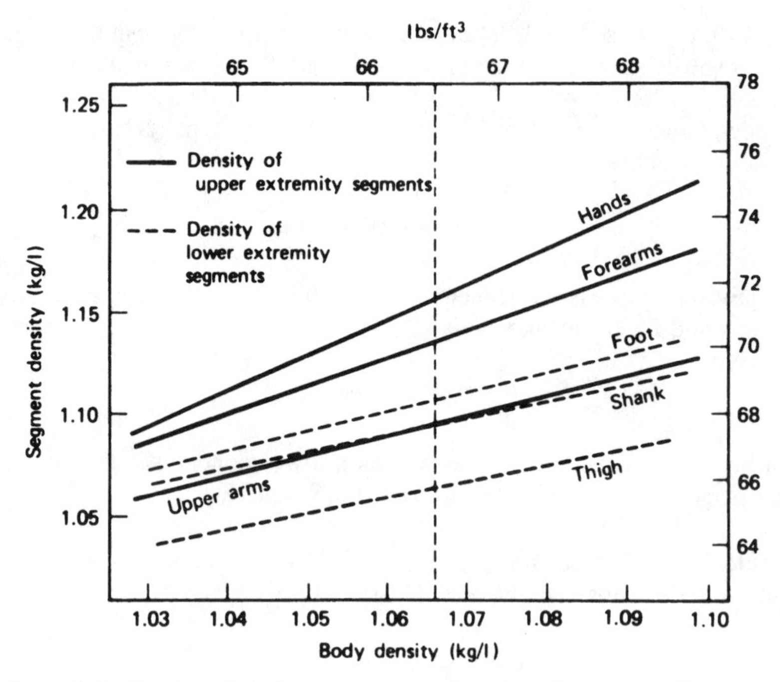

Figure 3.2 Density of limb segments as a function of average body density.

3.1.3 Segment Mass and Center of Mass

The terms *center of mass* and *center of gravity* are often used interchangeably. The more general term is center of mass, while the center of gravity refers to the center of mass in one axis only, that defined by the direction of gravity. In the two horizontal axes, the term center of mass must be used.

As the total body mass increases, so does the mass of each individual segment. Therefore, it is possible to express the mass of each segment as a percentage of the total body mass. Table 3.1 summarizes the compiled results of several investigators. These values are utilized throughout this text in subsequent kinetic and energy calculations. The location of the center of mass is also given as a percentage of the segment length from either the distal or the proximal end. In cadaver studies, it is quite simple to locate the center of mass by simply determining the center of balance of each segment. To calculate the center of mass in vivo, we need the profile of cross-sectional area and length. Figure 3.3 gives a hypothetical profile where the segment is broken into n sections, each with its mass indicated. The total mass M of the segment is

$$M = \sum_{i=1}^{n} m_i \tag{3.3}$$

where m_i is the mass of the ith section,

TABLE 3.1 Anthropometric Data

Segment	Definition	Segment Weight/ Total Body Weight	Center of Mass/ Segment Length		Radius of Gyration/ Segment Length			Density
			Proximal	Distal	C of G	Proximal	Distal	
Hand	Wrist axis/knuckle II middle finger	0.006 M	0.506	0.494 P	0.297	0.587	0.577 M	1.16
Forearm	Elbow axis/ulnar styloid	0.016 M	0.430	0.570 P	0.303	0.526	0.647 M	1.13
Upper arm	Glenohumeral axis/elbow axis	0.028 M	0.436	0.564 P	0.322	0.542	0.645 M	1.07
Forearm and hand	Elbow axis/ulnar styloid	0.022 M	0.682	0.318 P	0.468	0.827	0.565 P	1.14
Total arm	Glenohumeral joint/ulnar styloid	0.050 M	0.530	0.470 P	0.368	0.645	0.596 P	1.11
Foot	Lateral malleolus/head metatarsal II	0.0145 M	0.50	0.50 P	0.475	0.690	0.690 P	1.10
Leg	Femoral condyles/medial malleolus	0.0465 M	0.433	0.567 P	0.302	0.528	0.643 M	1.09
Thigh	Greater trochanter/femoral condyles	0.100 M	0.433	0.567 P	0.323	0.540	0.653 M	1.05
Foot and leg	Femoral condyles/medial malleolus	0.061 M	0.606	0.394 P	0.416	0.735	0.572 P	1.09
Total leg	Greater trochanter/medial malleolus	0.161 M	0.447	0.553 P	0.326	0.560	0.650 P	1.06

TABLE 3.1 (*Continued*)

Segment	Definition	Segment Weight/ Total Body Weight	Center of Mass/ Segment Length		Radius of Gyration/ Segment Length			Density
			Proximal	Distal	C of G	Proximal	Distal	
Head and neck	C7-T1 and 1st rib/ear canal	0.081 M	1.000	— PC	0.495	0.116	— PC	1.11
Shoulder mass	Sternoclavicular joint/ glenohumeral axis	—	0.712	0.288	—	—	—	1.04
Thorax	C7-T1/T12-L1 and diaphragm*	0.216 PC	0.82	0.18	—	—	—	0.92
Abdomen	T12-L1/L4-L5*	0.139 LC	0.44	0.56	—	—	—	—
Pelvis	L4-L5/greater trochanter*	0.142 LC	0.105	0.895	—	—	—	—
Thorax and abdomen	C7-T1/L4-L5*	0.355 LC	0.63	0.37	—	—	—	—
Abdomen and pelvis	T12-L1/greater trochanter*	0.281 PC	0.27	0.73	—	—	—	1.01
Trunk	Greater trochanter/ glenohumeral joint*	0.497 M	0.50	0.50	—	—	—	1.03
Trunk head neck	Greater trochanter/ glenohumeral joint*	0.578 MC	0.66	0.34 P	0.503	0.830	0.607 M	—
Head, arms, and trunk (HAT)	Greater trochanter/ glenohumeral joint*	0.678 MC	0.626	0.374 PC	0.496	0.798	0.621 PC	—
HAT	Greater trochanter/mid rib	0.678	1.142	—	0.903	1.456	—	—

*NOTE: These segments are presented relative to the length between the greater trochanter and the glenohumeral joint.

Source Codes: M, Dempster via Miller and Nelson; *Biomechanics of Sport,* Lea and Febiger, Philadelphia, 1973. P, Dempster via Plagenhoef; *Patterns of Human Motion,* Prentice-Hall, Inc. Englewood Cliffs, NJ, 1971. L, Dempster via Plagenhoef from living subjects; *Patterns of Human Motion,* Prentice-Hall, Inc., Englewood Cliffs, NJ, 1971. C, Calculated.

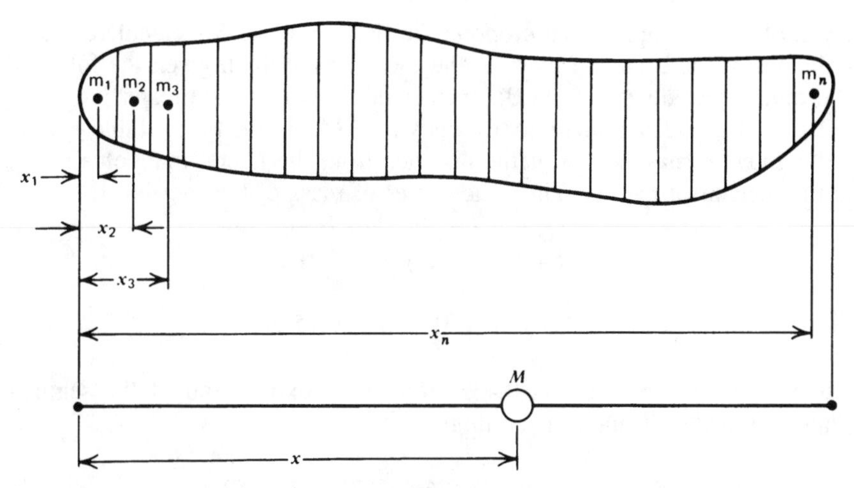

Figure 3.3 Location of the center of mass of a body segment relative to the distributed mass.

$$m_i = d_i V_i$$

where d_i = density of ith section
V_i = volume of ith section

If the density d is assumed to be uniform over the segment, then $m_i = dV_i$ and

$$M = d \sum_{i=1}^{n} V_i \tag{3.4}$$

The center of mass is such that it must create the same net gravitational moment of force about any point along the segment axis as did the original distributed mass. Consider the center of mass to be located a distance x from the left edge of the segment,

$$Mx = \sum_{i=1}^{n} m_i x_i$$

$$x = \frac{1}{M} \sum_{i=1}^{n} m_i x_i \tag{3.5}$$

We can now represent the complex distributed mass by a single mass M located at a distance x from one end of the segment.

Example 3.2. From the anthropometric data in Table 3.1 calculate the coordinates of the center of mass of the foot and the thigh given the following coordinates: ankle (84.9, 11.0), metatarsal (101.1, 1.3), greater trochanter (72.1, 92.8), and lateral femoral condyle (86.4, 54.9). From Table 3.1, the foot center of mass is 0.5 of the distance from the lateral malleolus (ankle) to the metatarsal marker. Thus, the center of mass of the foot is

$$x = (84.9 + 101.1) \div 2 = 93.0 \text{ cm}$$

$$y = (11.0 + 1.3) \div 2 = 6.15 \text{ cm}$$

The thigh center of mass is 0.433 from the proximal end of the segment. Thus, the center of mass of the thigh is

$$x = 72.1 + 0.433\ (86.4 - 72.1) = 78.3 \text{ cm}$$

$$y = 92.8 - 0.433\ (92.8 - 54.9) = 76.4 \text{ cm}$$

3.1.4 Center of Mass of a Multisegment System

With each body segment in motion, the center of mass of the total body is continuously changing with time. It is therefore necessary to recalculate it after each interval of time, and this requires a knowledge of the trajectories of the center of mass of each body segment. Consider at a particular point in time a three-segment system with the centers of mass as indicated in Figure 3.4. The center of mass of the total system is located at (x_0, y_0), and each of these coordinates can be calculated separately; $M = m_1 + m_2 + m_3$, and

$$x_0 = \frac{m_1x_1 + m_2x_2 + m_3x_3}{M} \tag{3.6}$$

$$y_0 = \frac{m_1y_1 + m_2y_2 + m_3y_3}{M} \tag{3.7}$$

The center of mass of the total body is a frequently calculated variable. Its usefulness in the assessment of human movement, however, is quite limited. Some researchers have used the time history center of mass to calculate the energy changes of the total body. Such a calculation is erroneous, because the center of mass does not account for energy changes related to reciprocal movements of the limb segments. Thus, the energy changes associated with the forward movement of one leg and the backward movement of another will not be detected in the center of mass, which may remain relatively unchanged. More about this will be said in Chapter 5. The major use of the body center of mass is in the analysis of sporting events, especially jumping

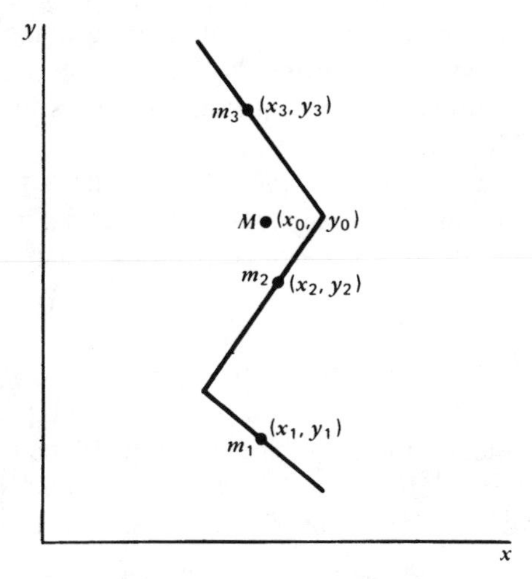

Figure 3.4 Center of mass of a three-segment system relative to the centers of mass of the individual segments.

events where the path of the center of mass is critical to the success of the event because its trajectory is decided immediately at takeoff. Also, in studies of body posture and balance, the center of mass is an essential calculation.

3.1.5 Mass Moment of Inertia and Radius of Gyration

The location of the center of mass of each segment is needed for an analysis of translational movement through space. If accelerations are involved, we need to know the inertial resistance to such movements. In the linear sense, $F = ma$ describes the relationship between a linear force F and the resultant linear acceleration a. In the rotational sense, $M = I\alpha$. M is the moment of force causing the angular acceleration α. Thus, I is the constant of proportionality that measures the ability of the segment to resist changes in angular velocity. M has units of N · m, α is in rad/s^2, and I is in kg · m^2. The value of I depends on the point about which the rotation is taking place and is a minimum when the rotation takes place about its center of mass. Consider a distributed mass segment as in Figure 3.3. The moment of inertia about the left end is

$$
\begin{aligned}
I &= m_1x_1^2 + m_2x_2^2 + \cdots + m_nx_n^2 \\
&= \sum_{i=1}^{n} m_i x_i^2 \qquad (3.8)
\end{aligned}
$$

It can be seen that the mass close to the center of rotation has very little influence on I, while the furthest mass has a considerable effect. This principle is used in industry to regulate the speed of rotating machines: the mass of a flywheel is concentrated at the perimeter of the wheel with as large a radius as possible. Its large moment of inertia resists changes in velocity, and therefore tends to keep the machine speed constant.

Consider the moment of inertia I_0 about the center of mass. In Figure 3.5 the mass has been broken into two equal point masses. The location of these two equal components is at a distance ρ_0 from the center such that

$$I_0 = m\rho_0^2 \tag{3.9}$$

ρ_0 is the radius of gyration and is such that the two equal masses shown in Figure 3.5 have the same moment of inertia in the plane of rotation about the center of mass as the original distributed segment did. Note that the center of mass of these two equal point masses is still the same as the original single mass.

3.1.6 Parallel-Axis Theorem

Most body segments do not rotate about their mass center, but rather about the joint at either end. In vivo measures of the moment of inertia can only be taken about a joint center. The relationship between this moment of inertia and that about the center of mass is given by the parallel-axis theorem. A short proof is now given.

$$\begin{aligned} I &= \frac{m}{2}(x - \rho_0)^2 + \frac{m}{2}(x + \rho_0)^2 \\ &= m\rho_0^2 + mx^2 \\ &= I_0 + mx^2 \end{aligned} \tag{3.10}$$

where I_0 = moment of inertia about center of mass
x = distance between center of mass and center of rotation
m = mass of segment

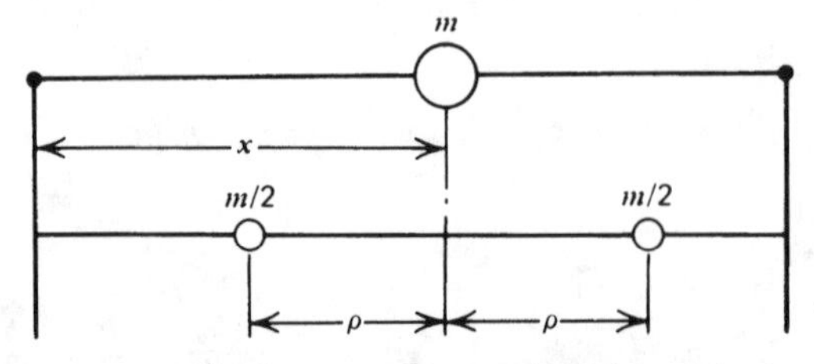

Figure 3.5 Radius of gyration of a limb segment relative to the location of the center of mass of the original system.

Actually, x can be any distance in either direction from the center of mass as long as it lies along the same axis as I_0 was calculated.

Example 3.3
(a) A prosthetic leg has a mass of 3 kg and a center of mass of 20 cm from the knee joint. The radius of gyration is 14.1 cm. Calculate I about the knee joint.

$$I_0 = m\rho_0^2 = 3(0.141)^2 = 0.06 \text{ kg} \cdot \text{m}^2$$

$$\begin{aligned} I &= I_0 + mx^2 \\ &= 0.06 + 3(0.2)^2 = 0.18 \text{ kg} \cdot \text{m}^2 \end{aligned}$$

(b) If the distance between the knee and hip joints is 42 cm, calculate I_h for this prothesis about the hip joint as the amputee swings through with a locked knee.

$$x = \text{distance from mass center to hip} = 20 + 42 = 62 \text{ cm}$$

$$\begin{aligned} I &= I_0 + mx^2 \\ &= 0.06 + 3(0.62)^2 = 1.21 \text{ kg} \cdot \text{m}^2 \end{aligned}$$

Note that I_h is about 20 times that calculated about the center of mass.

3.1.7 Use of Anthropometric Tables and Kinematic Data

Using Table 3.1 in conjunction with kinematic data we can calculate many variables needed for kinetic energy analyses (Chapters 4 and 5). This table gives the segment mass as a fraction of body mass and centers of mass as a fraction of their lengths from either the proximal or the distal end. The radius of gyration is also expressed as a fraction of the segment length about the center of mass, the proximal end, and the distal end.

3.1.7.1 Calculation of Segment Masses and Centers of Mass

Example 3.4. Calculate the mass of the foot, shank, thigh, and HAT and its location from the proximal or distal end assuming that the body mass of the subject is 80 kg. Using the mass fractions for each segment,

$$\begin{aligned}
\text{Mass of foot} &= 0.0145 \times 80 = 1.16 \text{ kg}\\
\text{Mass of leg} &= 0.0465 \times 80 = 3.72 \text{ kg}\\
\text{Mass of thigh} &= 0.10 \times 80 = 8.0 \text{ kg}\\
\text{Mass of HAT} &= 0.678 \times 80 = 54.24 \text{ kg}
\end{aligned}$$

Direct measures yielded the following segment lengths: foot = 0.195 m, leg = 0.435 m, thigh = 0.410 m, HAT = 0.295 m.

COM of foot = 0.50 × 0.195 = 0.098 m between ankle and metatarsal markers

COM of leg = 0.433 × 0.435 = 0.188 m below femoral condyle marker

COM of thigh = 0.433 × 0.410 = 0.178 m below greater trochanter marker

COM of HAT = 1.142 × 0.295 = 0.337 m above greater trochanter marker

where COM stands for center of mass.

3.1.7.2 Calculation of Total-Body Center of Mass. The calculation of the center of mass of the total body is a special case of Equations (3.6) and (3.7). For an n-segment body system, the center of mass in the X direction is

$$x = \frac{m_1x_1 + m_2x_2 + \cdots + m_nx_n}{m_1 + m_2 + \cdots + m_n} \tag{3.11}$$

where $m_1 + m_2 + \cdots + m_n = M$, the total body mass.

It is quite normal to know the values of $m_1 = f_1M$, $m_2 = f_2M$, and so on. Therefore,

$$x = \frac{f_1Mx_1 + f_2Mx_2 + \cdots + f_nMx_n}{M} = f_1x_1 + f_2x_2 + \cdots + f_nx_n \tag{3.12}$$

This equation is easier to use because all we require is a knowledge of the fraction of total body mass and the coordinates of each segment's center of mass. These fractions are given in Table 3.1.

It is not always possible to measure the center of mass of every segment, especially if it is not in full view of the camera. In the example data presented below, we have the kinematics of the right side of HAT and the right limb during walking. It may be possible to simulate data for the left side of HAT and the left limb. If we assume symmetry of gait, we can say that the trajectory of the left limb is the same as that of the right limb, but out of phase by half a stride. Thus, if we use data for the right limb one half stride later in time and shift them back in space one-half a stride length, we can simulate data for the left limb and left side of HAT.

Example 3.5. Calculate the total-body center of mass at a given frame 15. The time for one stride was 68 frames. Thus, the data from frame 15 become the data for the right lower limb and the right half of HAT, and the data one-half stride (34 frames) later become those for the left side of the body. All coordinates from frame 49 must now be shifted back in the x direction by a step length. An examination of the x coordinates of the heel during two successive periods of stance showed the stride length to be $264.2 - 122.8 = 141.4$ cm. Therefore, the step length is 70.7 cm = 0.707 m. Table 3.2 shows the coordinates of the body segments for both left and right halves of the body for this frame 15. The mass fractions for each segment are as follows: foot = 0.0145, leg = 0.0465, thigh = 0.10, 1/2 HAT = 0.339. The mass of HAT dominates the body center of mass, but the energy changes in the lower limbs will be seen to be dominant as far as walking is concerned (see Chapter 5).

Center of mass (COM) analyses in three dimensions are not an easy measure to make because every segment of the body must be identified with markers and tracked with a three-dimensional (3D) imaging system. In some studies of standing, the horizontal anterior/posterior displacement of a rod attached to the pelvis has been taken as an estimate of center of mass movement (Horak et al., 1992). However, in situations when a patient flexes the total body at the hip (called a "hip strategy") to defend against a forward fall, the pelvis moves posteriorly considerably more than the center of mass

TABLE 3.2 Coordinates for Body Segments, Example 3.5

	X (meters)		*Y* (meters)	
Segment	Right	Left	Right	Left
Foot	0.791	1.353 − 0.707 = 0.646	0.101	0.067
Leg	0.814	1.355 − 0.707 = 0.648	0.374	0.334
Thigh	0.787	1.402 − 0.707 = 0.695	0.708	0.691
½ HAT	0.721	1.424 − 0.707 = 0.717	1.124	1.122

$$x = 0.0145(0.791 + 0.645) + 0.0465(0.814 + 0.648) + 0.1(0.787 + 0.695) + 0.339(0.721 + 0.717) = 0.724 \text{ m}$$

$$y = 0.0145(0.101 + 0.067) + 0.0465(0.374 + 0.334) + 0.1(0.708 + 0.691) + 0.339(1.124 + 1.122) = 0.937 \text{ m}$$

(Horak and Nashner, 1986). In 3D assessments of COM displacements, the only technique is optical tracking of markers on all segments (or as many segments as possible). MacKinnon and Winter (1993) used a seven-segment total body estimate of the lower limbs and of the HAT to identify balance mechanisms in the frontal plane during level walking. Jian et al. (1993) reported a 3D analysis of a similar seven-segment estimate of the total body COM in conjunction with the center of pressure during initiation and termination of gait and identified the motor mechanisms responsible for that common movement.

The most complete measure of center of mass to date has been a 21-marker, 14-segment model that has been used to determine the mechanisms of balance during quiet standing (Winter et al., 1998). Figure 3.6 shows the location of the markers and the accompanying table gives the definition of each of the 14 segments, along with mass fraction of each segment. It is worth noting that most of the segments are fairly rigid segments (head, pelvis, upper and lower limbs). However, the trunk is not that rigid, and it required four separate segments to achieve a reliable estimate mainly because these trunk segments undergo internal mass shifts due to respiratory and cardiac functions. The validity of any COM estimate can be checked with the equation for the inverted pendulum model during the movement: $COP - COM = -K \cdot \ddot{COM}$ (Winter et al., 1998). COP is the center of pressure recorded from force plate data, $\ddot{COM}$ is the horizontal acceleration of COM in either the anterior/posterior or medial/lateral direction, and $K = I/Wh$, where I is the moment of inertia of the total body about the ankles, W is body weight, and h is the height of COM above the ankles.

3.1.7.3 Calculation of Moment of Inertia

Example 3.6. Calculate the moment of inertia of the leg about its center of mass, its distal end, and its proximal end. From Table 3.1, the mass of the leg is $0.0465 \times 80 = 3.72$ kg. The leg length is given as 0.435 m. The radius of gyration/segment length is 0.302 for the center of mass, 0.528 for the proximal end, and 0.643 for the distal end.

$$I_0 = 3.72(0.435 \times 0.302)^2 = 0.064 \text{ kg} \cdot \text{m}^2$$

About the proximal end,

$$I_p = 3.72(0.435 \times 0.528)^2 = 0.196 \text{ kg} \cdot \text{m}^2$$

About the distal end,

$$I_d = 3.72(0.435 \times 0.643)^2 = 0.291 \text{ kg} \cdot \text{m}^2$$

Note that the moment of inertia about either end could also have been calculated using the parallel-axis theorem. For example, the distance of the cen-

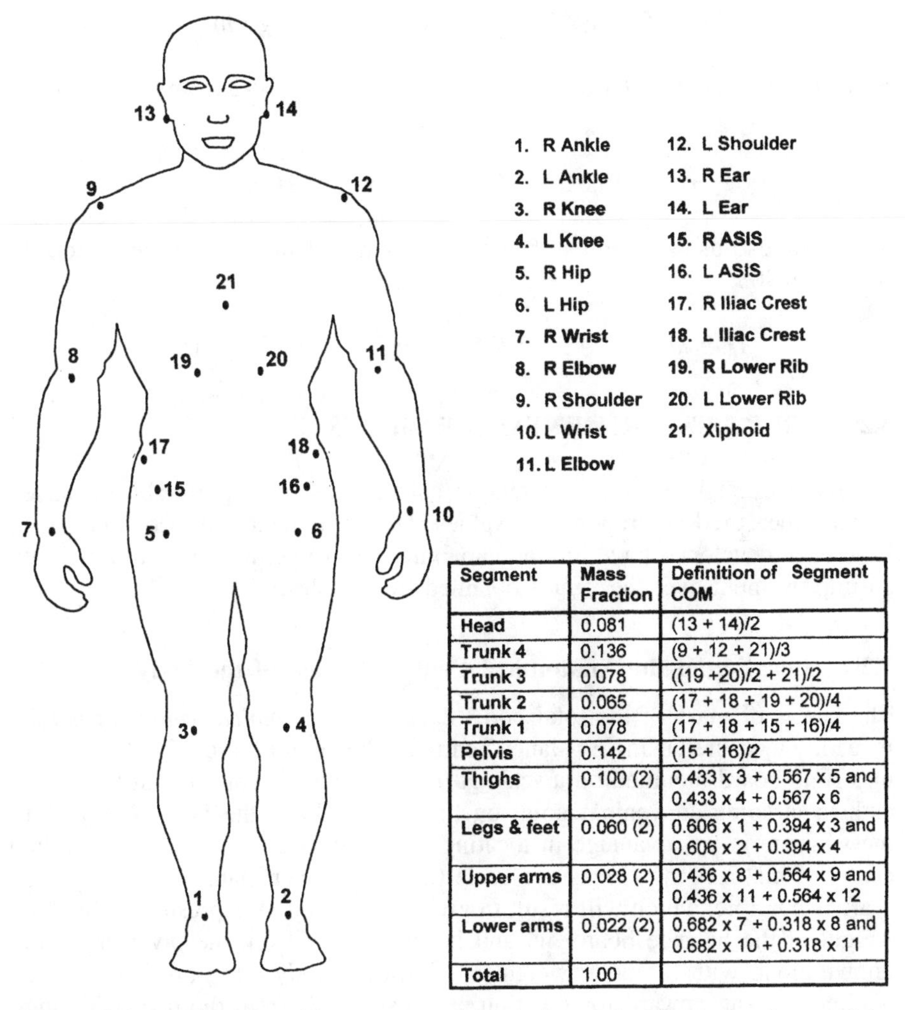

Segment	Mass Fraction	Definition of Segment COM
Head	0.081	(13 + 14)/2
Trunk 4	0.136	(9 + 12 + 21)/3
Trunk 3	0.078	((19 +20)/2 + 21)/2
Trunk 2	0.065	(17 + 18 + 19 + 20)/4
Trunk 1	0.078	(17 + 18 + 15 + 16)/4
Pelvis	0.142	(15 + 16)/2
Thighs	0.100 (2)	0.433 x 3 + 0.567 x 5 and 0.433 x 4 + 0.567 x 6
Legs & feet	0.060 (2)	0.606 x 1 + 0.394 x 3 and 0.606 x 2 + 0.394 x 4
Upper arms	0.028 (2)	0.436 x 8 + 0.564 x 9 and 0.436 x 11 + 0.564 x 12
Lower arms	0.022 (2)	0.682 x 7 + 0.318 x 8 and 0.682 x 10 + 0.318 x 11
Total	1.00	

Figure 3.6 A 21-marker, 14-segment model to estimate the 3D center of mass of the total body in balance control experiments. Four trunk segments were necessary to track the internal mass shifts of the thoracic/lumbar volumes.

ter of mass of the leg from the proximal end is $0.433 \times 0.435 = 0.188$ m, and

$$I_p = I_0 + mx^2 = 0.064 + 3.72(0.188)^2 = 0.196 \text{ kg} \cdot \text{m}^2$$

Example 3.7. Calculate the moment of inertia of HAT about its proximal end and about its center of mass. From Table 3.1, the mass of HAT is $0.678 \times 80 = 54.24$ kg. The HAT length is given as 0.295 m. The radius of gyration about the proximal end/segment length is 1.456.

$$I_p = 54.24(0.295 \times 1.456)^2 = 10.01 \text{ kg} \cdot \text{m}^2$$

From Table 3.1, the center of mass/segment length = 1.142 from the proximal end.

$$I_0 = I_p - mx^2 = 10.01 - 54.24(0.295 \times 1.142)^2 = 3.85 \text{ kg} \cdot \text{m}^2$$

We could also use the radius of gyration/segment length about the center of mass = 0.903.

$$I_0 = m\rho^2 = 54.24(0.295 \times 0.903)^2 = 3.85 \text{ kg} \cdot \text{m}^2$$

3.2 DIRECT EXPERIMENTAL MEASURES

For more exact kinematic and kinetic calculations, it is preferable to have directly measured anthropometric values. The equipment and techniques that have been developed have limited capability and sometimes are not much of an improvement over the values obtained from tables.

3.2.1 Location of the Anatomical Center of Mass of the Body

The center of mass of the total body, called the *anatomical center of mass,* is readily measured using a balance board, as shown in Figure 3.7*a*. It consists of a rigid board mounted on a scale at one end and a pivot point at the other end, or at some convenient point on the other side of the body's center of mass. There is an advantage in locating the pivot as close as possible to the center of mass. A more sensitive scale (0–5 kg) rather than a 50- or 100-kg scale is possible, which will result in greater accuracy. It is presumed that the weight of the balance board, w_1 and its location x_1 from the pivot are both known along with the body weight w_2. With the body lying prone the scale reading is S (an upward force acting at a distance x_3 from the pivot). Taking moments about the pivot

$$w_1 x_1 + w_2 x_2 = S x_3$$

$$x_2 = \frac{S x_3 - w_1 x_1}{w_2} \tag{3.13}$$

3.2.2 Calculation of the Mass of a Distal Segment

The mass or weight of a distal segment can be determined by the technique demonstrated in Figure 3.7*b*. The desired segment, here the leg and foot, is lifted to a vertical position so that its center of mass lies over the joint center. Prior to lifting, the center of mass was x_4 from the pivot point, with the scale

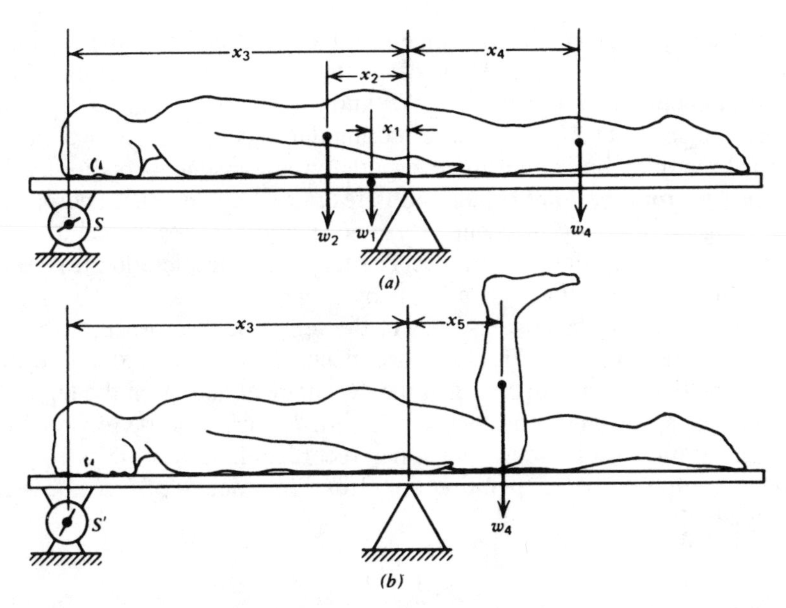

Figure 3.7 Balance board technique. (*a*) In vivo determination of mass of the location of the anatomical center of mass of the body. (*b*) Mass of a distal segment. See text for details.

reading S. After lifting, the leg center of mass is x_5 from the pivot, and the scale reading has increased to S^1. The decrease in the clockwise moment due to the leg movement is equal to the increase in the scale reaction force moment about the pivot point,

$$w_4(x_4 - x_5) = (S^1 - S)x_3$$

$$w_4 = \frac{(S^1 - S)x_3}{(x_4 - x_5)} \tag{3.14}$$

The major error in this calculation is due to errors in x_4, usually obtained from anthropometric tables. To get the mass of the total limb, this experiment can be repeated with the subject lying on his back and the limb flexed at an angle of 90°. From the mass of the total limb we can now subtract that of the leg and foot to get the thigh mass.

3.2.3 Moment of Inertia of a Distal Segment

The equation for the moment of inertia, described in Section 3.1.5, can be used to calculate I at a given joint center of rotation. I is the constant of proportionality that relates the joint moment to the segment's angular acceleration, assuming the proximal segment is fixed. A method called the *quick release experiment* can be used to calculate I directly and requires the arrangement pictured in Figure 3.8. We know that $I = M/\alpha$, so if we can measure the moment M that causes an angular acceleration α, we can calculate I directly. A horizontal force F pulls on a convenient rope or cable at a distance y_1 from the joint center and is restrained by an equal and opposite force acting on a release mechanism. An accelerometer is attached to the leg at a distance y_2 from the joint center. The tangential acceleration a is related to the angular acceleration of the leg α by $a = y_2\alpha$.

With the forces in balance as shown, the leg is held in a neutral position and no acceleration occurs. If the release mechanism is actuated, the restraining force suddenly drops to zero and the net moment acting on the leg is Fy_1, which causes an instantaneous acceleration α. F and a can be recorded on a dual-beam storage oscilloscope; most pen recorders have too low a frequency response to capture the acceleration impulse. The moment of inertia can now be calculated,

$$I = \frac{M}{\alpha} = \frac{Fy_1y_2}{a} \tag{3.15}$$

Figure 3.8 shows the sudden burst of acceleration accompanied by a rapid decrease in the applied force F. This force drops after the peak of acceleration and does so because the forward displacement of the limb causes the tension to drop in the pulling cable. A convenient release mechanism can be achieved by suddenly cutting the cable or rope that holds back the leg. The sudden accelerometer burst can also be used to trigger the oscilloscope sweep so that the rapidly changing force and acceleration can be captured.

More sophisticated experiments have been devised to measure more than one parameter simultaneously. Such techniques were developed by Hatze

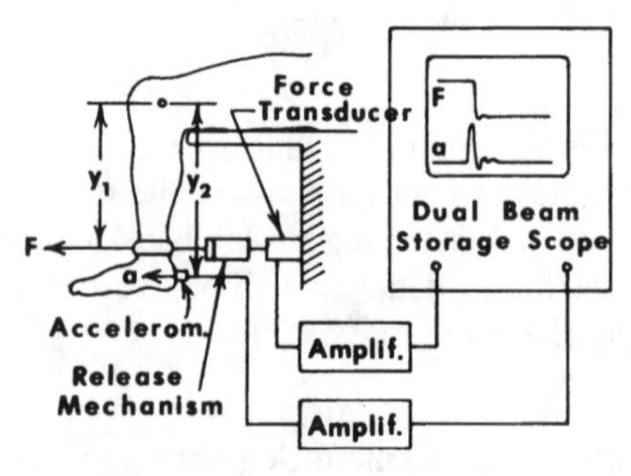

Figure 3.8 Quick-release technique for the determination of the mass moment of inertia of a distal segment. Force F applied horizontally results, after release of the segment, in an initial acceleration a. Moment of inertia can then be calculated from F, y_1, y_2.

(1975) and are capable of determining the moment of inertia, the location of the center of mass, and the damping coefficient simultaneously.

3.2.4 Joint Axes of Rotation

Markers attached to the body are usually placed to represent our best estimate of a joint center. However, because of anatomical constraints, our location can be somewhat in error. The lateral malleolus, for example, is a common location for ankle joint markers. However, the articulation of the tibial/talus surfaces is such that the distal end of the tibia (and the fibula) move in a small arc over the talus. The true axis of rotation is actually a few centimeters distal of the lateral malleolus. Even more drastic differences are evident at some other joints. The hip joint is often identified in the sagittal plane by a marker on the upper border of the greater trochanter. However, it is quite evident that the marker is somewhat more lateral than the center of the hip joint such that internal and external rotations of the thigh relative to the pelvis may cause considerable errors, as will abduction/adduction at that joint.

Thus, it is important that the true axes of rotation be identified relative to anatomical markers that we have placed on the skin. Several techniques have been developed to calculate the instantaneous axis of rotation of any joint based on the displacement histories of markers on the two adjacent segments. Figure 3.9 shows two segments in a planar movement. First, they must be

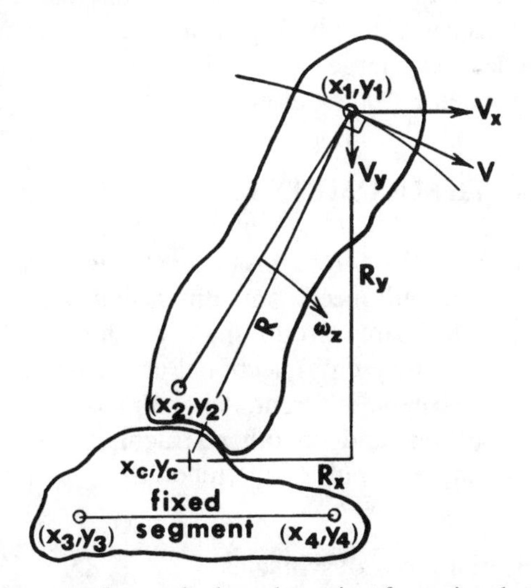

Figure 3.9 Technique used to calculate the axis of rotation between two adjacent segments x_c, y_c. Each segment must have two markers in the plane of movement. After data collection, the segments are rotated and translated so that one segment is fixed in space. Thus, the moving-segment kinematics reflects the relative movement between the two segments and the axis of rotation can be located relative to the anatomical location of markers on that fixed segment. See text for complete details.

translated and rotated in space so that one segment is fixed in space and the second rotates as shown. At any given instant in time, the true axis of rotation is at (x_c, y_c) within the fixed segment, and we are interested in the location of (x_c, y_c) relative to anatomical coordinates (x_3, y_3) and (x_4, y_4) of that segment. Markers (x_1, y_1) and (x_2, y_2) are located as shown; (x_1, y_1) has an instantaneous tangential velocity $\overline{V}$ and is located at a radius $\overline{R}$ from the axis of rotation. From the line joining (x_1, y_1) to (x_2, y_2) we calculate the angular velocity of the rotating segment $\overline{\omega}_z$. With one segment fixed in space, $\overline{\omega}_z$ is nothing more than the joint angular velocity,

$$\overline{V} = \overline{\omega}_z \times \overline{R} \tag{3.16a}$$

or, in Cartesian coordinates,

$$V_x\hat{\imath} + V_y\hat{\jmath} = (R_y\omega_z)\hat{\imath} - (R_x\omega_z)\hat{\jmath}$$

Therefore,

$$V_x = R_y\omega_z \qquad \text{and} \qquad V_y = -R_x\omega_z \tag{3.16b}$$

Since V_x, V_y, and ω_z can be calculated from the marker trajectory data, R_y and R_x can be determined. Since x_1, y_1 is known, the axis of rotation x_c, y_c can be calculated. Care must be taken when ω_z approaches 0 or reverses its polarity, because R, as calculated by Equation (3.16*a*), becomes indeterminate or falsely approaches very large values. In practice, we have found errors become significant when ω_z falls below 0.5 r/s.

3.3 MUSCLE ANTHROPOMETRY

Before we can calculate the forces produced by individual muscles during normal movement, we usually need some dimensions from the muscles themselves. If muscles of the same group share the load, they probably do so proportionally to their relative cross-sectional areas. Also, the mechanical advantage of each muscle can be different, depending on the moment arm length at its origin and insertion, and on other structures beneath the muscle or tendon that alter the angle of pull of the tendon.

3.3.1 Cross-Sectional Area of Muscles

The functional or physiologic cross-sectional area (PCA) of a muscle is a measure of the number of sarcomeres in parallel with the angle of pull of the muscles. In pennate muscles, the fibers act at an angle from the long axis and therefore are not as effective as fibers in a parallel-fibered muscle. The angle between the long axis of the muscle and the fiber angle is called *pennation angle.* In parallel-fibered muscle, the PCA is

$$\text{PCA} = \frac{m}{dl}\ \text{cm}^2 \tag{3.17}$$

where m = mass of muscle fibers, grams
d = density of muscle, g/cm^3, = 1.056 g/cm^3
l = length of muscle fibers, centimeters

In pennate muscles, the physiological cross-sectional area becomes

$$\text{PCA} = \frac{m \cos \theta}{dl}\ \text{cm}^2 \tag{3.18}$$

where θ is the pennation angle, which increases as the muscle shortens.

Wickiewicz et al. (1983), using data from three cadavers, measured muscle mass, fiber lengths, and pennation angle for 27 muscles of the lower extremity. Representative values are given in Table 3.3. The PCA as a percentage of the total cross-sectional area of all muscles crossing a given joint is presented in Table 3.4. In this way, the relative potential contribution of a group of agonist muscles can be determined, assuming that each is generating the same stress. Note that a double joint muscle, such as the gastrocnemius, may represent different percentages at different joints because of the different total PCA of all muscles crossing each joint.

3.3.2 Change in Muscle Length During Movement

A few studies have investigated the changes in the length of muscles as a function of the angles of the joints they cross. Grieve and colleagues (1978), in a study on eight cadavers, reported percentage length changes of the gas-

TABLE 3.3 Mass, Length, and PCA of Some Muscles

Muscle	Mass (g)	Fiber Length (cm)	PCA (cm^2)	Pennation Angle (deg)
Sartorius	75	38	1.9	0
Biceps femoris (long)	150	9	15.8	0
Semitendinosus	75	16	4.4	0
Soleus	215	3.0	58	30
Gastrocnemius	158	4.8	30	15
Tibialis posterior	55	2.4	21	15
Tibialis anterior	70	7.3	9.1	5
Rectus femoris	90	6.8	12.5	5
Vastus lateralis	210	6.7	30	5
Vastus medialis	200	7.2	26	5
Vastus intermedius	180	6.8	25	5

TABLE 3.4 Percent PCA of Muscles Crossing Ankle, Knee, and Hip Joints

Ankle		Knee		Hip	
Muscle	%PCA	Muscle	%PCA	Muscle	%PCA
Soleus	41	Gastrocnemius	19	Iliopsoas	9
Gastrocnemius	22	Biceps Femoris (small)	3	Sartorius	1
Flexor Hallucis Longus	6	Biceps Femoris (long)	7	Pectineus	1
Flexor Digitorum Longus	3	Semitendinosus	3	Rectus Femoris	7
Tibialis Posterior	10	Semimembranosus	10	Gluteus Maximus	16
Peroneus Brevis	9	Vastus Lateralis	20	Gluteus Medius	12
Tibialis Anterior	5	Vastus Medialis	15	Gluteus Minimus	6
Extensor Digitorum Longus	3	Vastus Intermedius	13	Adductor Magnus	11
Extensor Hallucis Longus	1	Rectus Femoris	8	Adductor Longus	3
		Sartorius	1	Adductor Brevis	3
		Gracilis	1	Tensor Fasciae Latae	1
				Biceps Femoris (long)	6
				Semitendinosus	3
				Semimembranosus	8
				Piriformis	2
				Lateral Rotators	13

trocnemius muscle as a function of the knee and ankle angle. The resting length of the gastrocs was assumed to be when the knee was flexed 90° and the ankle was in an intermediate position, neither plantarflexed nor dorsiflexed. With 40° plantarflexion, the muscle shortened 8.5% and linearly changed its length to a 4% increase at 20° dorsiflexion. An almost linear curve described the changes at the knee: 6.5% at full extension to a 3% decrease at 150° flexion.

3.3.3 Force per Unit Cross-Sectional Area (Stress)

A wide range of stress values for skeletal muscles has been reported (Haxton, 1944; Alexander and Vernon, 1975; Maughan et al., 1983). Most of these stress values were measured during isometric conditions and range from 20 to 100 N/cm^2. These higher values were recorded in pennate muscles, which are those whose fibers lie at an angle from the main axis of the muscle. Such an orientation effectively increases the cross-sectional area above that measured and used in the stress calculation. Haxton (1944) related force to stress in two pennate muscles (gastrocs and soleus) and found stresses as high as 38 N/cm^2. Dynamic stresses have been calculated in the quadriceps during running and jumping to be about 70 N/cm^2 (based on a peak knee extensor moment of 210 $N \cdot m$ in adult males) and about 100 N/cm^2 in isometric maximum voluntary contractions (MVCs) (Maughan et al., 1983).

3.3.4 Mechanical Advantage of Muscle

The origin and insertion of each muscle defines the angle of pull of the tendon on the bone, and therefore the mechanical leverage it has at the joint center. Each muscle has its unique moment arm length, which is the length of a line normal to the muscle passing through the joint center. This moment arm length changes with the joint angle. One of the few studies done in this area (Smidt, 1973) reports the average moment arm length (26 subjects) for the knee extensors and for the hamstrings acting at the knee. Both these muscle groups showed an increase in the moment length as the knee was flexed, reaching a peak at 45°, then decreasing again as flexion increased to 90°. Wilkie (1950) has also documented the moments and lengths for elbow flexors.

3.3.5 Multijoint Muscles

A large number of the muscles in the human body pass over more than one joint. In the lower limbs, the hamstrings are extensors of the hip and flexors of the knee, the rectus femoris is a combined hip flexor and knee extensor, and the gastrocnemius are knee flexors and ankle plantarflexors. The fiber length of many of these muscles may be insufficient to allow a complete range of movement of both joints involved. Elftman (1966) has suggested that many normal movements require lengthening at one joint simultaneously with shortening at the other. Consider the action of the rectus femoris, for example, during early swing in running. This muscle shortens as a result of hip flexion and lengthens at the knee as the leg swings backward in preparation for swing. The tension in the rectus femoris simultaneously creates a flexor hip moment (positive work) and an extensor knee moment to decelerate the swinging leg (negative work) and start accelerating it forward. In this way, the net change in muscle length is reduced compared with two equivalent single-joint muscles, and excessive positive and negative work within the muscle can be reduced. A double-joint muscle could even be totally isometric in such situations and would effectively be transferring energy from the leg to the pelvis in the example just described. In running during the critical push-off phase, when the plantarflexors are generating energy at a high rate, the knee is continuing to extend. Thus, the gastrocnemii may be essentially isometric (they may appear to be shortening at the distal end and lengthening at the proximal end). Similarly, toward the end of swing in running, the knee is rapidly extending while the hip has reached full flexion and is beginning to reverse (i.e., it has an extensor velocity). Thus, the hamstrings appear to be rapidly lengthening at the distal end and shortening at the proximal end, with the net result that they may be lengthening at a slower rate than a single joint would.

It is also critical to understand the role of the major biarticulate muscles of the lower limb during stance phase of walking or running. Figure 3.10 shows the gastrocnemii, hamstrings, and rectus femoris and their moment-arm lengths at their respective proximal and distal ends. The hamstrings have a 5-cm moment-arm at the ankle and 3.5-cm moment-arm at the knee. Thus, when they are active during stance, their contribution to the ankle extensor moment is about 50% greater than their contribution to the knee flexor moment. The net effect of these two contributions is to cause the leg to rotate posteriorly and prevent the knee from collapsing. The hamstrings, with the exception of the short head of the biceps femoris, have moment-arms of 6–7 cm at the hip but only 3.5 cm at the knee. Thus, when these muscles are active during stance, their contribution to hip extension is about twice their contribution to knee flexion. The net effect of these two actions is to cause the thigh to rotate posteriorly and prevent the knee from collapsing. Finally, the rectus femoris is the only biarticulate muscle of the large quadriceps group, and its moment-arm at the hip is slightly larger than at the knee. However, the quadriceps activate as a group, and because the uniarticulate quadriceps comprise 84% of the PCA of the quadriceps (see Table 3.3), the dominant action is knee extension. Thus, the net effect of the major biarticulate muscles of the lower limb is extension at all three joints, and therefore they contribute, along with all the uniarticulate extensors, to defending against a gravity-induced collapse. The algebraic summation of all three moments during stance phase of gait has been calculated and been found to be dominantly extensor (Winter, 1984). This summation has been labeled the support moment, and is discussed further in Section 4.2.5.

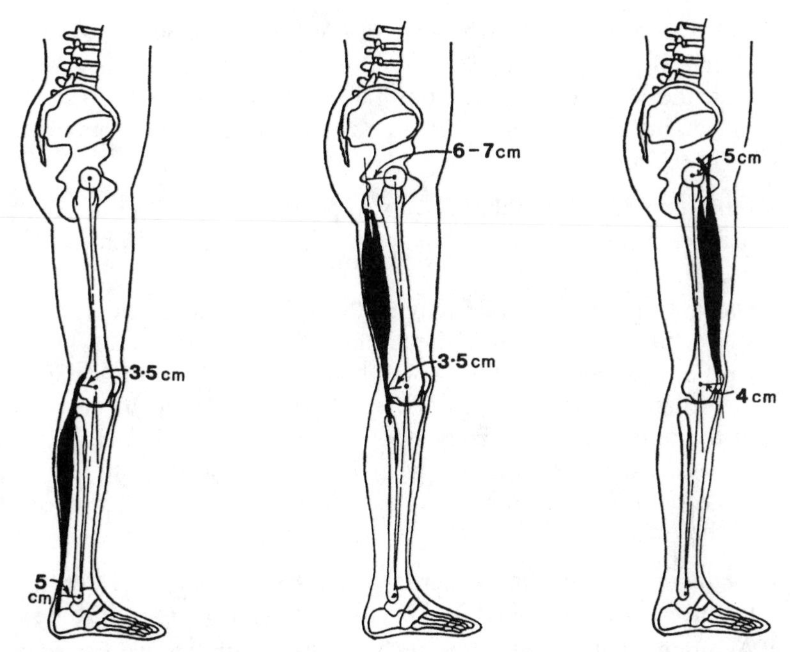

Figure 3.10 Three major biarticulate muscles of the lower limb. Shown are the gastrocnemii, hamstrings, and rectus femoris, and their moment-arm lengths at their proximal and distal ends. These moment-arms are critical to the functional role of these muscle groups during weight bearing; see text for details.

3.4 PROBLEMS BASED ON ANTHROPOMETRIC DATA

1. (a) Calculate the average body density of a young adult whose height is 1.68 m and whose mass is 68.5 kg. *Answer:* 1.059 kg/l.

(b) For the adult in (a), determine the density of the forearm and use it to estimate the mass of the forearm that measures 24.0 cm from the ulnar styloid to the elbow axis. Circumference measures (in cm) taken at 1-cm intervals starting at the wrist are 20.1, 20.3, 20.5, 20.7, 20.9, 21.2, 21.5, 21.9, 22.5, 23.2, 23.9, 24.6, 25.1, 25.7, 26.4, 27.0, 27.5, 27.9, 28.2, 28.4, 28.4, 28.3, 28.2, 28.0. Assuming the forearm to have a circular cross-sectional area over its entire length, calculate the volume of the forearm and its mass. Compare the mass as calculated with that estimated using averaged anthropometric data (Table 3.1). *Answer:* Forearm density = 1.13 kg/l; volume = 1.174 l; mass = 1.33 kg. Mass calculated from Table 3.1 = 1.10 kg.

(c) Calculate the location of the center of mass of the forearm along its long axis and give its distance from the elbow axis. Compare that with the center of mass as determined from Table 3.1. *Answer:* COM = 10.34 cm from the elbow; from Table 3.1, COM = 10.32 cm from the elbow.

(d) Calculate the moment of inertia of the forearm about the elbow axis. Then calculate its radius of gyration about the elbow and compare with the value calculated from Table 3.1. *Answer:* $I_p = 0.0201$ kg · m^2; radius of gyration = 12.27 cm; radius of gyration from Table 3.1 = 12.62 cm.

2. (a) From data listed in Table A.3 in Appendix A, calculate the center of mass of the lower limb for frame 70. *Answer:* $x = 1.755$ m; $y = 0.522$ m.

(b) Using your data of stride length (Problem 2(f) in Section 2.8) and a stride time of 68 frames, create an estimate (assuming symmetrical gait) of the coordinates of the left half of the body for frame 30. From the segment centers of mass (Table A.4), calculate the center of mass of the right half of the body (foot + leg + high + 1/2 HAT), and of the left half of the body (using segment data suitably shifted in time and space). Average the two centers of mass to get the center of mass of the total body for frame 30. *Answer:* x 1.025 m, $y = 0.904$ m.

3. (a) Calculate the moment of inertia of the HAT about its center of mass for the subject described in Appendix A. *Answer:* $I_0 = 1.96$ kg · m^2.

(b) Assuming that the subject is standing erect with the two feet together, calculate the moment of inertia of HAT about the hip joint, the knee joint, and the ankle joint. What does the relative size of these moments of inertia tell us about the relative magnitude of the joint moments required to control the inertial load of HAT. *Answer:* $I_h = 5.09$ kg · m^2, $I_k = 15.78$ kg · m^2, $I_a = 42.31$ kg · m^2.

(c) Assuming that the center of mass of the head is 1.65 m from the ankle, what percentage does it contribute to the moment of inertia of HAT about the ankle? *Answer:* I_{head} about ankle = 12.50 kg · m^2, which is 29.6% of I_{hat} about the ankle.

4. (a) Calculate the moment of inertia of the lower limb of the subject in Appendix A about the hip joint. Assume the knee is not flexed and the foot is a point mass located 6 cm distal to the ankle. *Answer:* $I_{\text{lower limb}}$ about hip = 1.39 kg · m^2.

(b) Calculate the increase in the moment of inertia as calculated in (a) when a ski boot is worn. The mass of the ski boot is 1.8 kg, and assume it to be a point mass located 1 cm distal to the ankle. *Answer:* I_{boot} about hip = 1.01 kg · m^2.

3.5 REFERENCES

Alexander, R. McN., and A. Vernon. "The Dimensions of Knee and Ankle Muscles and the Forces They Exert," *J. Human Movement Studies* **1**:115–123, 1975.

Contini, R. "Body Segment Parameters, Part II," *Artificial Limbs* **16**:1–19, 1972.

Dempster, W. T. "Space Requirements of the Seated Operator," WADC-TR-55-159, Wright Patterson Air Force Base, 1955.

Dempster, W. T., W. C. Gabel, and W. J. L. Felts. "The Anthropometry of Manual Work Space for the Seated Subjects," *Am. J. Phys. Anthrop.* **17**:289–317, 1959.

Drillis, R., and R. Contini. "Body Segment Parameters," Rep. 1163-03, Office of Vocational Rehabilitation, Department of Health, Education, and Welfare, New York, 1966.

Elftman, H. "Biomechanics of Muscle, with Particular Application to Studies of Gait," *J. Bone Joint Surg.* **48-A**:363–377, 1966.

Grieve, D. W., P. R. Cavanagh, and S. Pheasant. "Prediction of Gastrocnemius Length from Knee and Ankle Joint Posture," *Biomechanics, Vol. VI-A,* E. Asmussen and K. Jorgensen, Eds. (University Park Press, Baltimore, MD, 1978), pp. 405–412.

Haxton, H. A. "Absolute Muscle Force in the Ankle Flexors of Man," *J. Physiol.* **103**: 267–273, 1944.

Hatze, H. "A New Method for the Simultaneous Measurement of the Moment of Inertia, the Damping Coefficient and the Location of the Center of Mass of a Body Segment in situ," *Eur. J. Appl. Physiol.* **34**:217–266, 1975.

Horak, F. B. and L. M. Nashner. "Central Programming of Postural Movements: Adaptation to Altered Support Surface Configurations," *J. Neurophysiol.* **55**:1369–1381, 1986.

Horak, F. B., J. G. Nutt, and L. M. Nashner. "Postural Inflexibility in Parkinsonian Subjects," *J. Neurol. Sci.* **111**:46–58, 1992.

Jian, Y., D. A. Winter, M. G. Ishac, and L. Gilchrist. "Trajectory of the Body COG and COP During Initiation and Termination of Gait," *Gait and Posture* **1**:9–22, 1993.

MacKinnon, C. D. and D. A. Winter. "Control of Whole Body Balance and Posture in the Frontal Plane During Walking," *J. Biomech.* **26**:633–644, 1993.

Maughan, R. J., J. S. Watson, and J. Weir. "Strength and Cross-Sectional Area of Human Skeletal Muscle," *J. Physiol.* **338**:37–49, 1983.

Smidt, G. L. "Biomechanical Analysis of Knee Flexion and Extension," *J. Biomech.,* **6**:79–92, 1973.

Wickiewcz, T. L., R. R. Roy, P. L. Powell, and V. R., Edgerton. "Muscle Architecture of the Human Lower Limb," *Clin. Orthop. Rel. Rs.* **179**:275–283, 1983.

Wilkie, D. R. "The Relation between Force and Velocity in Human Muscle," *J. Physiol.* **110**:249–280, 1950.

Winter, D. A. "Kinematic and Kinetic Patterns in Human Gait: Variability and Compensating Effects," *Hum. Movement Sci.* **3**:51–76, 1984.

Winter, D. A., A. E. Patla, F. Prince, M. G. Ishac, and K. Gielo-Perczak. "Stiffness Control of Balance in Quiet Standing," *J. Neurophysiol.* **80**:1211–1221, 1998.

4

KINETICS: FORCES AND MOMENTS OF FORCE

4.0 BIOMECHANICAL MODELS

Chapter 2 has dealt at length with the movement itself, without regard to the forces that cause the movement. The study of these forces and the resultant energetics is called *kinetics.* Knowledge of the patterns of the forces is necessary for an understanding of the cause of any movement.

Transducers have been developed that can be implanted surgically to measure the force exerted by a muscle at the tendon. However, such techniques have applications only in animal experiments, and even then only to a limited extent. It therefore remains that we attempt to calculate these forces indirectly, using readily available kinematic and anthropometric data. The process by which the reaction forces and muscle moments are calculated is called *link-segment modeling.* Such a process is depicted in Figure 4.1. If we have a full kinematic description, accurate anthropometric measures, and the external forces, we can calculate the joint reaction forces and muscle moments. This prediction is called an *inverse solution* and is a very powerful tool in gaining insight into the net summation of all muscle activity at each joint. Such information is very useful to the coach, surgeon, therapist, and kinesiologist in their diagnostic assessments. The effect of training, therapy, or surgery is extremely evident at this level of assessment, although it is often obscured in the original kinematics.

4.0.1 Link-Segment Model Development

The validity of any assessment is only as good as the model itself. Accurate measures of segment masses, centers of mass (COMs), joint centers, and

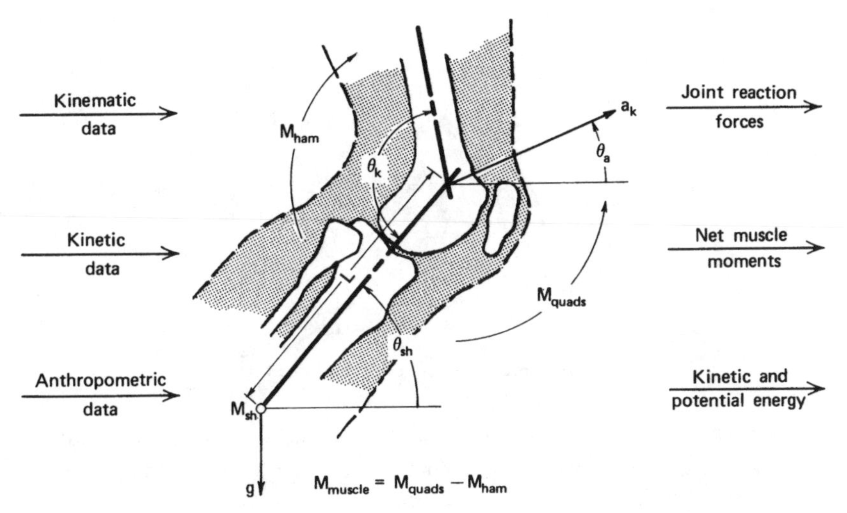

Figure 4.1 Schematic diagram of the relationship between kinematic, kinetic, and anthropometric data and the calculated forces, moments, energies, and powers using an inverse solution of a link-segment model.

moments of inertia are required. Such data can be obtained from statistical tables based on the person's height, weight, and, sometimes, sex, as was detailed in Chapter 3. A limited number of these variables can be measured directly, but some of the techniques are time consuming and have limited accuracy. Regardless of the source of the anthropometric data, the following assumptions are made with respect to the model:

1. Each segment has a fixed mass located as a point mass at its COM (which will be the center of gravity in the vertical direction).
2. The location of each segment's COM remains fixed during the movement.
3. The joints are considered to be hinge (or ball and socket) joints.
4. The mass moment of inertia of each segment about its mass center (or about either proximal or distal joints) is constant during the movement.
5. The length of each segment remains constant during the movement (e.g., the distance between hinge or ball and socket joints remains constant).

Figure 4.2 shows the equivalence between the anatomical and the link-segment models for the lower limb. The segment masses m_1, m_2, and m_3 are considered to be concentrated at points. The distance from the proximal joint to the mass centers is considered to be fixed, as are the length of the segments and each segment's moment of inertia I_1, I_2, and I_3 about each COM.

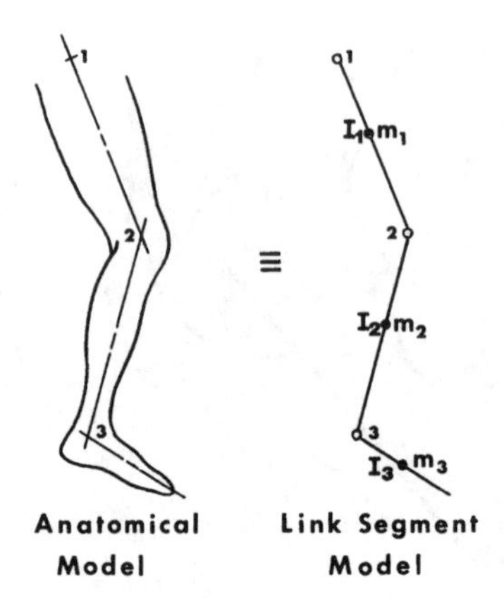

Figure 4.2 Relationship between anatomical and link-segment models. Joints are replaced by hinge (pin) joints and segments are replaced by masses and moments of inertia located at each segment's center of mass.

4.0.2 Forces Acting on the Link-Segment Model

1. *Gravitational Forces.* The forces of gravity act downward through the COMs of each segment and are equal to the magnitude of the mass times acceleration due to gravity (normally 9.8 m/s^2).

2. *Ground Reaction or External Forces.* Any external forces must be measured by a force transducer. Such forces are distributed over an area of the body (such as the ground reaction forces under the area of the foot). In order to represent such forces as vectors, they must be considered to act at a point that is usually called the center of pressure (COP). A suitably constructed force plate, for example, yields signals from which the COP can be calculated.

3. *Muscle and Ligament Forces.* The net effect of muscle activity at a joint can be calculated in terms of net muscle moments. If a cocontraction is taking place at a given joint, the analysis yields only the net effect of both agonist and antagonistic muscles. Also, any friction effects at the joints or within the muscle cannot be separated from this net value. Increased friction merely reduces the effective "muscle" moment; the muscle contractile elements themselves are actually creating moments higher than that analyzed at the tendon. However, the error at low and moderate speeds of movement is usually only a few percent. At the extreme range of movement of any joint, passive structures such as ligaments come into play to contain the range. The moments generated by these tissues will add to or subtract from those generated by the muscles. Thus, unless the muscle is silent, it is impossible to determine the contribution of these passive structures.

4.0.3 Joint Reaction Forces and Bone-on-Bone Forces*

The three forces described in the preceding sections constitute all the forces acting on the total body system itself. However, our analysis examines the segments one at a time and therefore must calculate the reaction between segments. A free-body diagram of each segment is required, as shown in Figure 4.3. Here the original link-segment model is broken into its segmental parts. For convenience, we make the break at the joints and the forces that act across each joint must be shown on the resultant free-body diagram. This procedure now permits us to look at each segment and calculate all unknown joint reaction forces. In accordance with Newton's third law, there is an equal and opposite force acting at each hinge joint in our model. For example, when a leg is held off the ground in a static condition, the foot is exerting a downward force on the tendons and ligaments crossing the ankle joint. This is seen as a downward force acting on the leg equal to the weight of the foot. Like-

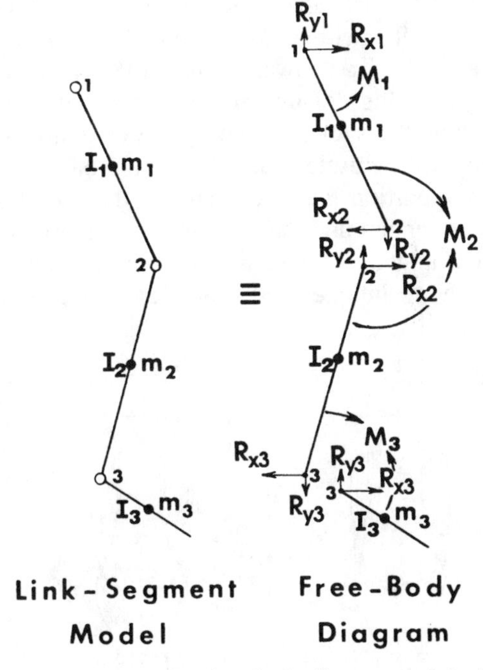

Figure 4.3 Relationship between the free-body diagram and the link-segment model. Each segment is "broken" at the joints, and the reaction forces and moments of force acting at each joint are indicated.

*Representative paper: Paul, 1966.

wise, the leg is exerting an equal upward force on the foot through the same connective tissue.

Considerable confusion exists regarding the relationship between joint reaction forces and joint bone-on-bone forces. The latter forces are the actual forces seen across the articulating surfaces and include the effect of muscle activity. Actively contracting muscles pull the articulating surfaces together, creating compressive forces and, sometimes, shear forces. In the simplest situation, the bone-on-bone force equals the active compressive force due to muscle plus joint reaction forces. Figure 4.4 illustrates these differences. Case 1 has the lower segment with a weight of 100 N hanging passively from the muscles originating in the upper segment. The two muscles are not contracting but are, assisted by ligamentous tissue, pulling upward with an equal and opposite 100 N. The link-segment model shows these equal and opposite reaction forces. The bone-on-bone force is zero, indicating that the joint articulating surfaces are under neither tension nor compression. In case 2, there is an active contraction of the muscles, so that the total upward force is now 170 N. The bone-on-bone force is 70 N compression. This means that a force of 70 N exists across the articulating surfaces. As far as the lower segment is concerned, there is still a net reaction force of 100 N upward (170 N upward through muscles and 70 N downward across the articulating surfaces). The lower segment is still acting downward with a force of 100 N; thus, the free-body diagram remains the same. Generally the anatomy is not as simple as depicted. More than one muscle is usually active on each side of the joint, so it is difficult to apportion the forces among the muscles. Also, the exact angle of pull of each tendon and the geometry of the articulating surfaces are not always readily available. Thus, a much more complex free-body diagram must be used, and the techniques and examples are described in Section 4.3.

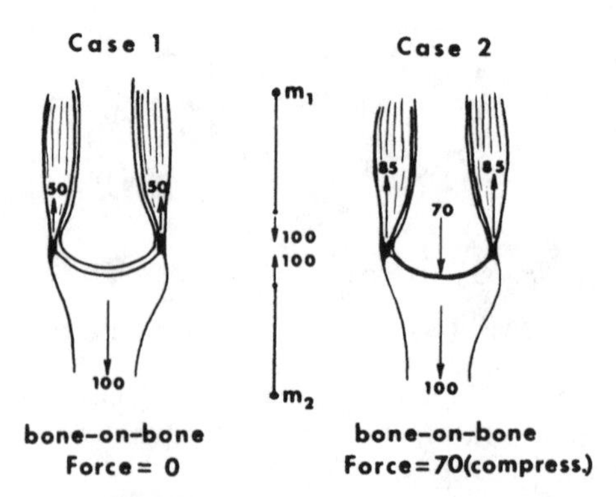

Figure 4.4 Diagrams to illustrate the difference between joint reaction forces and bone-on-bone forces. In both cases, the reaction force is 100 N acting upward on M_2 and downward on M_1. With no muscle activity (case 1), the bone-on-bone force is zero; with muscle activity (case 2) it is 70 N.

4.1 BASIC LINK-SEGMENT EQUATIONS— THE FREE-BODY DIAGRAM*

Each body segment acts independently under the influence of reaction forces and muscle moments, which act at either end, plus the forces due to gravity. Consider the planar movement of a segment in which the kinematics, anthropometrics, and reaction forces at the distal end are known (Figure 4.5).

Known

a_x, a_y = acceleration of segment COM
θ = angle of segment in plane of movement
α = angular acceleration of segment in plane of movement
R_{xd}, R_{yd} = reaction forces acting at distal end of segment, usually determined from a prior analysis of the proximal forces acting on distal segment
M_d = net muscle moment acting at distal joint, usually determined from an analysis of the proximal muscle acting on distal segment

Unknown

R_{xp}, R_{yp} = reaction forces acting at proximal joint
M_p = net muscle moment acting on segment at proximal joint

Equations

1. $\Sigma F_x = ma_x$

$$R_{xp} - R_{xd} = ma_x \tag{4.1}$$

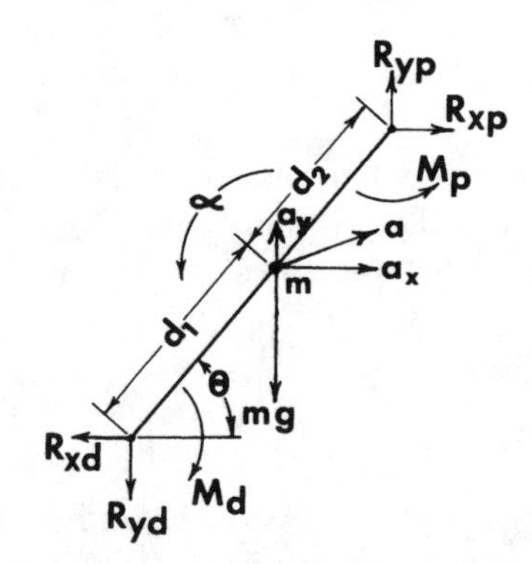

Figure 4.5 Complete free-body diagram of a single segment, showing reaction and gravitational forces, net moments of force, and all linear and angular accelerations.

*Representative paper: Bresler and Frankel, 1950.

2. $\Sigma F_y = ma_y$

$$R_{yp} - R_{yd} - mg = ma_y \tag{4.2}$$

3. About the segment COM, $\Sigma M = I_0\alpha$ (4.3)

Note that the muscle moment at the proximal end cannot be calculated until the proximal reaction forces R_{xp} and R_{yp} have first been calculated.

Example 4.1 (Figure 4.6). In a static situation, a person is standing on one foot on a force plate. The ground reaction force is found to act 4 cm anterior to the ankle joint. Note that convention has the ground reaction force R_{y1} always acting upward. We also show the horizontal reaction force R_{x1} to be acting in the positive direction (to the right). If this force actually acts to the left, it will be recorded as a negative number. The subject's mass is 60 kg, and the mass of the foot is 0.9 kg. Calculate the joint reaction forces and net muscle moment at the ankle. R_{y1} = body weight $= 60 \times 9.8 = 588$ N.

1. $\Sigma F_x = ma_x$,

   $$R_{x2} + R_{x1} = ma_x = 0$$

   Note that this is a redundant calculation in static conditions.

2. $\Sigma F_y = ma_y$,

   $$R_{y2} + R_{y1} - mg = ma_y$$

   $$R_{y2} + 588 - 0.9 \times 9.8 = 0$$

   $$R_{y2} = -579.2 \text{ N}$$

   The negative sign means that the force acting on the foot at the ankle joint acts *downward*. This is not surprising because the entire body weight, less that of the foot, must be acting downward on the ankle joint.

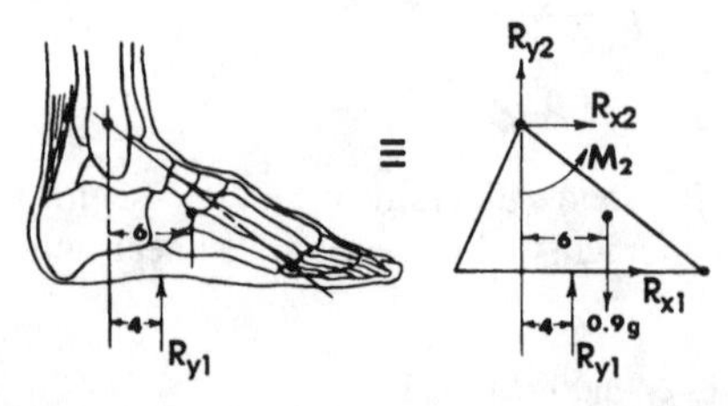

Figure 4.6 Anatomical and free-body diagram of foot during weight bearing.

3. About the COM, $\Sigma M = I_0\alpha$,

$$M_2 - R_{y1} \times 0.02 - R_{y2} \times 0.06 = 0$$

$$M_2 = 588 \times 0.02 + (-579.2 \times 0.06) = -22.99 \text{ N} \cdot \text{m}$$

The negative sign means that the real direction of the muscle moment acting on the foot at the ankle joint is clockwise, which means that the plantar-flexors are active at the ankle joint to maintain the static position. These muscles have created an action force that resulted in the ground reaction force that was measured, and whose center of pressure was 4 cm anterior to the ankle joint.

Example 4.2 (Figure 4.7). From the data collected during the swing of the foot, calculate the muscle moment and reaction forces at the ankle. The subject's mass was 80 kg and the ankle-metatarsal length was 20.0 cm. From Table 3.1, the inertial characteristics of the foot are calculated:

$$m = 0.0145 \times 80 = 1.16 \text{ kg}$$

$$\rho_0 = 0.475 \times 0.20 = 0.095 \text{ m}$$

$$I_0 = 1.16(0.095)^2 = 0.0105 \text{ kg} \cdot \text{m}^2$$

$$\alpha = 21.69 \text{ rad/s}^2$$

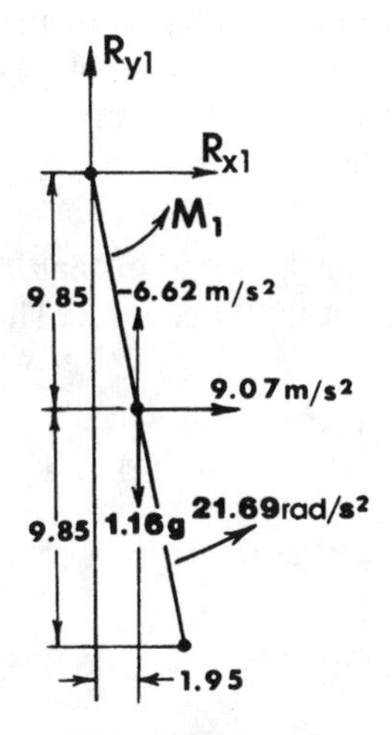

Figure 4.7 Free-body diagram of foot during swing showing the linear accelerations of the center of mass and the angular acceleration of the segment. Distances are in centimeters. Three unknowns, R_{x1}, R_{y1}, and M_1, are to be calculated assuming a positive direction as shown.

1. $\Sigma F_x = ma_x$,

$$R_{x1} = 1.16 \times 9.07 = 10.52 \text{ N}$$

2. $\Sigma F_y = ma_y$,

$$R_{y1} - 1.16g = m(-6.62)$$

$$R_{y1} = 1.16 \times 9.8 - 1.16 \times 6.62 = 3.69 \text{ N}$$

3. At the COM of the foot, $\Sigma M = I_0\alpha$,

$$M_1 - R_{x1} \times 0.0985 - R_{y1} \times 0.0195 = 0.0105 \times 21.69$$

$$M_1 = 0.0105 \times 21.69 + 10.52 \times 0.0985 + 3.69 \times 0.0195$$
$$= 0.23 + 1.04 + 0.07 = 1.34 \text{ N} \cdot \text{m}$$

Discussion

1. The horizontal reaction force of 10.52 N at the ankle is the cause of the horizontal acceleration that we calculated for the foot.
2. The foot is decelerating its upward rise at the end of lift-off. Thus, the vertical reaction force at the ankle is somewhat less than the static gravitational force.
3. The ankle muscle moment is positive, indicating net dorsiflexor activity (tibialis anterior), and most of this moment (1.04 out of 1.34 N · m) is required to cause the horizontal acceleration of the foot's center of gravity, with very little needed (0.23 N · m) to angularly accelerate the low moment of inertia of the foot.

Example 4.3 (Figure 4.8). For the same instant in time, calculate the muscle moments and reaction forces at the knee joint. The leg segment was 43.5 cm long.

$$m = 0.0465 \times 80 = 3.72 \text{ kg}$$

$$\rho_0 = 0.302 \times 0.435 = 0.131 \text{ m}$$

$$I_0 = 3.72(0.131)^2 = 0.0638 \text{ kg} \cdot \text{m}^2$$

$$\alpha = 36.9 \text{ rad/s}^2$$

From Example 4.2, $R_{x1} = 10.52$ N, $R_{y1} = 3.69$ N, and $M_1 = 1.34$ N · m.

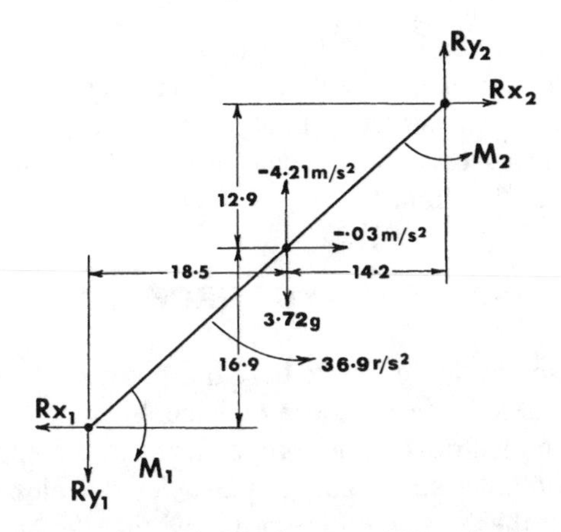

Figure 4.8 Free-body diagram of leg at the same instant in time as the foot in Figure 4.7. Linear and angular accelerations are as shown. Distances are in centimeters. The distal end and reaction forces and moments have been reversed, recognizing Newton's third law. Again the three unknowns, R_{x2}, R_{y2}, and M_2, are assumed to be positive with directions as indicated.

1. $\Sigma F_x = ma_x$,

$$R_{x2} - R_{x1} = ma_x$$

$$R_{x2} = 10.52 + 3.72(-0.03) = 10.41 \text{ N}$$

2. $\Sigma F_y = ma_y$,

$$R_{y2} - R_{y1} - mg = ma_y$$

$$R_{y2} = 3.69 + 3.72 \times 9.8 + 3.72(-4.21) = 24.48 \text{ N}$$

3. About the COM of the leg, $\Sigma M = I\alpha$,

$$M_2 - M_1 - 0.169R_{x1} + 0.185R_{y1} - 0.129R_{x2} + 0.142R_{y2} = I\alpha$$

$$M_2 = 1.34 + 0.169 \times 10.52 - 0.185 \times 3.69 + 0.129 \times 10.41$$
$$-0.142 \times 24.48 + 0.0638 \times 36.9$$
$$= 1.34 + 1.78 - 0.68 + 1.34 - 3.48 + 2.35 = 2.65 \text{ N} \cdot \text{m}$$

Discussion

1. M_2 is positive. This represents a counterclockwise (extensor) moment acting at the knee. The quadricep muscles at this time are rapidly extending the swinging leg.

2. The angular acceleration of the leg is the net result of two reaction forces and one muscle moment acting at each end of the segment. Thus, there may not be a single primary force causing the movement we observe. In this case, each force and moment had a significant influence on the final acceleration.

4.2 FORCE TRANSDUCERS AND FORCE PLATES

In order to measure the force exerted by the body on an external body or load, we need a suitable force-measuring device. Such a device, called a *force transducer,* gives an electrical signal proportional to the applied force. There are many kinds available: strain gauge, piezoelectric, piezoresistive, capacitive, and others. All these work on the principle that the applied force causes a certain amount of strain within the transducer. For the strain gauge type, a calibrated metal plate or beam within the transducer undergoes a very small change (strain) in one of its dimensions. This mechanical deflection, usually a fraction of 1%, causes a change in resistances connected as a bridge circuit (see Section 2.2.2), resulting in an unbalance of voltages proportional to the force. Piezoelectric and piezoresistive types require minute deformations of the atomic structure within a block of special crystalline material. Quartz, for example, is a naturally found piezoelectrical material, and deformation of its crystalline structure changes the electrical characteristics such that the electrical charge across appropriate surfaces of the block is altered and can be translated via suitable electronics to a signal proportional to the applied force. Piezoresistive types exhibit a change in resistance which, like the strain gauge, upset the balance of a bridge circuit.

4.2.1 Multidirectional Force Transducers

In order to measure forces in two or more directions, it is necessary to use a bi- or tri-directional force transducer. Such a device is nothing more than two or more force transducers mounted at right angles to each other. The major problem is to ensure that the applied force acts through the central axis of each of the individual transducers.

4.2.2 Force Plates*

The most common force acting on the body is the ground reaction force, which acts on the foot during standing, walking, or running. This force vector is three-dimensional and consists of a vertical component plus two shear components acting along the force plate surface. These shear forces are usually resolved into anterior–posterior and medial–lateral directions.

*Representative paper: Elftman, 1939.

The fourth variable needed is the location of the center of pressure of this ground reaction vector. The foot is supported over a varying surface area with different pressures at each part. Even if we knew the individual pressures under every part of the foot, we would be faced with the expensive problem of calculating the net effect of all these pressures as they change with time. Some attempts have been made to develop suitable pressure-measuring shoes, but they have been very expensive and are limited to vertical forces only. It is therefore necessary to use a force plate to give us all the forces necessary for a complete inverse solution.

Two common types of force plates are now explained. The first is a flat plate supported by four triaxial transducers, depicted in Figure 4.9. Consider the coordinates of each transducer to be (0, 0), (0, Z), (X, 0), and (X, Z). The location of the center of pressure is determined by the relative vertical forces seen at each of these corner transducers. If we designate the vertical forces as F_{00}, F_{X0}, F_{0Z}, and F_{XZ}, the total vertical force is $F_Y = F_{00} + F_{X0} + F_{0Z} + F_{XZ}$. If all four forces are equal, the COP is at the exact center of the force plate, at ($X/2$, $Z/2$). In general,

$$x = \frac{X}{2}\left[1 + \frac{(F_{X0} + F_{XZ}) - (F_{00} + F_{0Z})}{F_Y}\right] \tag{4.4}$$

$$z = \frac{Z}{2}\left[1 + \frac{(F_{0Z} + F_{XZ}) - (F_{00} + F_{X0})}{F_Y}\right] \tag{4.5}$$

A second type of force plate has one centrally instrumented pillar that supports an upper flat plate. Figure 4.10 shows the forces that act on this instrumented support. The action force of the foot F_y acts downward, and the anterior–posterior shear force can act either forward or backward. Consider a reverse shear force F_x, as shown. If we sum the moments acting about the central axis of the support, we get

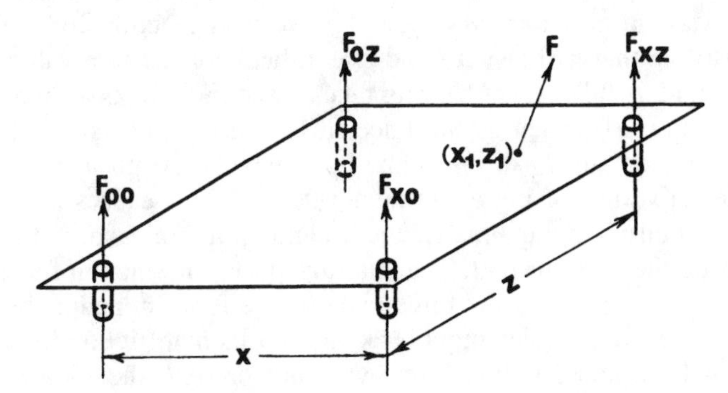

Figure 4.9 Force plate with force transducers in the four corners. Magnitude and location of ground reaction force F can be determined from the signals from the load cells in each of the support bases.

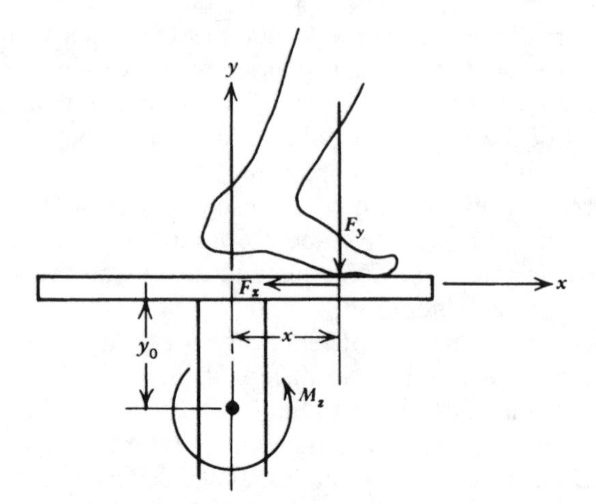

Figure 4.10 Central support type force plate, showing the location of the center of pressure of the foot and the forces and moments involved.

$$M_z - F_y \cdot x + F_x \cdot y_0 = 0$$

$$x = \frac{F_x \cdot y_0 + M_z}{F_y} \tag{4.6}$$

where M_z = bending moment about axis of rotation of support
y_0 = distance from support axis to force plate surface

Since F_x, F_y, and M_z continuously change with time, x can be calculated to show how the COP moves across the force plate. Caution is required for both types of force platforms when F_y is very low (<2% of body weight). During this time, a small error in F_y represents a large percentage error in x and z, as calculated by Equations (4.4), (4.5), and (4.6).

Typical force plate data are shown in Figure 4.11 plotted against time for a subject walking at a normal speed. The vertical reaction force F_y is very characteristic in that it shows a rapid rise at heel contact to a value in excess of body weight as full weight bearing takes place (i.e., the body mass is being accelerated upward). Then as the knee flexes during midstance, the plate is partially unloaded and F_y drops below body weight. At push-off the plantarflexors are active in causing ankle plantarflexion. This causes a second peak greater than body weight. Finally, the weight drops to zero as the opposite limb takes up the body weight. The anterior–posterior reaction force F_x is the horizontal force exerted by the force plate on the foot. Immediately after heel contact, it is negative, indicating a backward horizontal friction force between the floor and the shoe. If this force were not present, the foot would slide forward, as sometimes happens when walking on icy or slippery surfaces.

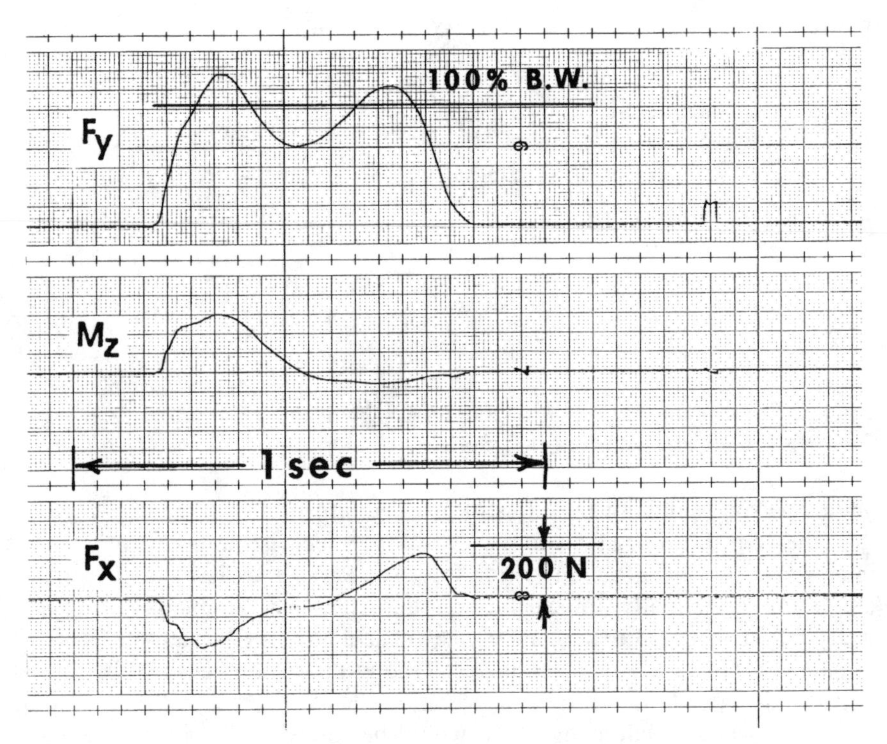

Figure 4.11 Force plate record obtained during gait, using a central support type as shown in Figure 4.10.

Near midstance, F_x goes positive, indicating that the force plate reaction is acting forward as the muscle forces (mainly plantarflexor) cause the foot to push back against the plate.

The COP starts at the heel, assuming that initial contact is made by the heel, and then progresses forward toward the ball and toe. The position of the COP relative to the foot cannot be obtained from the force plate data themselves. We must first know where the foot is, relative to the midline of the plate. The type of force plate used to collect these data is the centrally instrumented pillar type depicted in Figure 4.10. The student should note that M_z is positive at heel contact and then becomes negative as the body weight moves forward. The centers of pressure X_{cp} and Y_{cp} are calculated in absolute coordinates to match those given in the kinematics listing. Y_{cp} was set at 0 to indicate ground level.

It is very important to realize that these reaction forces are merely the algebraic summation of all mass–acceleration products of all body segments. In other words, for an N-segment system the reaction force in the x direction is

$$F_x = \sum_{i=1}^{N} m_i a_{xi} \tag{4.7}$$

where m_i = mass of ith segment
a_{xi} = acceleration of ith-segment COM in the x direction

Similarly, in the vertical direction,

$$F_y = \sum_{i=1}^{N} m_i(a_{yi} + g) \tag{4.8}$$

where a_{yi} = acceleration of ith-segment COM in the y direction
g = acceleration due to gravity

With a person standing perfectly still on the platform, F_y will be nothing more than the sum of all segment masses times g, which is the body weight. The interpretation of the ground reaction forces as far as what individual segments are doing is virtually impossible. In the algebraic summation of mass–acceleration products, there can be many cancellations. For example, F_Y could remain constant while one arm is accelerating upward while the other has an equal and opposite downward acceleration. Thus, to interpret the complete meaning of any reaction force, we would be forced to determine the accelerations of the COMs of all segments and carry out the summations seen in Equations (4.7) and (4.8).

4.2.3 Special Pressure Measuring Sensory Systems

The COP measured by the force plate is a weighted average of the distributed COPs under the two feet if both are in contact with the ground or under the one foot that is in contact. However, the COP signal tells us nothing about the pressure at any of the contact points under the foot. For example, during midstance in walking or running, there are two main pressure areas: the ball of the foot and the heel, but the force plate records the COP as being under the arch of the foot, where there may in fact be negligible pressure. To get some insight into the distributed pressures around all contact points, a number of special pressure measurement systems have been developed. A typical pressure measurement system was introduced a number of years ago by Tekscan and is schematically shown in Figure 4.12. It is a flexible sheet of tactile force sensors that is cut in the shape of a shoe insole and that can be trimmed down to any shoe size. Each sensel shown in the exploded view is about 0.2 in. × 0.2 in. (5 mm × 5 mm), giving 25 sensels/sq. in. Two thin flexible polyester sheets have electrically conductive electrodes deposited in rows and columns. A thin semiconductive coating is applied between the conductive rows and columns, and its electrical resistance changes with the pressure

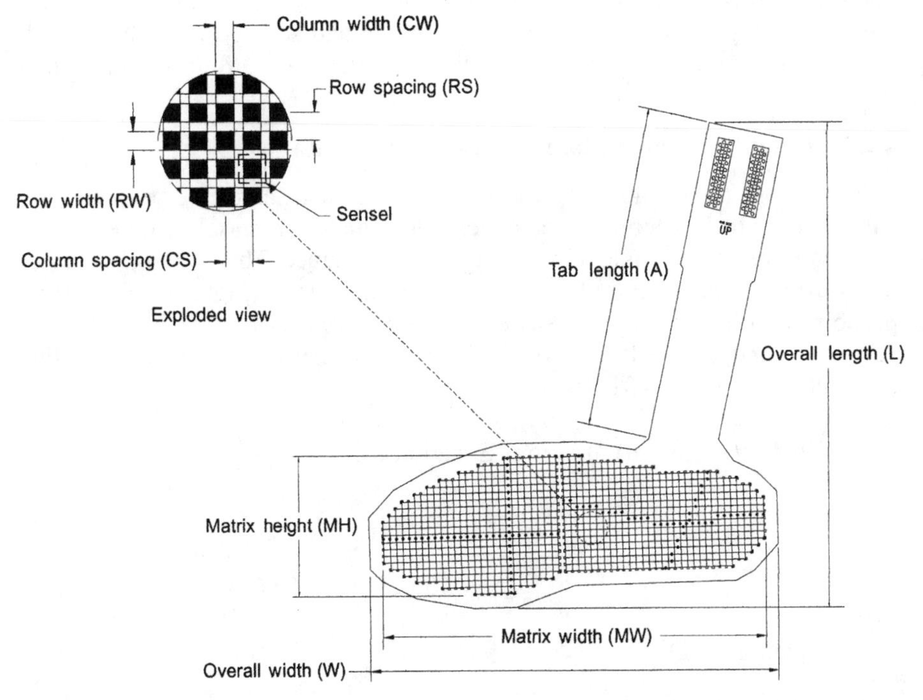

Figure 4.12 Typical foot pressure measurement system introduced by Tekscan, Inc. It is a shoe insole insert with an array of pressure sensors that vary their resistance with applied pressure. See text for detailed operation of sensor and computer display system. (Reproduced with permission of Tekscan, Inc.)

applied. The matrix of sensels results in a matrix of voltage changes that are computer displayed as different colors for different pressure levels. Colors range from blue through green, yellow, orange, and red (1–125 psi). Thus, the high- and low-pressure areas under the foot are visually evident as the pressure moves from the heel to the toes over the stance period (Hsiao et al., 2002). Such devices have been valuable in identifying high pressure points in various foot deformities and in diabetic feet (Pitie et al., 1999). The results of surgery and shoe orthotics to relieve the pressure points are also immediately available.

4.2.4 Synchronization of Force Plate and Kinematic Data

Because kinematic data are coming from a completely separate system, there may be problems in time synchronization with the ground reaction data. Most optoelectric systems have synchronizing pulses that must be recorded simultaneously with the force records. Similarly, TV systems must generate a pulse for each TV field that can be used to synchronize with the force signals. The major imaging system that has problems is cine. Here, the movie cameras must have a sync pulse generated every frame, and somehow the number of the pulse must be made available to the person doing the film digitization.

4.2.5 Combined Force Plate and Kinematic Data

It is valuable to see how the reaction force data from the force plate are combined with the segment kinematics to calculate the muscle moments and reaction forces at the ankle joint during dynamic stance. This is best illustrated in an example calculation. For a subject in late stance (Figure 4.13) during pushoff the following foot accelerations were recorded: $a_x = 3.25$ m/s², $a_y = 1.78$ m/s², and $\alpha = -45.35$ rad/s². The mass of the foot is 1.12 kg, and the moment of inertia is 0.01 kg · m².

Example 4.4 (Figure 4.13). From Equation (4.1),

$$F_{ax} + F_x = ma_x$$

$$F_{ax} = 1.12 \times 3.25 - 160.25 = -156.6 \text{ N}$$

From Equation (4.2),

$$F_{ay} + F_y - mg = ma_y$$

$$F_{ay} = 1.12 \times 1.78 - 765.96 + 1.12 \times 9.81 = -753.0 \text{ N}$$

From Equation (4.3), about the center of mass of the foot, $\Sigma M = I\alpha$,

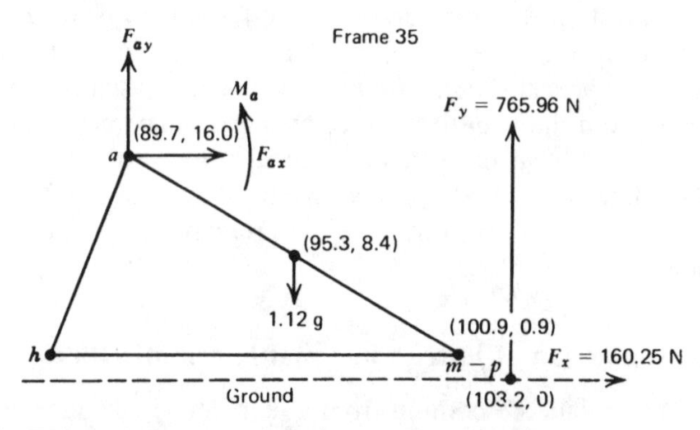

Figure 4.13 Free-body diagram of foot during weight bearing with the ground reaction forces F_x and F_y shown as being located at the COP.

$$M_a + F_x \times 0.084 + F_y \times 0.079 - F_{ay} \times 0.056 - F_{ax} \times 0.076$$
$$= 0.01(-45.35)$$

$$M_a = -0.01 \times 45.35 - 0.084 \times 160.25 - 0.079 \times 765.96 - 0.056$$
$$\times 753.0 - 0.076 \times 156.6 = -128.5 \text{ N} \cdot \text{m}$$

The polarity and the magnitude of this ankle moment indicate strong plantarflexor activity acting to push off the foot and cause it to rotate clockwise about the metatarsophalangeal joint.

4.2.6 Interpretation of Moment-of-Force Curves*

A complete link-segment analysis yields the net muscle moment at every joint during the time course of movement. As an example, we discuss in detail the ankle moments of force of a hip joint replacement patient as shown in Figure 4.14. Three repeat trials are plotted. The convention for the moments is shown to the right of the plot; counterclockwise moments acting on a segment distal to the joint are positive, clockwise moments are negative. Thus, a plantarflexor moment (acting on its distal segment) is negative, a knee extensor moment is shown to be positive, and a hip extensor moment is negative.

The moments are plotted during stance, with heel contact at 0 and toe-off at 680 ms. The ankle muscles generate a positive (dorsiflexor) moment for the first 80 ms of stance as the pretibial muscles act eccentrically to lower the foot to the ground. Then the plantarflexors increase their activity during mid and late stance. During midstance they act to control the amount of forward rotation of the leg over the foot, which is flat on the ground. As the plantarflexors generate their peak of about 60 N · m, they cause the foot to plantarflex and create a push-off. This concentric action results in a major generation of energy in normals (Winter, 1983), but in this patient it was somewhat reduced because of the pathology related to the patient's hip joint replacement. Prior to toe-off, the plantarflexor moment drops to 0 because that limb is now unloaded due to the fact that the patient is now weight bearing on his good limb and, for the last 90 ms prior to toe-off, the toe is just touching the ground with a light force. At this same point in time, the hip flexors are active to pull this limb upward and forward as a first phase of lower limb swing.

The knee muscles effectively show one pattern during all of stance. The quadriceps are active to generate an extensor moment, which acts to control the amount of knee flexion during early stance and also extends the knee during midstance. Even during push-off, when the knee starts flexing in prep-

*Representative paper: Pedotti, 1977.

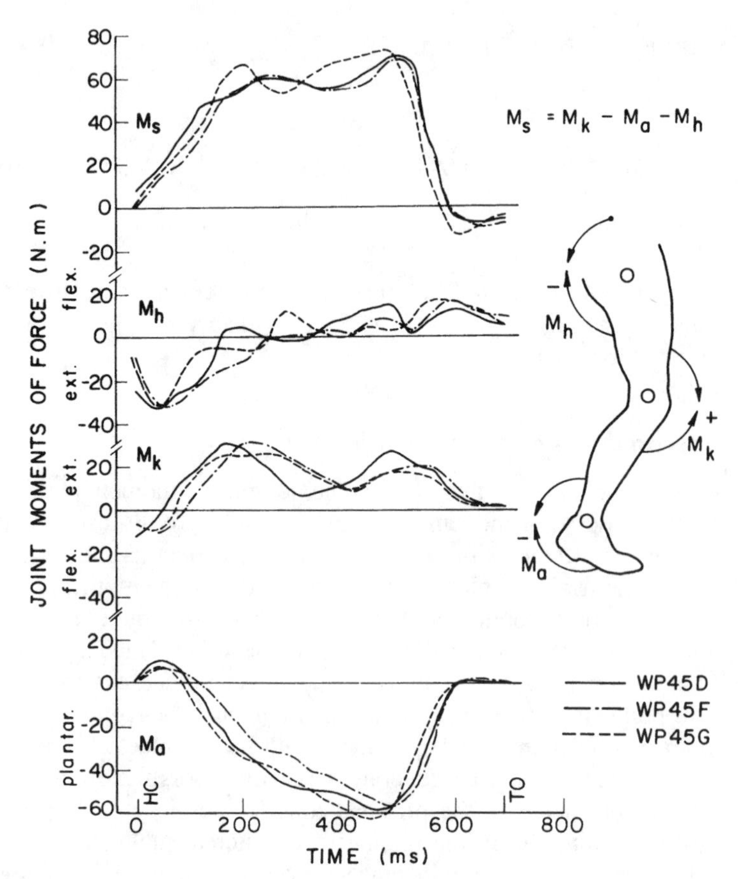

Figure 4.14 Joint moment-of-force profiles from three repeat trials of a patient fitted with a total hip replacement. See text for detailed discussion.

aration for swing, the quadriceps act eccentrically to control the amount of knee flexion. At heel contact, the hip moment is negative (extensor) and remains so until midstance. Such activity has two functions. First, the hip muscles act on the thigh to assist the quadriceps in controlling the amount of knee flexion. Second, the hip extensors act to control the forward rotation of the upper body as it attempts to rotate forward over the hip joint (the reaction force at the hip has a backward component during the first half of stance). Then, during the latter half of stance, the hip moment becomes positive (flexor), initially to reverse the backward rotating thigh and then, as described earlier, to pull the thigh forward and upward.

The fourth curve, M_s, bears some explanation. It is the net summation of the moments at all three joints, such that extensor moments are positive. It is called the *support moment* (Winter, 1980) because it represents a total limb pattern to push away from the ground. In scores of walking and running trials on normal subjects and patients, this synergy has been shown to be consistently positive during single support, in spite of considerable variability at individual joints (Winter, 1984). This latter curve is presented as an example to demonstrate that moment-of-force curves should not be looked at in isolation, but rather as part of a total integrated synergy in a given movement task.

4.2.7 A Note About the Wrong Way to Analyze Moments of Force

In the 1980s, an erroneous technique has evolved and is still in use in the clinical area in spite of technical notes and letters to the editor. This technique is referred as the FRFV (floor reaction force vector) approach and calculates the moments as depicted in Figure 4.15, which shows the ground reaction vector directed posteriorly during early stance. The FRFV approach calculates the moment to be equal to the magnitude of the vector times the perpendicular distance between the joint center and the vector. Thus, the ankle moment

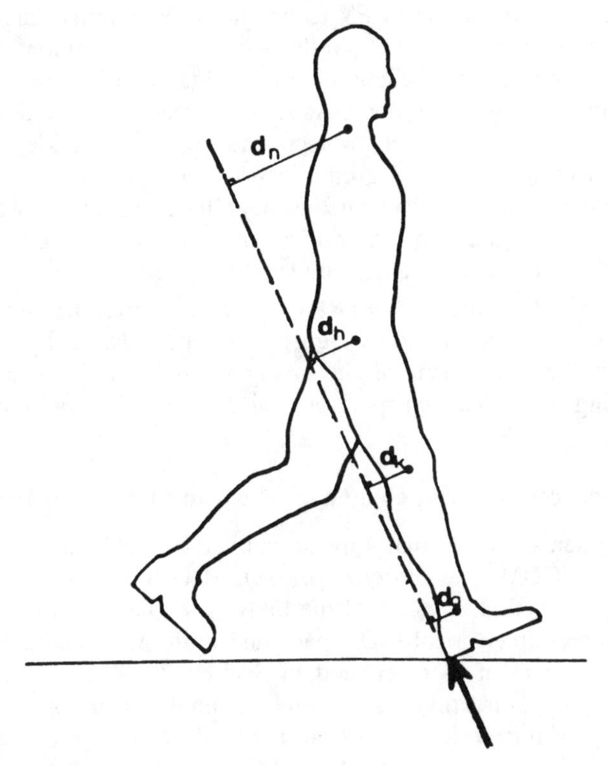

Figure 4.15 Projection of the ground reaction force vector that is used erroneously to predict the joint moments of force. The distance of the projection from the relevant joint center to the vector times the magnitude of that vector gives the net joint moment using FRFV approach. (Reproduced with permission of R. Wells.)

would be described as plantarflexor, at the knee it would be flexor, and it would be extensor at the hip. The folly of this technique becomes obvious if we extend the vector to calculate the moment of force at the neck. We would estimate an extensor neck moment several times larger than any moment calculated at the ankle, knee, or hip. There are three errors or shortcomings in the technique.

1. Numerically, the magnitude of the moment is incorrect because the mass–acceleration products and the moment of inertia–angular acceleration products of the stance limb are not accounted for. As was discussed in Section 4.2.2, the ground reaction vectors represent the algebraic summation of the mass–acceleration products of all segments. Using the correct technique, those effects must be subtracted as our calculation moves from the ground upward. Thus, by the time we reach the hip joint calculation, we have taken into account all inertial forces in the thigh, leg, and foot. The FRFV approach does not do this. Numerically the error has been shown to be negligible at the ankle, small but significant at the knee, and large at the hip (Wells, 1981).
2. The second error in the FRFV technique is the polarity that is attributed to the moment at the joint, and this reflects basic misconceptions about action and reaction. The FRFV is a reaction to the gravitational (external) and muscle (internal) forces acting on the body. As such the reaction force does not cause the moment of force, it merely reflects the cause. An ankle plantarflexor moment, for example, causes a ground action vector to move forward of the ankle joint, and this is reflected in the magnitude and the forward location (COP) of the ground reaction vector. If the ankle muscles were not generating any force, the FRFV would remain under the ankle joint.
3. A final shortcoming of the FRFV approach is that it is not capable of calculating moments during non-weight-bearing periods. Such a deficiency is significant in the assessment of the swing phase of walking and more so during running, and it is even more critical during athletic jumping events.

4.2.8 Differences Between Center of Mass and Center of Pressure

For many students and for many in the applied and clinical areas the terms *center of mass* (COM) and *center of pressure* (COP) are often misinterpreted or even interchanged. The COM of the body is the net location of the center of mass in three-dimensional (3D) space and is the weighted average of the COM of each segment as calculated in Section 3.1.4. The location of the COM in the vertical direction is sometimes called the *center of gravity* (COG). The trajectory of this vertical line from the COM to the ground allows us to compare the trajectories of the COM and COP. The trajectory of the COP is totally independent of the COM, and it is the location of the vertical ground reaction force vector from a single force platform, assuming that all body contact points are on that platform. The vertical ground reaction force is a weighted average of the location of all downward (action) forces acting on the force plate. These forces depend on the foot placement and the motor control of the ankle muscles. Thus, the COP is the neuromuscular response to the imbalances of the body's COM. The major misuse of the COP comes from researchers who refer to COP as "sway," thereby inferring it to be the kinematic measure COM.

The difference between COM and COP is demonstrated in Figure 4.16. Here we see a subject swaying back and forth while standing erect on a force plate. Each figure shows the changing situation at one of five different points in time. Time 1 has the body's COM (shown by the vertical body weight vector, W) to be ahead of the COP (shown by the vertical ground reaction vector, R). This "parallelogram of forces" acts at distances g and p, respectively, from the ankle joint. The magnitudes of W and R are equal and constant during quiet standing. Assuming the body to be pivoting about the ankles and neglecting the small mass of the feet, a counterclockwise moment equal to Rp and a clockwise moment equal to Wg will be acting. At Time 1, $Wg > Rp$ and the body will experience a clockwise angular acceleration α. It will also have a clockwise angular velocity ω. In order to correct this forward imbalance, the subject will increase his or her plantarflexor activity, which

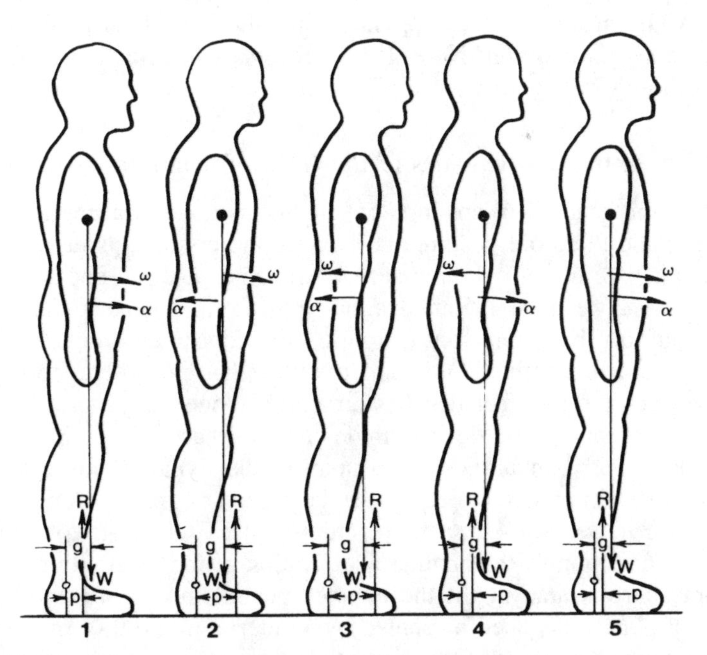

Figure 4.16 A subject swaying back and forth while standing on a force platform. Five different points in time are described, showing the COM and the COP locations along with the associated angular accelerations and velocities of the body. See text for detailed description.

will increase the COP such that at Time 2 the COP will be anterior of the COM. Now $Rp > Wg$. Thus, α will reverse and will start to decrease ω until, at Time 3, the time integral of α will result in a reversal of ω. Now both ω and α are counterclockwise, and the body will be experiencing a backward sway. The subject's response to the backward sway at Time 4 is to decrease his or her COP by reduced plantarflexor activation. Now $Wg > Rp$ and α will reverse, and after a period of time ω will decrease and reverse, and the body will return to the original conditions, as seen at Time 5. From this sequence of events relating COP to COM, it is evident that the plantarflexors/dorsiflexors vary the net ankle moment to control the COP and thereby regulate the body's COM. However, it is apparent that the COP must be continuously moving anterior and posterior of the COM; thus, the dynamic range of the COP must be somewhat greater than that of the COM. If the COM were allowed to move within a few centimeters of the toes, it is possible that a corrective movement of the COP to the extremes of the toes might not be adequate to reverse ω. Here the subject would be forced to take a step forward to arrest a forward fall.

Figure 4.17 is a typical 40-s record of the center of pressure (COP_x) and center of mass (COM_x) in the anterior/posterior direction of an adult subject standing quietly. Note that both signals are virtually in phase and that COP is slightly greater than COM. As was seen in the discussion regarding Figure 4.16, the COP must move ahead of and behind the COM in order to decelerate it and reverse its direction. Note that all reversals of direction of COM coincide with an overshoot of the COP signal.

4.2.9 Kinematics and Kinetics of the Inverted Pendulum Model

All human movement (except in space flights) is done in a gravitational environment, and therefore posture and balance are continuous tasks that must be accomplished. In normal daily activity at home, at work, and in our sports and recreation, we must maintain a safe posture and balance. The base of support can vary from one foot (running) to a four-point support (football), and it is essential that the COM remain within that base of support or move safely between the two feet if it lies temporarily outside the base of support (as it does in running and during single support phase of walking). There is a common model that allows us to analyze the dynamics of balance: the inverted pendulum model, which relates the trajectories of the COP and COM. As was seen in Section 4.2.8, the position of the COP relative to the COM decides the direction of the angular acceleration of the inverted pendulum. A full biomechanical analysis of the inverted pendulum model in both sagittal and frontal planes has been presented by Winter et al. (1998). In the sagittal plane, assuming the body swayed about the ankles, it was shown that:

$$COP_x - COM_x = -\frac{I_S \ddot{COM}_x}{Wh} \tag{4.9}$$

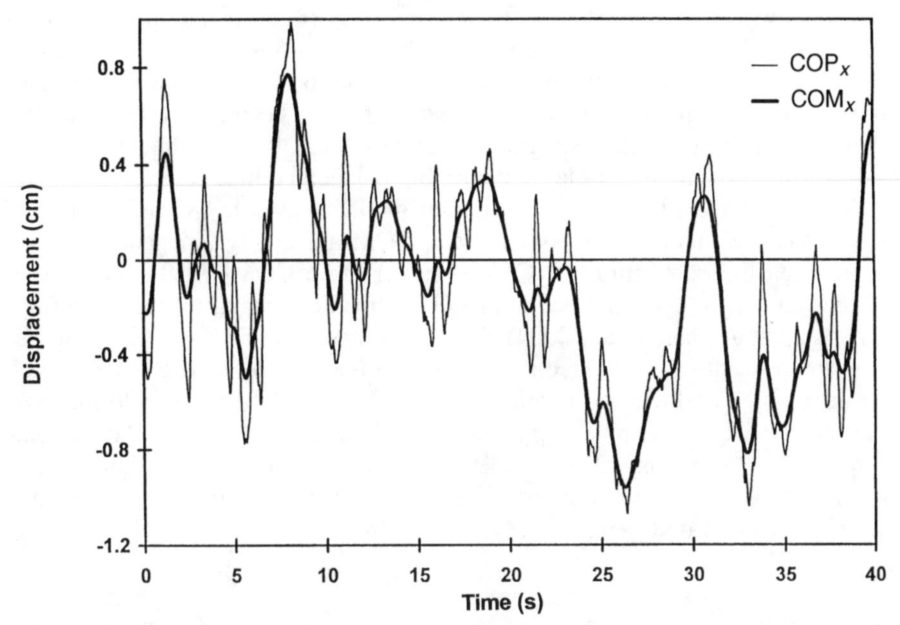

Figure 4.17 Typical 40-s record of the total body center of mass (COM_x) and center of pressure (COP_x) in the anterior/posterior direction during quiet standing. The COP amplitude exceeds that of the COM, and the reversals of direction of the COM (i.e., high +*ve* or −*ve* accelerations) are caused by overshoots of COP, as predicted by the discussion related to Figure 4.16.

where I_s = moment of inertia of body about ankles in sagittal plane
$\ddot{COM}_x$ = forward acceleration of COM
W = body weight above ankles
h = height of COM above ankles

In the frontal plane, the balance equation is virtually the same:

$$COP_z - COM_z = -\frac{I_f \ddot{COM}_z}{Wh} \tag{4.10}$$

where I_f = moment of inertia of body about ankles in frontal plane
$\ddot{COM}_z$ = medial/lateral acceleration of COM

The major difference between Equations (4.9) and (4.10) is in the muscle groups that control the COP. In Equation (4.9), $COP_x = M_a/R$ where M_a is the sum of the right and left plantarflexor moments and R is the total vertical reaction force at the ankles. In Equation (4.10), $COP_z = M_t/R$ where M_t =

$M_{al} + M_{ar} + M_{hl} + M_{hr}$. M_{al} with M_{ar} being the left and right frontal ankle moments while M_{hl} and M_{hr} are the left and right frontal plane hip moments. Thus, frontal plane balance is controlled by four torque motors, one at each corner of the closed-link parallelogram consisting of the two lower limbs and the pelvis. The hip abductor/adductor moments have been shown to be totally dominant in side-by-side standing (Winter et al., 1996), while the ankle invertor/evertor moments play a negligible balance control role.

The validity of the inverted pendulum model is evident in the validity of Equations (4.9) and (4.10). As COP and COM are totally independent measures, then the correlation of (COP − COM) with $\ddot{\text{COM}}$ will be a measure of the validity of this simplified model. Recent validations of the model during quiet standing (Gage et al., 2004) showed a correlation of $r = -0.954$ in the anterior/posterior (A/P) direction and $r = -0.85$ in the medial/lateral (M/L) direction. The lower correlation in the M/L direction was due to the fact that the M/L COM displacement was about 45% of that in the A/P direction. Similar validations have been made to justify the inverted pendulum model during initiation and termination of gait (Jian et al., 1993); correlations averaged −0.94 in both A/P and M/L directions.

4.3 BONE-ON-BONE FORCES DURING DYNAMIC CONDITIONS

Link-segment models assume that each joint is a hinge joint and that the moment of force is generated by a torque motor. In such a model, the reaction force calculated at each joint would be the same as the force across the surface of the hinge joint (i.e., the bone-on-bone forces). However, our muscles are not torque motors; rather, they are linear motors that produce additional compressive and shear forces across the joint surfaces. Thus, we must overlay on the free-body diagram these additional muscle-induced forces. In the extreme range of joint movement we would also have to consider the forces due to the ligaments and anatomical constraints. However, for the purpose of this text, we will limit the analyses to estimated muscle forces.

4.3.1 Indeterminacy in Muscle Force Estimates

Estimating muscle force is a major problem, even if we have good estimates of the moment of force at each joint. The solution is indeterminate as initially described in Section 1.3.4. Figure 4.18 demonstrates the number of major muscles responsible for the sagittal plane joint moments of force in the lower limb. At the knee, for example, there are nine muscles whose forces create the net moment from our inverse solution. The line of action of each of these muscles is different and continuously changes with time. Thus, the moment arms are also dynamic variables. Thus, the extensor moment is a net algebraic sum of the cross product of all force vectors and moment arm vectors,

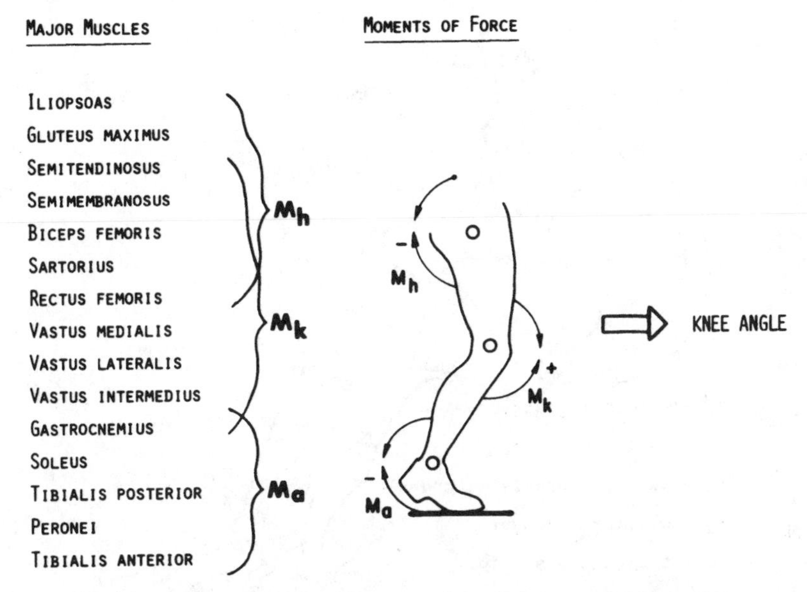

Figure 4.18 Fifteen major muscles responsible for the sagittal plane moments of force at the ankle, knee, and hip joints. During weight bearing, all three moments control the knee angle. Thus, there is considerable indeterminacy when relating knee angle changes to any single moment pattern or to any unique combination of muscle activity.

$$M_j(t) = \sum_{i=1}^{N_e} F_{ei}(t) \times d_{ei}(t) - \sum_{i=1}^{Nf} F_{fi}(t) \times d_{fi}(t) \quad (4.11)$$

where N_e = number of extensor muscles
N_f = number of flexor muscles
$F_{ei}(t)$ = force in ith extensor muscle at time t
$d_{ei}(t)$ = moment arm of ith extensor muscle at time t

Thus, a first major step is to make valid estimates of individual muscle forces and to combine them with a detailed kinematic/anatomical model of lines of pull of each muscle relative to the joint center (or our best estimate of it). Thus, a separate model must be developed for each joint, and a number of simplifying assumptions are necessary in order to resolve the indeterminacy problem. An example is now presented of a runner during the rapid push-off phase when the plantarflexors are dominant.

4.3.2 Example Problem (Scott and Winter, 1990)

During late stance, a runner's foot and ankle are shown (Figure 4.19) along with the direction of pull of each of the plantarflexors. The indeterminacy

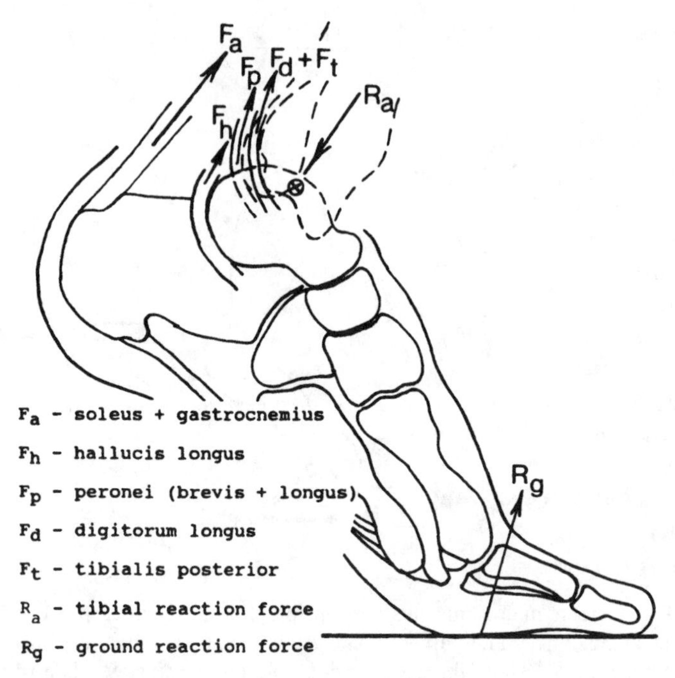

Figure 4.19 Anatomical drawing of foot and ankle during the push-off phase of a runner. The tendons for the major plantarflexors as they cross the ankle joint are shown, along with ankle and ground reaction forces.

problem is solved assuming that there is no cocontraction and that each active muscle's stress is equal [i.e., its force is proportional to its physiological cross-sectional area (PCA)]. Thus, Equation (4.11) can be modified as follows for the five major muscles of the extensor group acting at the ankle:

$$M_a(t) = \sum_{i=1}^{5} PCA_i \times S_{ei}(t) \times d_{ei}(t) \tag{4.12}$$

Since the PCA for each muscle is known and d_{ei} can be calculated for each point in time, the stress, $S_{ei}(t)$ can be estimated.

For this model, the ankle joint center is assumed to remain fixed and is located as shown in Figure 4.20. The location of each muscle origin and insertion is defined from the anatomical markers on each segment using polar coordinates. The attachment point of the ith muscle is defined by a distance R_i from the joint center and an angle θ_{mi} between the segment's neutral axis and the line joining the joint center to the attachment point. Thus, the coordinates of any origin or insertion at any instant in time are given by

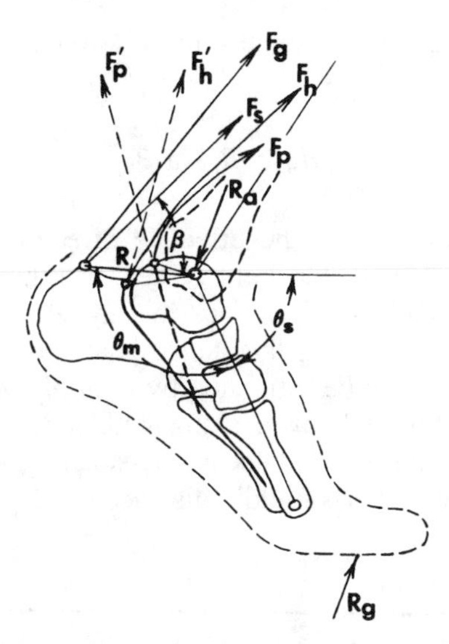

Figure 4.20 Free-body diagram of foot segment showing the actual and effective lines of pull of four of the plantarflexor muscles. Also shown are the reaction forces at the ankle and the ground as they act on the foot segment.

$$X_{mi}(t) = X_i(t) + R_i \cos\left[\theta_{mi}(t) + \theta_s(t)\right] \tag{4.13a}$$

$$Y_{mi}(t) = Y_i(t) + R_i \sin\left[\theta_{mi}(t) + \theta_s(t)\right] \tag{4.13b}$$

where $\theta_s(t)$ is the angle of the foot segment in the spatial coordinate system. In Figure 4.20, the free-body diagram of the foot is presented showing the internal anatomy to demonstrate the problems of defining the effective insertion point of the muscles and the effective line of pull of each muscle. Four muscle forces are shown here: soleus F_s, gastrocnemius F_g, flexor hallucis longus F_h, and peronei F_p. The ankle joint center is defined by a marker on the lateral malleolus. The insertion of the Achilles tendon is at a distance R with an angle θ_m from the foot segment (defined by a line joining the ankle to the fifth-metatarsal–phalangeal joint). The foot angular position is θ_s in the plane of movement. The angle of pull for the soleus and gastrocnemius muscles from this insertion is rather straightforward. However, for F_h and F_p the situation is quite different. The effective angle of pull is the direction of the muscle force as it leaves the foot segment. The flexor hallucis longus tendon curves under the talus hone and inserts on the distal phalanx of the big toe. As this tendon leaves the foot, it is rounding the pulleylike groove in the talus. Thus, its effective direction of pull is F_h'. Similarly, the peronei tendon curves around the distal end of the lateral malleolus. However, its effective direction of pull is F_p'.

The moment arm length d_{ei} for any muscle required by Equation (4.12), can now he calculated,

$$d_{ei} = R_i \sin \beta_i \tag{4.14}$$

where β_i is the angle between the effective direction of pull and the line joining the insertion point to the joint center. In Figure 4.20, the β for the soleus only is shown. Thus, it is now possible to calculate $S_{ei}(t)$ over the time that the plantarflexors act during the stance phase of a running cycle. Thus, each muscle force $F_{ei}(t)$ can be estimated by multiplying $S_{ei}(t)$ by each PCA_i. With all five muscle forces known, along with R_g and R_a, it is possible to estimate the total compressive and shear forces acting at the ankle joint. Figure 4.21 plots these forces for a middle-distance runner over the stance period

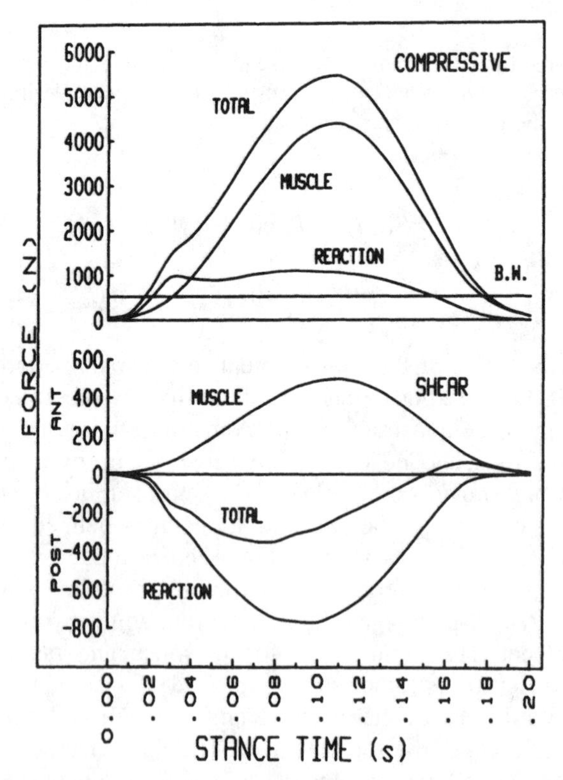

Figure 4.21 Compressive and shear forces at the ankle calculated during the stance phase of a middle-distance runner. The reaction force accounts for less than 20% of the total compressive force. The direction of pull of the major plantarflexors is such as to cause an anterior shear force (the talus is shearing anteriorly with respect to the tibia), which is opposite to that caused by the reaction forces. Thus, the muscle action can be classified as an antishear mechanism.

of 0.22 s. The compressive forces reach a peak of more than 5500 N, which is in excess of 11 times this runner's body weight. It is interesting to note that the ground reaction force accounts for only 1000 N of the total force, hut the muscle forces themselves account for over 4500 N. The shear forces (at right angles to the long axis of the tibia) are actually reduced by the direction of the muscle forces. The reaction shear force is such as to cause the talus to move posteriorly under the tibia. However, the line of pull of the major plantarflexors (soleus and gastrocnemius) is such as to pull the foot in an anterior direction. Thus, the total shear force is drastically reduced from 800 N to about 300 N, and the muscle action has been classified as an antishear mechanism.

4.4 PROBLEMS BASED ON KINETIC AND KINEMATIC DATA

1. **(a)** Plot the vertical ground reaction force (Table A.5*a* in the Appendix A) over the stance period (frames 28–69) and draw a line indicating the body weight. Discuss the reasons why this force exceeds the body weight during early stance and again during late stance. During the period in midstance, when this force is less than the body weight, what can we say about the acceleration of the center of mass of the body?
   (b) At heel contact right (HCR), the horizontal ground reaction force is positive. What does this tell you about the way the heel made contact with the ground?
   (c) If this subject were walking at a perfectly constant speed, how would this be indicated in the horizontal ground reaction force curve? Was this subject speeding up, slowing down, or at a constant speed?
2. **(a)** Calculate the reaction forces at the ankle in the x and y directions for frames 20 and 60 and compare your answers with those listed in Table A.5*a*.
   (b) Calculate the ankle moment of force for frame 72 using appropriate kinematic data plus data listed in Table A.5*a*. Discuss your answer in terms of ankle function during early swing.
   (c) Repeat Problem 2*b* for frame 60. Discuss your answer in terms of the role of the ankle muscles during late stance.
3. **(a)** Given the calculated ankle reaction forces, calculate the reaction forces at the knee in the x and y directions for frames 20 and 60 and compare your answers with those listed in Table A.5*a*.
   (b) Given the calculated ankle reaction forces and moments of force, calculate the knee moment of force for frame 90 using appropriate kinematic data plus data listed in Table A.5*a*. Discuss your answer in terms of the role of knee muscles during late swing.

(c) Repeat Problem 3*b* for frame 35. Discuss your answer in terms of the role of knee muscles during the weight acceptance phase of stance.

4. (a) Given the calculated knee reaction forces, calculate the reaction forces at the hip in the x and y directions for frames 20 and 60 and compare your answers with those listed in Table A.5*b*.
   (b) Given the calculated knee reaction forces and moment of force, calculate the hip moment of force from frame 90 using appropriate kinematic data plus data listed in Table A.5*b*. Discuss your answer in terms of hip muscle activity during late swing.
   (c) Repeat Problem 4*b* for frame 30. Discuss your answer in terms of the role of hip muscles during early weight acceptance.

5. (a) Calculate and plot the support moment for the total lower limb during stance (every second frame commencing with frame 28). The support moment M_s is the total extensor pattern at all three joints and $M_s = -M_a + M_k - M_h$. Note that the negative sign for M_a and M_h is introduced to make the clockwise extensor moments at those two joints positive, and vice versa when the moments are flexor. Discuss total lower limb synergy as seen in this total limb pattern.
   (b) Scan the polarity and magnitude of the ankle moments during the entire stride commencing at heel contact right (HCR) (frame 28) to the next HCR (frame 97) and describe the function of the ankle muscles over the stride period.
   (c) Repeat Problem 5*b* for the knee moments.
   (d) Repeat Problem 5*b* for the hip moments.

4.5 REFERENCES

Bresler, B. and J. P. Frankel. "The Forces and Moments in the Leg during Level Walking," *Trans. ASME* **72**:27–36, 1950.

Elftman, H. "Forces and Energy Changes in the Leg during Walking," *Am. J. Physiol.* **125**:339–356, 1939.

Gage, W. G., D. A. Winter, J. S. Frank, and A. L. Adkin. "Kinematic and Kinetic Validation of Inverted Pendulum Model in Quiet Standing," *Gait & Posture* **19**: 124–132, 2004.

Hsiao, H., J. Guan, and M. Wetherly. "Accuracy and Precision of Two In-Shoe Pressure Measurement Systems," *Ergonomics* **45**:537–555, 2002.

Jian, Y., D. A. Winter, M. G. Ishac, and L. Gilchrist. "Trajectory of the Body COG and COP during Initiation and Termination of Gait," *Gait and Posture* **1**:9–22, 1993.

Paul, J. P. "Forces Transmitted by Joints in the Human Body," *Proc. Inst. Mech. Eng.* **18**(3):8–15, 1966.

Pedotti, A. "A Study of Motor Coordination and Neuromuscular Activities in Human Locomotion," *Biol. Cybern.* **26**:53–62, 1977.

Pitie, D. L., M. Lord, A. Foster, S. Wilson, P. Watkins, and M. E. Edmonds. "Plantar Pressures Are Elevated in the Neuroischemic and Neuropathic Diabetic Foot," *Diabetic Care* **23**:1966–1970, 1999.

Scott, S. H. and D. A. Winter. "Internal Forces at Chronic Running Injury Sites," *Med. Sci. in Sports & Exercise* **22**:357–369, 1990.

Wells, R. P. "The Projection of Ground Reaction Force as a Predictor of Internal Joint Moments," *Bull. Prosthet. Res.* **18**:15–19, 1981.

Winter, D. A. "Overall Principle of Lower Limb Support during Stance Phase of Gait," *J. Biomech.* **13**:923–927, 1980.

Winter, D. A. "Energy Generation and Absorption at the Ankle and Knee during Fast, Natural and Slow Cadences," *Clin. Orthop. Rel. Res.* **197**:147–154, 1983.

Winter, D. A. "Kinematic and Kinetic Patterns in Human Gait: Variability and Compensating Effects," *Human Movement Sci.* **3**:51–76. 1984.

Winter, D. A., F. Prince, J. S. Frank, C. Powell, and K. F. Zabjek. "A Unified Theory Regarding A/P and M/L Balance During Quiet Standing," *J. Neurophysiol.* **75**: 2334–2343, 1996.

Winter, D. A., A. E. Patla, F. Prince, M. G. Ishac, and K. Gielo-Perczak. "Stiffness Control of Balance in Quiet Standing," *J. Neurophysiol.* **80**:1211–1221, 1998.

5

MECHANICAL WORK, ENERGY, AND POWER

5.0 INTRODUCTION

If we were to choose which biomechanical variable contains the most information, we would be forced to look at a variable that relates to the energetics. Without that knowledge, we would know nothing about the energy flows that *cause* the movement we are observing; and no movement would take place without those flows. Diagnostically, we have found joint mechanical powers to be the most discriminating in all our assessment of pathological gait. Without them, we could have made erroneous or incomplete assessments that would not have been detected by EMG or moment-of-force analyses alone. Also, valid mechanical work calculations are essential to any efficiency assessments that are made in sports and work-related tasks.

Before proceeding, the student should have certain terms and laws relating to mechanical energy, work, and power clearly in mind, and these are now reviewed.

5.0.1 Mechanical Energy and Work

Mechanical energy and work have the same units (joules) but have different meanings. Mechanical energy is a measure of the state of a body *at an instant in time* as to its ability to do work. For example, a body which has 200 J kinetic energy and 150 J potential energy is capable of doing 350 J of work (on another body). Work, on the other hand, is the measure of energy flow from one body to another, and time must elapse for that work to be done. If energy flows from body A to body B, we say that body A does work on body

B; or muscle *A* can do work on segment *B* if energy flows from the muscle to the segment.

5.0.2 Law of Conservation of Energy

At all points in the body at all instants in time, the law of conservation of energy applies. For example, any body segment will change its energy only if there is a flow of energy or out of any adjacent structure (tendons, ligaments, or joint contact surfaces). Figure 5.1 depicts a segment which is in contact at the proximal and distal ends and has four muscle attachments. In this situation, there are six possible routes for energy flow. Figure 5.1*a* shows the work (in joules) done at each of these points over a short period of time Δt. The law of conservation of energy states that the algebraic sum of all the energy flows must equal the energy change of that segment. In the case shown, $\Delta E_s = 4 + 2.4 + 5.3 - 1.7 - 0.2 - 3.8 = 6.0$ J. Thus, if we are able to calculate each of the individual energy flows at the six attachment points, we should be able to confirm ΔE_s through an independent analysis of mechanical energy of that segment. The balance will not be perfect because of measurement errors and because our link-segment model does not perfectly satisfy the assumptions inherent in link-segment analyses. A second way to look at the energy balance is through a *power balance,* which is shown in Figure 5.1*b* and effectively looks at the rate of flow of energy into and out

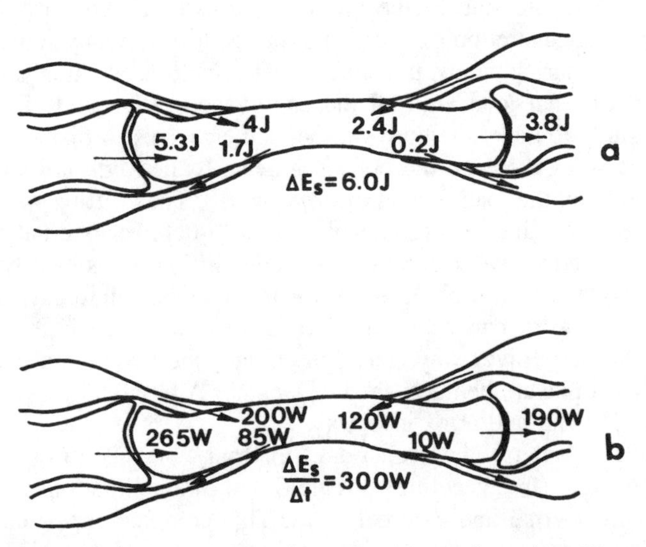

Figure 5.1 (*a*) Flow of energy into and out of a segment from the adjacent connective tissue and joint contacts over a period of time Δt. (*b*) Rate of flow of energy (power) for the same segment and same point in time as (*a*). A "power balance" can be calculated; see text for discussion.

of the segment and equates that to the rate of change of energy of the segment. Thus if, Δt is 20 ms, $\Delta E_s/\Delta t = 200 + 120 + 265 - 85 - 10 - 190 = 300$ W.

Another aspect of energy conservation takes place within each segment. Energy storage within each segment takes the form of potential and kinetic energy (translational and rotational). Thus, the segment energy E_s at any given point in time could be made up of any combination of potential and kinetic energies, quite independent of the energy flows into or out of the segment. In Section 5.3.1, the analysis of these components and the determination of the amount of conservation that occurs within a segment over any given period of time are demonstrated.

5.0.3 Internal versus External Work

The only source of mechanical energy generation in the human body is muscles, and the major site of energy absorption is also muscles. A very small fraction of energy is dissipated into heat as a result of joint friction and viscosity in the connective tissue. Thus, mechanical energy is continuously flowing into and out of muscles and from segment to segment. To reach an external load, there may be many energy changes in the intervening segments between the source and the external load. In a lifting task (Figure 5.2), the work rate on the external load might be 200 W, but the work rate to increase the energy of the total body by the source muscles of the lower limb might be 400 W. Thus, the sum of the internal and external work rates would be 600 W, and this generation of energy might result from many source muscles, as shown. Or, during many movement tasks such as walking and running, there is no external load, and all the energy generation and absorption are required simply to move the body segments themselves. A distinction is made between the work done on the body segments (called *internal* work) versus the work done on the load (called *external* work). Thus, lifting weights, pushing a car, or bicycling an ergometer have well-defined external loads. One exception to external work definition includes lifting one's own body weight to a new height. Thus, running up a hill involves both internal and external work. External work can be negative if an external force is exerted on the body and the body gives way. Thus, in contact sports, external work is regularly done on players being pushed or tackled. A baseball does work on the catcher as his hand and arms give way.

In bicycle ergometry, the cyclist does internal work just to move his limbs through the cycle (freewheeling). Figure 5.3 shows a situation where the cyclist did both internal and external work. This complex experiment has one bicycle ergometer connected via the chain to a second bicycle. Thus, one cyclist can bicycle in the forward direction (positive work), while the other cycles backwards (negative work). The assumption made by the researches who introduced this novel idea was that each cyclist was doing equal amounts of work (Abbot et al., 1952). This is not true because the positive-work cyclist

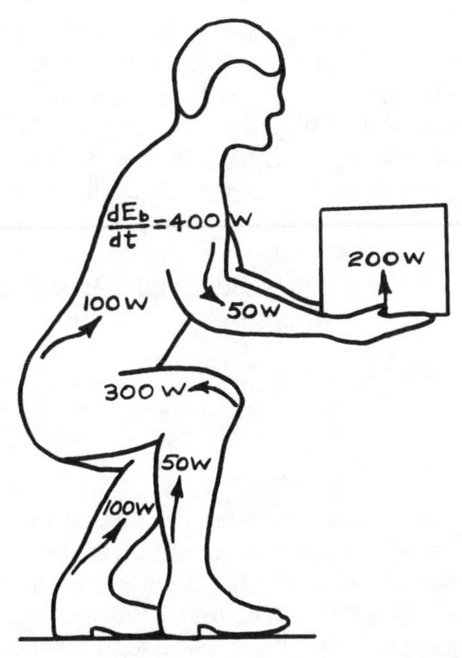

Figure 5.2 Lifting task showing the power generation from a number of muscles and the combined rate of change of energy of the body (internal work) and rate of energy flow to the load (external work).

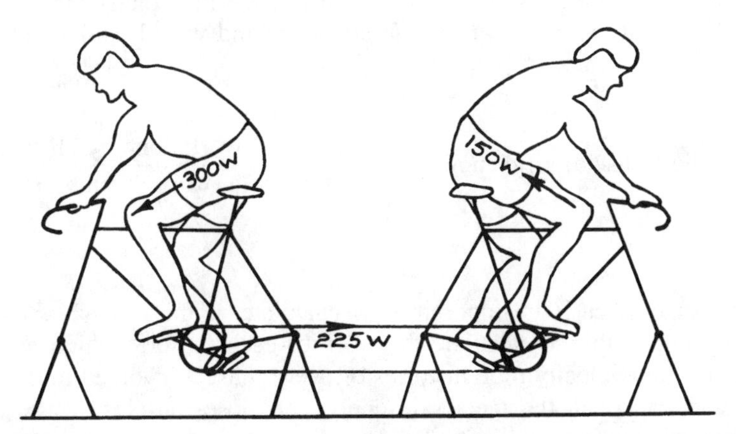

Figure 5.3 Bicycle ergometry situation where one subject (left) cycles in a forward direction and does work on a second subject who cycles in the reverse direction. The positive-work cyclist not only does external work on the negative-work cyclist but also does the internal work to move the limbs of both cyclists. Contrary to common interpretation, both cyclists are not performing equal magnitudes of mechanical work.

must do his or her own internal work plus the internal work on the negative-work cyclist plus any additional negative work of that cyclist. Thus, if the internal work of each cyclist was 75 W, the positive-work cyclist would have to do mechanical work at 150-W rate just to "freewheel" both cyclists. Then as the negative-work cyclist contracted his or her muscles, an additional load would be added. Thus, if the negative-work cyclist worked at 150 W, the positive-work cyclists would not be loaded to 300 W. It is no wonder that the negative-work cyclist can cycle with ease while the positive-work cyclist rapidly fatigues. They are simply not working at the same mechanical work rate, plus the metabolic demands of positive work far exceeds that of negative work.

5.1 EFFICIENCY

The term *efficiency* is probably the most abused and misunderstood term in human movement energetics. Confusion and error result from an improper definition of both the numerator and the denominator of the efficiency equation (Gaesser and Brooks, 1975; Whipp and Wasserman, 1969). In the next section, four causes of inefficiency are discussed in detail, and these mechanisms must be recognized in whatever formula evolves. Overlaid on these four mechanisms are two fundamental reasons for inefficiency: inefficiency in the conversion of metabolic energy to mechanical energy, and neurological inefficiency in the control of that energy. Metabolic energy is converted to mechanical energy at the tendon, and the metabolic efficiency depends on the conditioning of each muscle, the metabolic (fatigue) state of muscle, the subject's diet, and any possible metabolic disorder. This conversion of energy would be called *metabolic* or *muscle efficiency* and would be defined as follows:

$$\text{metabolic (muscle) efficiency} = \frac{\Sigma \text{ mechanical work done by all muscles}}{\text{metabolic work of muscles}} \tag{5.1}$$

Such an efficiency is impossible to calculate at this time because it is currently impossible to calculate the work of each muscle (which would require force and velocity time histories of every muscle involved in the movement) and to isolate the metabolic energy of those muscles. Thus, we are forced to compromise and calculate an efficiency based on segmental work and to correct the metabolic cost by subtracting estimates of overhead costs not associated with the actual mechanical work involved. Thus, an efficiency would be defined as

$$\text{mechanical efficiency} = \frac{\text{mechanical work (internal + external)}}{\text{metabolic cost} - \text{resting metabolic cost}} \tag{5.2}$$

The resting metabolic cost in bicycling, for example, could be the cost associated with sitting still on the bicycle.

A further modification is work efficiency, which is defined as

$$\text{work efficiency} = \frac{\text{external mechanical work}}{\text{metabolic cost} - \text{zero-work metabolic cost}} \tag{5.3}$$

The zero-work cost would be the cost measured with the cyclist freewheeling.

In all of the efficiency calculations described, there are varying amounts of positive and negative work. The metabolic cost of positive work exceeds that of equal levels of negative work. However, negative work is not negligible in most activities. Level gait has equal amounts of positive and negative work. Running uphill has more positive work than negative work, and vice versa for downhill locomotion. Thus, any of the efficiency calculations yield numbers that are strongly influenced by the relative percentages of positive and negative work. An equation that gets around this problem is

$$\text{metabolic cost (positive work)} + \text{metabolic cost (negative work)} = \text{metabolic cost}$$

or

$$\frac{\text{positive work}}{\eta_+} + \frac{\text{negative work}}{\eta_-} = \text{metabolic cost} \tag{5.4}$$

where η_+ and η_- are the positive and negative work efficiencies, respectively.

The interpretation of efficiency is faulty if it is assumed to be simply a measure of how well the metabolic system converts biochemical energy into mechanical energy, rather than a measure of how well the neural system is performing to control the conversion of that energy. An example will demonstrate the anomaly that results. A normal healthy adult walks with 100 J mechanical work per stride (half positive, half negative). The metabolic cost is 300 J per stride, and this would yield an efficiency of 33%. A neurologically disabled adult would do considerably more mechanical work because of his or her jerky gait pattern, say 200 J per stride. Metabolically, the cost might be 500 J per stride, which would given an efficiency of 40%. Obviously, the healthy adult is a more efficient walker, but our efficiency calculation does not reflect that fact. Neurologically, the disabled person is quite inefficient because he or she is not generating an effective and smooth neural pattern. However, the disabled is quite efficient in the actual conversion of metabolic energy to mechanical energy (at the tendon), and that is all that is reflected in the higher efficiency score.

5.1.1 Positive Work of Muscles

Positive work is work done during a concentric contraction, when the muscle moment acts in the same direction as the angular velocity of the joint. If a flexor muscle is causing a shortening, we can consider the flexor moment to be positive and the angular velocity to be positive. The product of muscle moment and angular velocity is positive; thus, power is positive, as depicted in Figure 5.4*a*. Conversely, if an extensor muscle moment is negative and an extensor angular velocity is negative, the product is still positive, as shown in Figure 5.4*b*. The integral of the power over the time of the contraction is the net work done by the muscle and represents generated energy transferred from the muscles to the limbs.

5.1.2 Negative Work of Muscles

Negative work is work done during an eccentric contraction when the muscle moment acts in the opposite direction to the movement of the joint. This usually happens when an external force, F_{ext}, acts on the segment and is such that it creates a joint moment greater than the muscle moment. The external force could include gravitational or ground reaction forces. Using the polarity convention as described, we can see in Figure 5.5*a* that we have a flexor moment (positive) with an extensor angular velocity (negative). The product yields a negative power, so that the work done during this angular change is negative. Similarly, when there is an extensor moment (negative) during a

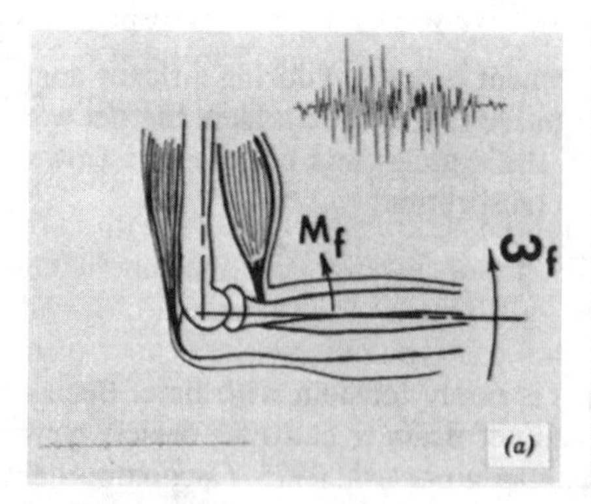

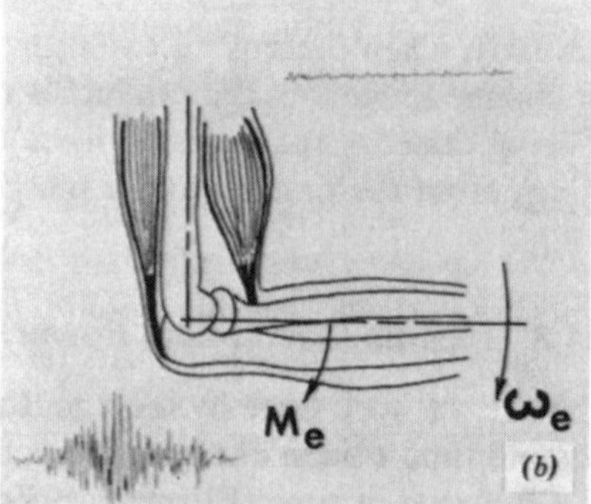

Figure 5.4 Positive power as defined by the net muscle moment and angular velocity. (*a*) A flexion moment acts while the forearm is flexing. (*b*) An extension moment acts during and extensor angular velocity. (Reproduced by permission of *Physiotherapy Canada.*)

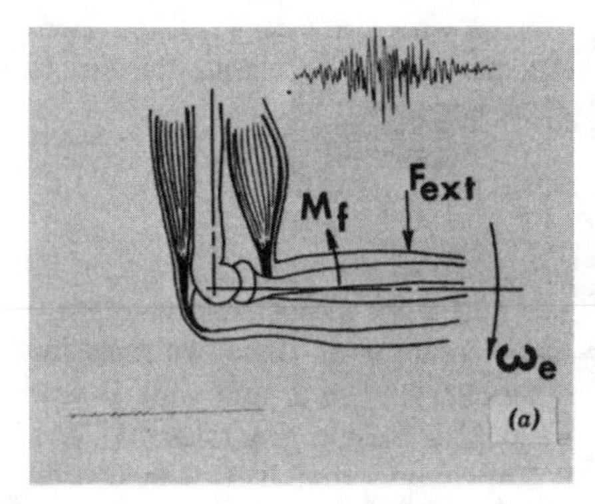

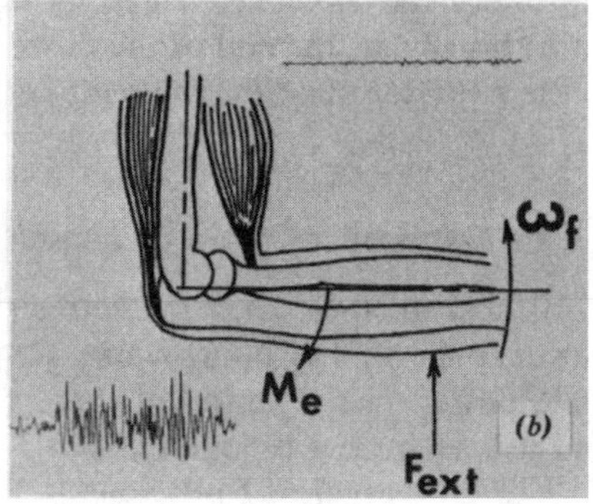

Figure 5.5 Negative power as defined by net muscle moment and angular velocity. (*a*) An external force causes extension when the flexors are active. (*b*) An external force causes flexion in the presence of an extensor muscle moment. (Reproduced by permission of *Physiotherapy Canada.*)

flexor angular change (positive), the product is negative (Figure 5.5*b*). Here, the net work is being done by the external force on the muscles and represents a flow of energy from the limbs into the muscles (absorption).

5.1.3 Muscle Mechanical Power

The rate of work done by most muscles is rarely constant with time. Because of rapid time-course changes, it has been necessary to calculate muscle power as a function of time (Elftman, 1939; Quanbury et al., 1975; Cappozzo et al., 1976; Winter and Robertson, 1978). At a given joint, muscle power is the product of the net muscle moment and angular velocity,

$$P_m = M_j \omega_j \quad \text{W} \tag{5.5}$$

where P_m = muscle power, watts
M_j = net muscle moment, N · m
ω_j = joint angular velocity, rad/s

As has been described in the previous sections, P_m can be either positive or negative. During even the simplest movements, the power will reverse sign several times. Figure 5.6 depicts the muscle moment, angular velocities, and muscle power as a function of time during a simple extension and flexion of the forearm. As can be seen, the time courses of M_j and ω_j are roughly out of phase by 90°. During the initial extension, there is an extensor moment and an extensor angular velocity as the triceps do positive work on the forearm. During the latter extension phase, the forearm is decelerated by the biceps (flexor moment). Here the biceps are doing negative work (absorbing

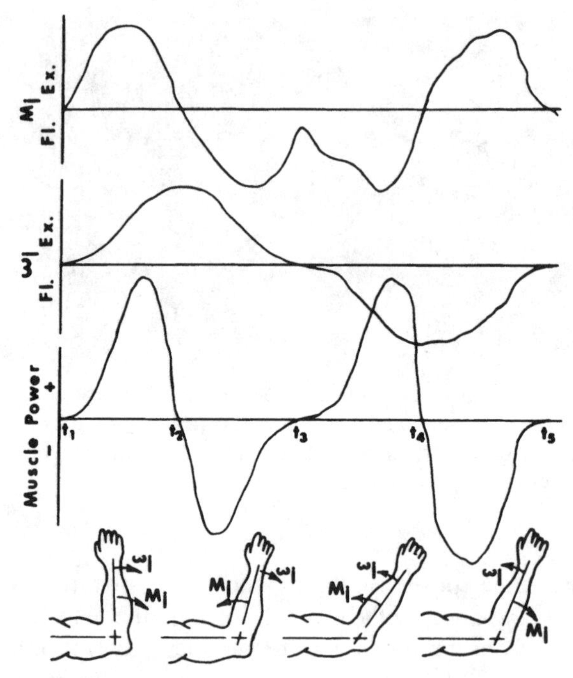

Figure 5.6 Sequence of events during simple extension and flexion of forearm. Muscle power shows two positive bursts alternating with two negative bursts.

mechanical energy). Once the forearm is stopped, it starts accelerating in a flexor direction still under the moment created by the biceps, which are now doing positive work. Finally, at the end of the movement, the triceps decelerate the forearm as the extensor muscles lengthen; here, P_m is negative.

5.1.4 Mechanical Work of Muscles

Until now we have used the terms *power* and *work* almost interchangeably. Power is the rate of doing work. Thus, to calculate work done, we must integrate power over a period of time. The product of power and time is work, and it is measured in joules (1 J = 1 W · s). If a muscle generates 100 W for 0.1 s, the mechanical work done is 10 J. This means that 10 J of mechanical energy has been transferred from the muscle to the limb segments. As the example of Figure 5.6 shows, power is continuously changing with time. Thus, the mechanical work done must be calculated from the time integral of the power curve. The work done by a muscle during a period t_1 to t_2 is

$$W_m = \int_{t_1}^{t_2} P_m \, dt \quad \text{J} \tag{5.6}$$

In the example described, the work done from t_1 to t_2 is positive, from t_2 to t_3 it is negative, from t_3 to t_4 it is positive again, and during t_4 to t_5 it is negative. If the forearm returns to the starting position, the net mechanical work done by the muscles is zero, meaning that the time integral of P_m from t_1 to t_5 is zero. It is therefore critical to know the exact times when P_m is reversing polarities in order to calculate the total negative and the total positive work done during the event.

5.1.5 Mechanical Work Done on an External Load

When any part of the body exerts a force on an adjacent segment or on an external body, it can only do work if there is movement. In this case, work is defined as the product of the force acting on a body and the displacement of the body in the direction of the applied force. The work dW done when a force causes an infinitesimal displacement ds is

$$dW = F \, ds \tag{5.7}$$

Or the work done when F acts over a distance S_1 is

$$W = \int_0^{S_1} F \, ds = FS_1 \tag{5.8}$$

If the force is not constant (which is most often the case), then we have two variables that change with time. Therefore, it is necessary to calculate the power as a function of time and integrate the power curve with respect to time to yield the work done. Power is the rate of doing work, or dW/dt.

$$\begin{aligned} P &= \frac{dW}{dt} = F \frac{ds}{dt} \\ &= \overline{F} \cdot \overline{V} \end{aligned} \tag{5.9}$$

where P = instantaneous power, watts
$\overline{F}$ = force, newtons
$\overline{V}$ = velocity, m/s

Since both force and velocity are vectors, we must take the dot product, or the product of the force and the component of the velocity that is in the same direction as the force. This will yield

$$P = FV \cos \theta = F_x V_x + F_y V_y \tag{5.10}$$

where θ = angle between force and velocity vectors in the plane defined by those vectors
F_x and F_y = forces in x and y directions
V_x and V_y = velocities in x and y directions

For the purpose of this initial discussion, let us assume that the force and the velocity are always in the same direction. Therefore, $\cos \theta = 1$ and

$$P = FV \quad \text{W}$$

$$W = \int_0^t P\, dt = \int_0^t FV\, dt \quad \text{J} \tag{5.11}$$

Example 5.1. A baseball is thrown with a constant accelerating force of 100 N for a period of 180 ms. The mass of the baseball is 1.0 kg, and it starts from rest. Calculate the work done on the baseball during the time of force application.

Solution

$$S_1 = ut + \tfrac{1}{2}at^2$$

$$u = 0$$

$$a = F/m = 100/1.0 = 100 \text{ m/s}^2$$

$$S_1 = \tfrac{1}{2} \times 100(0.18)^2 = 1.62 \text{ m}$$

$$W = \int_0^{S_1} F\, ds = FS_1 = 100 \times 1.62 = 162 \text{ J}$$

Example 5.2. A baseball of mass 1 kg is thrown with a force that varies with time, as indicated in Figure 5.7. The velocity of the baseball in the direction of the force is also plotted on the same time base and was calculated from the time integral of the acceleration curve (which has the same numerical value as the force curve because the mass of the baseball is 1 kg). Calculate the instantaneous power to the baseball and the total work done on the baseball during the throwing period.

The peak power calculated here may be considered quite high, but it should be noted that this peak has a short duration. The average power for the throwing period is less than 500 W. In real-life situations, it is highly unlikely that the force will ever be constant; thus, instantaneous power must always be

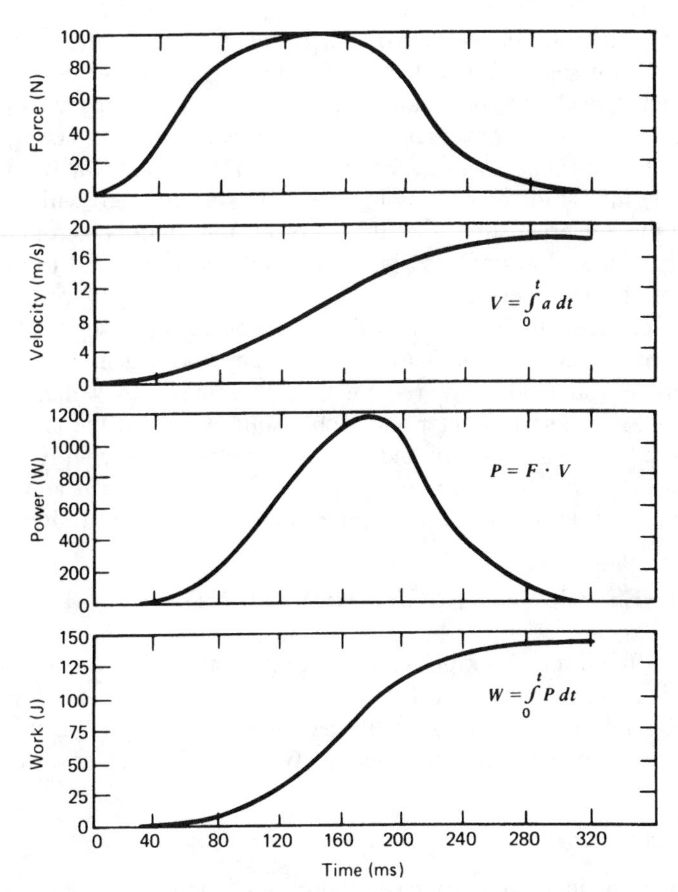

Figure 5.7 Forces, velocity, mechanical power, and work done on a baseball while being thrown. See text for details.

calculated. When the baseball is caught, the force of the hand still acts against the baseball, but the velocity is reversed. The force and velocity vectors are now in opposite directions. Thus, the power is negative and the work done is also negative, indicating that the baseball is doing work on the body.

5.1.6 Mechanical Energy Transfer Between Segments

Each body segment exerts forces on its neighboring segments, and if there is a translational movement of the joints, there is a mechanical energy transfer between segments. In other words, one segment can do work on an adjacent segment by a force displacement through the joint center (Quanbury et al., 1975). This work is in addition to the muscular work described in Sections 5.1.1 to 5.1.4. Equations (5.9) and (5.10) can be used to calculate the rate of energy transfer (i.e., power) across the joint center. Consider the situation in

Figure 5.8 at the joint between two adjacent segments. F_{j1}, the reaction force of segment 2 on segment 1, acts at an angle θ_1 from the velocity vector V_j. The product of $F_{j1}V_j \cos \theta_1$ is positive, indicating that energy is being transferred into segment 1. Conversely, $F_{j2}V_j \cos \theta_2$ is negative, denoting a rate of energy outflow from segment 2. Since $P_{j1} = -P_{j2}$, the outflow from segment 2 equals the inflow to segment 1. In an n-joint system there will be n power flows, but the algebraic sum of all those power flows will be zero, reinforcing the fact that these flows are passive and therefore do not add to or subtract from the total body energy.

This mechanism of energy transfer between adjacent segments is quite important in the conservation of energy of any movement because it is a passive process and does not require muscle activity. In walking, this has been analyzed in detail (Winter and Robertson, 1978). At the end of swing, for example, the swinging foot and leg lose much of their energy by transfer upward through the thigh to the trunk, where it is conserved and converted to kinetic energy to accelerate the upper body in the forward direction.

5.2 CAUSES OF INEFFICIENT MOVEMENT

It is often difficult for a therapist or coach to concentrate directly on efficiency. Rather, it is more reasonable to focus on the individual causes of inefficiency, and thereby automatically improve the efficiency of the movement. The four major causes of mechanical inefficiency (Winter, 1978) are now described.

5.2.1 Cocontractions

Obviously it is inefficient to have muscles cocontract because they fight against each other without producing a net movement. Suppose a certain

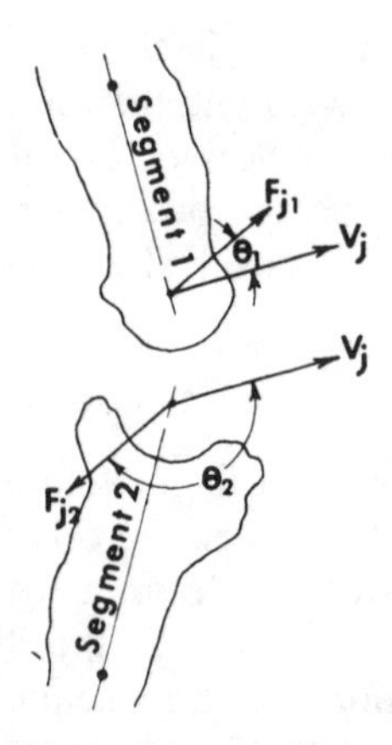

Figure 5.8 Reaction forces and velocities at a joint center during dynamic activity. The dot product of the force and velocity vectors is the mechanical power (rate of mechanical energy transfer) across the joint.

movement can be accomplished with a flexor moment of 30 N · m. The most efficient way to do this is with flexor activity only. However, the same movement can be achieved with 40 N · m flexion and 10 N · m extension, or with 50 N · m flexion and 20 N · m extension. In the latter case, there is an unnecessary 20 N · m moment in both the extensors and the flexors. Another way to look at this situation is that the flexors are doing unnecessary positive work to overcome the negative work of the extensors.

Cocontractions occur in many pathologies, notably hemiplegia and spastic cerebral palsy. They also occur to a limited extent during normal movement when it is necessary to stabilize a joint, especially if heavy weights are being lifted or at the ankle joint during walking or running. At present, the measurement of unnecessary cocontractions is only possible by monitoring the EMG activity of the antagonistic muscles. Without an exact EMG calibration versus tension for each muscle, it is impossible to arrive at a quantitative measure of cocontraction. Falconer and Winter (1985) presented a formula by which cocontraction can be quantified,

$$\%\text{COCON} = 2 \times \frac{M_{\text{antag}}}{M_{\text{agon}} + M_{\text{antag}}} \times 100\% \tag{5.12}$$

where M_{antag} and M_{agon} are the moments of force of antagonists and agonists, respectively.

In the example reported, the antagonist activity results in an equal increase in agonist activity; thus, the unnecessary activity must be twice that of the antagonist alone. If the antagonists created a 20 N · m extensor moment and the agonists generated a 50 N · m flexor moment, %COCON would be 40/70 × 100% = 57%. However, most movements involve continuously changing muscle forces; thus, an agonist muscle at the beginning of the movement will likely reverse its role and become an antagonist later on in the movement. Joint moments of force, as seen in many graphs in Chapter 4, reverse their polarities many times; thus, a modification of Equation (5.12) is needed to cope with these time-varying changes. Figure 5.9 demonstrates the profile of activity of two antagonistic muscles during a given movement. The cross-hatching of muscles *A* and *B* shows a common area of activity that indicates the cocontraction area. Thus, the percent cocontraction is defined as

$$\%\text{COCON} = 2 \times \frac{\text{common area } A\&B}{\text{area } A + \text{area } B} \times 100\% \tag{5.13}$$

If EMG is the primary measure of relative tension in the muscle, we can suitably process the raw EMG to yield a tension-related activation profile (Milner-Brown et al., 1973; Winter, 1976). The activity profiles of many common muscles look very much like those portrayed in Figure 5.9. In this case, muscle *A* is the tibialis anterior and muscle *B* is the soleus as seen over one

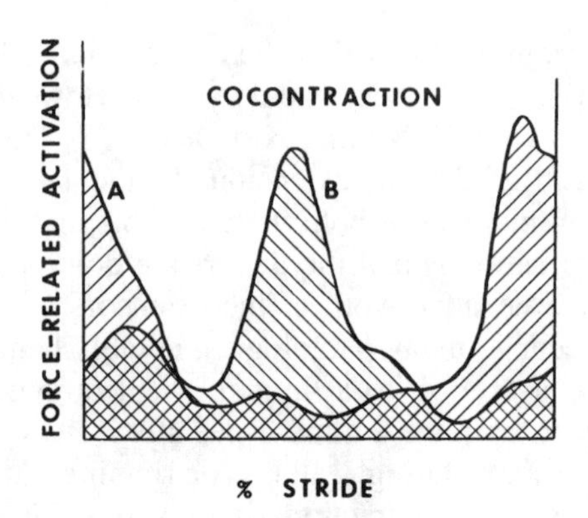

Figure 5.9 Profiles of activity of two antagonist muscles, with cross-hatched area representing the cocontraction. See text for detailed discussion and analysis.

walking stride. Using Equation (5.13) for the profiles in Figure 5.9, %COCON was calculated to be 24%.

5.2.2 Isometric Contractions Against Gravity

In normal dynamic movement, there is minimal muscle activity that can be attributed to holding limb segments against the forces of gravity. This is because the momentum of the body and limb segments allows for a smooth interchange of energy. However, in many pathologies the movement is so slow that there are extended periods of time when limb segments or the trunk are being held in near-isometric contractions. Spastic cerebral palsy patients often crouch with their knee flexed, requiring excessive quadriceps activity to keep them from falling down. Or, as seen in Figure 5.10, the crutch-walking cerebral palsy child holds her leg off the ground for a period of time prior to swing-through.

At the present time, it is impossible to quantify work against gravity because there is no movement involved. The only possible technique that might be used is the EMG, and each muscle's EMG would have to be calibrated against the extra metabolism required to contract that muscle. At present, no valid technique has been developed to separate the metabolic cost of this inefficiency.

5.2.3 Generation of Energy at One Joint and Absorption at Another

The least known and understood cause of inefficiency occurs when one muscle group at one joint does positive work at the same time as negative work is being done at others. Such an occurrence is really an extension of what occurs during a cocontraction (e.g., positive work being canceled out by negative work). It is quite difficult to visualize when this happens. During normal

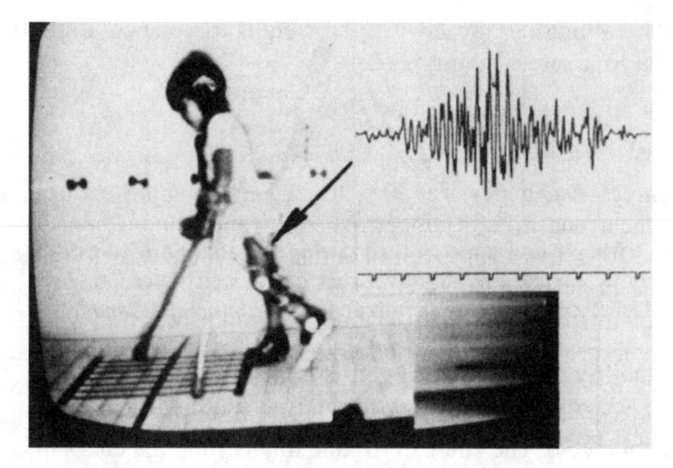

Figure 5.10 Example of "work" against gravity, one of the causes of inefficient movement. Here a cerebral palsy child holds her leg against gravity for an extended period prior to swinging through. (Reproduced by permission of *Physiotherapy Canada.*)

walking, it occurs during double support when the energy increase of the push-off leg takes place at the same time as the weight-accepting leg absorbs energy. Figure 5.11 shows this point in gait: the left leg push-off (positive work) is due primarily to plantarflexors; the right leg energy absorption (negative work) takes place in the quadriceps and tibialis anterior. There is no doubt that the instability of pathological gait is a major cause of this type of inefficient muscle activity. The only way to analyze such inefficiencies is to calculate the muscle power at each joint separately and to quantify the overlap of simultaneous phases of positive and negative work.

In spite of the inefficiencies inferred by such events, it must be remembered that many complex movements such as walking or running require that several functional tasks be performed at the same time. The example in Figure 5.11 illustrates such a situation: the plantarflexors are completing push-off while

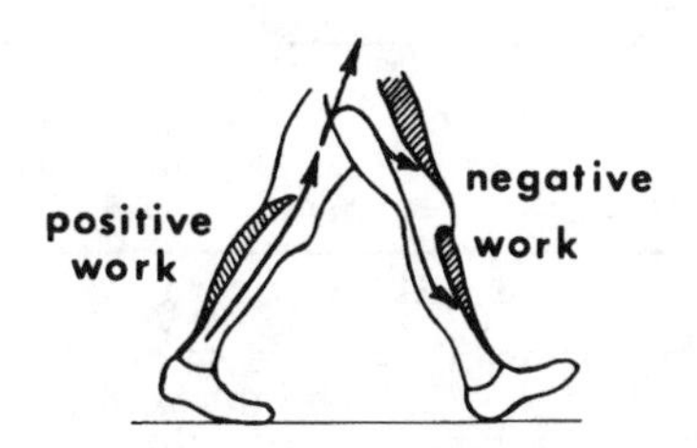

Figure 5.11 Example of a point in time during gait that positive work by the push-off muscles can be canceled by negative work of the weight-accepting muscles of the contralateral leg. (Reproduced by permission of *Physiotherapy Canada.*)

the contralateral muscles are involved in weight acceptance. Both these events are essential to a safe walking pattern.

5.2.4 Jerky Movements

Efficient energy exchanges are characterized by smooth-looking movements. A ballet dancer and a high jumper execute smooth movements for different reasons, one for artistic purposes, the other for efficient performance. Energy added to the body by positive work at one point in time is conserved, and very little of this energy is lost by muscles doing negative work. The jerky gait of a cerebral palsy child is quite the opposite. Energy added at one time is removed a fraction of a second later. The movement has a steady succession of stops and starts, and each of these bursts of positive and negative work has a metabolic cost. The energy cost due to jerky movements can be assessed in two ways: by work analysis based on a segment-by-segment energy analysis or by a joint-by-joint power analysis. Both of these techniques are described later.

5.2.5 Summary of Energy Flows

It is valuable to summarize the flows of energy from the metabolic level through to an external load. Figure 5.12 depicts this process schematically. Metabolic energy cannot be measured directly, but can be calculated indirectly from the amount of O_2 required or by the CO_2 expired. The details of these calculations and their interpretation are the subject of many textbooks and are beyond the scope of this book.

At the basal level (resting, lying down), the muscles are relaxed but still require metabolic energy to keep them alive. The measure of this energy level is called *maintenance heat.* Then as a muscle contracts, it requires energy, which shows up as additional heat called *activation heat.* It has been shown to be associated with the rate of buildup of tension within the muscle and is accompanied by an internal shortening of the muscle contractile elements. Stable heat is the heat which measures the energy required to maintain tension

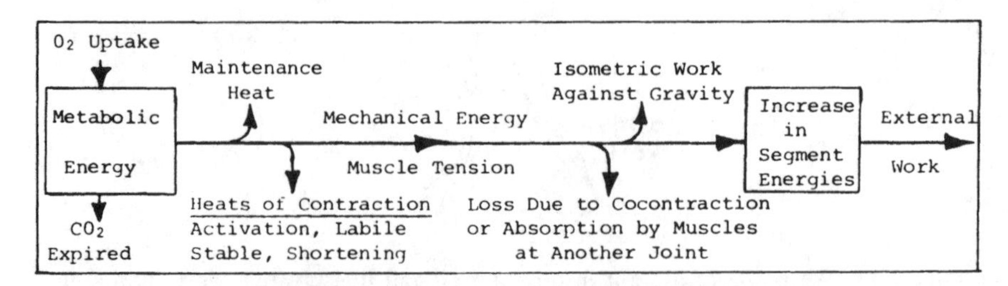

Figure 5.12 Flow of energy from metabolic level of external mechanical work. Energy is lost as heat associated with the contractile process or by inefficiencies after energy has been converted to mechanical energy.

within the muscle. *Labile heat* is a third heat seen in isometric contractions, and is the heat not accounted for by either tension or rate of tension generation. The final heat loss is *shortening heat,* that associated with the actual shortening of the muscle under load. Students are referred to the excellent review by Hill (1960).

Finally, at the tendon we see the energy in mechanical form. The muscle tension can contribute to four types of mechanical loads. It can be involved with a cocontraction, isometric "work" against gravity, or a simultaneous absorption by another muscle, or it can cause a net change in the body energy. In the latter case, if positive work is done, the net body energy will increase; if negative work is done, there will be a decrease in total body energy. Finally, if the body exerts forces on an external body, some of the energy may be transferred as the body performs external work.

5.3 FORMS OF ENERGY STORAGE

1. Potential Energy. Potential energy (PE) is the energy due to gravity and, therefore, increases with the height of the body above ground or above some other suitable reference datum,

$$\text{PE} = mgh \quad \text{J} \tag{5.14}$$

where m = mass, kg
g = gravitational acceleration, = 9.8 m/s^2
h = height of center of mass, meters

With $h = 0$, the potential energy decreases to zero. However, the ground reference datum should be carefully chosen to fit the problem in question. Normally it is taken as the lowest point the body takes during the given movement. For a diver, it could be the water level; for a person walking, it would be the lowest point in the pathway.

2. Kinetic Energy. There are two forms of kinetic energy (KE), that due to translational velocity and that due to rotational velocity,

$$\text{translational KE} = \tfrac{1}{2}mv^2 \quad \text{J} \tag{5.15}$$

where v = velocity of center of mass, m/s

$$\text{rotational KE} = \tfrac{1}{2}I\omega^2 \quad \text{J} \tag{5.16}$$

where I = rotational moment of inertia, $\text{kg} \cdot \text{m}^2$
ω = rotational velocity of segment, rad/s

Note that these two energies increase as the velocity squared. The polarity of direction of the velocity is unimportant because velocity squared is always positive. The lowest level of kinetic energy is therefore zero when a body is at rest.

3. Total Energy and Exchange within a Segment. As mentioned previously, energy of a body exists in three forms so that the total energy of a body is

$$\begin{aligned} E_s &= \text{PE} + \text{translational KE} + \text{rotational KE} \\ &= mgh + \tfrac{1}{2}mv^2 + \tfrac{1}{2}I\omega^2 \quad \text{J} \end{aligned} \tag{5.17}$$

It is possible for a body to exchange energy within itself and still maintain a constant total energy.

Example 5.3. Suppose the baseball in Example 5.1 is thrown vertically. Calculate the potential and kinetic energies at the time of release, at maximum height, and when its reaches the ground. Assume that it is released at a height of 2 m above the ground, and that the vertical accelerating force of 100 N is in excess of gravitational force. At release,

$$a = 100 \text{ m/s}^2 \qquad \text{(as calculated previously)}$$

$$v = \int_0^{t_1} a\, dt = at_1 = 100t_1$$

$$t_1 = 180 \text{ ms}$$

$$v = 18 \text{ m/s}$$

$$\text{translational KE} = \tfrac{1}{2}mv^2 = \tfrac{1}{2} \times 1 \times 18^2 = 162 \text{ J}$$

Note that this 162 J is equal to the work done on the baseball prior to release.

$$\begin{aligned} \text{total energy} &= \text{PE}(t_1) + \text{translational KE}(t_1) \\ &= 19.6 + 162 = 181.6 \text{ J} \end{aligned}$$

If we ignore air resistance, the total energy remains constant during the flight of the baseball, such that at t_2, when the maximum height is reached, all the energy is potential energy and KE $= 0$. Therefore, PE$(t_2) = 181.6$ J. This means that the baseball reaches a height such that $mgh_2 = 181.6$ J, and

$$h_2 = \frac{181.6}{1.0 \times 9.8} = 18.5 \text{ m}$$

At t_3 the baseball strikes the ground, and $h = 0$. Thus, $\text{PE}(t_3) = 0$ and $\text{KE}(t_3) = 181.6$ J. This means that the velocity of the baseball is such that $\frac{1}{2}mv^2 = 181.6$ J, or

$$v = 19.1 \text{ m/s}$$

This velocity is slightly higher than the release velocity of 18 m/s because the ball was released from 2 m above ground level.

5.3.1 Energy of a Body Segment and Exchanges of Energy Within the Segment

Most body segments contain all three energies in various combinations at any point in time during a given movement. A diver at the top of a dive has considerable potential energy, and during the dive, converts it to kinetic energy. Similarly, a boomerang when released has rotational and translational kinetic energy, and at peak height some of the translational kinetic energy has been converted to potential energy. At the end of its travel, the boomerang will have regained most of its translational kinetic energy.

In a multisegment system, such as the human body, the exchange of energy can be considerably more complex. There can be exchanges within a segment or between adjacent segments. A good example of energy exchange within a segment is during normal gait. The upper part of the body [head, arms, and trunk (HAT)] has two peaks of potential energy each stride—during midstance of each leg. At this time, HAT has slowed its forward velocity to a minimum. Then, as the body falls forward to the double support position, HAT picks up velocity at the expense of a loss in height. Evidence of energy exchange should be seen from a plot of the horizontal velocity and the vertical displacement of the center of gravity of HAT (Figure 5.13). The potential energy, which varies with height, changes roughly as a sinusoidal wave, with a minimum during double support and reaching a maximum during midstance. The forward velocity is almost completely out of phase, with peaks approximately during double support and minima during midstance.

Exchanges of energy within a segment are characterized by opposite changes of the potential and kinetic energy components. Figure 5.14 shows what would happen if a perfect exchange took place, as in a swinging frictionless pendulum. The total energy would remain constant over time in the presence of large changes of potential and kinetic energy.

Consider the other extreme, in which no energy exchange takes place. Such a situation would be characterized by totally in-phase energy components, not necessarily of equal magnitude, as depicted in Figure 5.15

5.3.1.1 Approximate Formula for Energy Exchanges Within a Segment. An approximation of energy exchange can be calculated if we know the peak-to-peak change in each of the energy components over a period of time:

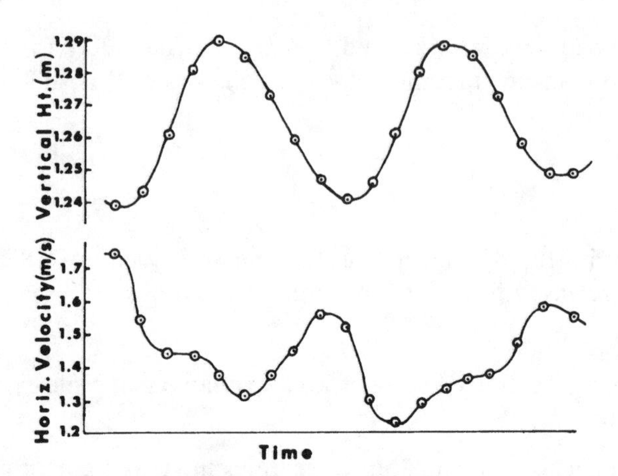

Figure 5.13 Plot of vertical displacement and horizontal velocity of HAT shows evidence of energy exchange within the upper part of the body during gait.

$$E_{ex} = \Delta E_p + \Delta E_{kt} + \Delta E_{kr} - \Delta E_s \tag{5.18}$$

If there is no exchange, $\Delta E_p + \Delta E_{kt} + \Delta E_{kr} = \Delta E_s$. If there is 100% exchange, $\Delta E_s = 0$.

Example 5.4. A visual scan of the energies of the leg segment during walking yields the following maximum and minimum energies on the stride period: E_s (max) = 29.30 J, E_s (min) = 13.14 J, E_p (max) = 15.18 J, E_p (min) = 13.02 J, E_{kt} (max) = 13.63 J, E_{kt} (min) = 0.09 J, E_{kr} (max) = 0.95 J, E_{kr} (min) = 0 J. Thus, $\Delta E_s = 29.30 - 13.14 = 16.16$ J, $\Delta E_p = 15.18 - 13.02 = 2.16$ J, $\Delta E_{kt} = 13.63 - 0.09 = 13.54$ J, $\Delta E_{kr} = 0.95 - 0 = 0.95$ J.

Since $\Delta E_p + \Delta E_{kt} + \Delta E_{kr} = 16.65$ J, it can be said that $16.65 - 16.16 = 0.49$ J exchanged during the stride. Thus, the leg is a highly nonconservative system.

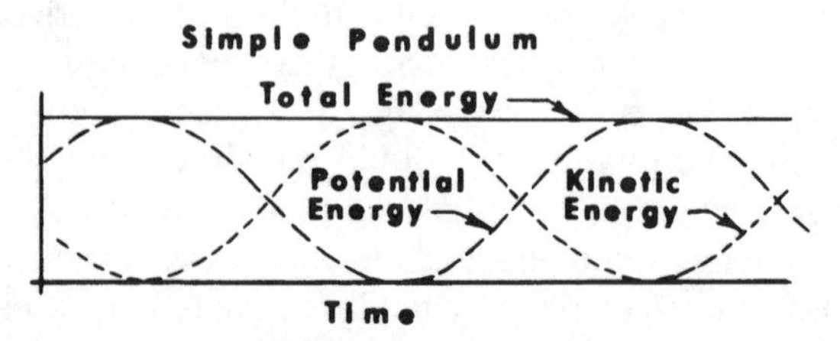

Figure 5.14 Exchange of kinetic and potential energies in a swinging frictionless pendulum. Total energy of the system is constant, indicating that no energy is being added or lost. (Reproduced by permission of *Physiotherapy Canada.*)

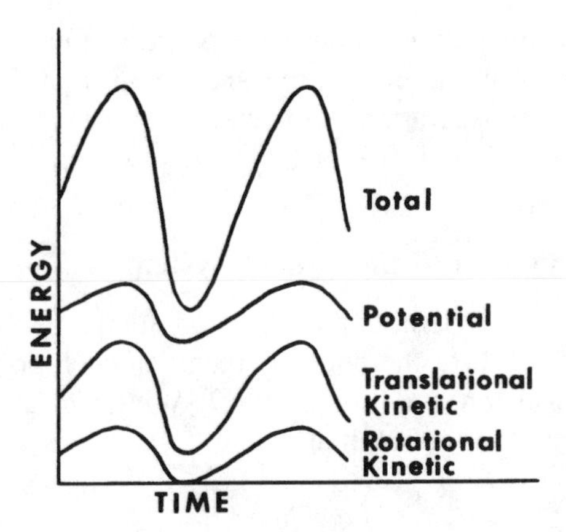

Figure 5.15 Energy patterns for a segment in which no exchanges are taking place. All energy components are perfectly in phase.

5.3.1.2 Exact Formula for Energy Exchange within Segments. The example just discussed illustrates a simple situation where only one minimum and maximum occur over the period of interest. If individual energy components have several maxima and minima, we must calculate the sum of the absolute energy changes over the time period. The work W_s done on and by a segment during N sample periods is

$$W_s = \sum_{i=1}^{N} |\Delta E_s| \quad \text{J} \tag{5.19}$$

Assuming no energy exchanges between any of the three components (Normal et al., 1976), the work done by the segment during the N sample periods is

$$W_s' = \sum_{i=1}^{N} (|\Delta E_p| + |\Delta E_{kt}| + |\Delta E_{kr}|) \quad \text{J} \tag{5.20}$$

Therefore, the energy W_c conserved within the segment during the time is

$$W_c = W_s' - W_s \quad \text{J} \tag{5.21}$$

The percentage energy conservation C_s during the time of this event is

$$C_s = \frac{W_c}{W_s'} \times 100\% \tag{5.22}$$

If $W_s' = W_s$, all three energy components are in phase (they have exactly the same shape and have their minima and maxima at the same time), and there is no energy conservation. Conversely, as demonstrated by an ideal pendulum, if $W_s = 0$, then 100% of the energy is being conserved.

5.3.2 Total Energy of a Multisegment System

As we proceed with the calculation of the total energy of the body, we merely sum the energies of each of the body segments at each point in time (Bresler and Berry, 1951; Ralston and Lukin, 1969; Winter et al., 1976). Thus, the total body energy E_b at a given time is

$$E_b = \sum_{i=1}^{B} E_{si} \quad \text{J} \tag{5.23}$$

where E_{si} = total energy of ith segment at that point in time
B = number of body segments

The individual segment energies continuously change with time, so it is not surprising that the sum of these energies will also change with time. However, the interpretation of changes in E_b must be done with caution when one considers the potential for transfer of energy between segments and the number of possible generators and absorbers of energy at each joint. For example, transfers of energy between segments (see Section 5.1.6) will not result in either an increase or a loss in total body energy. However, in other than the simplest movements, there may be several simultaneous concentric and eccentric contractions. Thus, over a period of time, two muscle groups may generate 30 J, while one other muscle group may absorb 20 J. The net change in body energy over that time would be an increase of 10 J. Only through a detailed analysis of mechanical power at the joints (see Section 5.4.1.4) are we able to assess the extent of such a cancellation. Such events are obviously inefficient, but are necessary to accomplish many desired movement patterns.

Consider now a simple muscular system that can be represented by a pendulum mass with a pair of antagonistic muscle groups m_1 and m_2, crossing a simple hinge joint. Figure 5.16 shows such an arrangement along with a time history of the total energy of the system. At t_1, the segment is rotating counterclockwise at ω_1 rad/s. No muscle activity occurs until t_2, when m_2 contracts. Between t_1 and t_2, the normal pendulum energy exchange takes place and the total energy remains constant. However, between t_2 and t_3, muscle m_2 causes an increase in both kinetic and potential energies of the segment. The muscle moment has been in the same direction as the direction of rotation, so positive work has been done by the muscle on the limb segment, and the total energy of the limb has increased. Between t_3 and t_4, both muscles are inactive, and the total energy remains at the higher but constant level. At t_4, muscle m_1

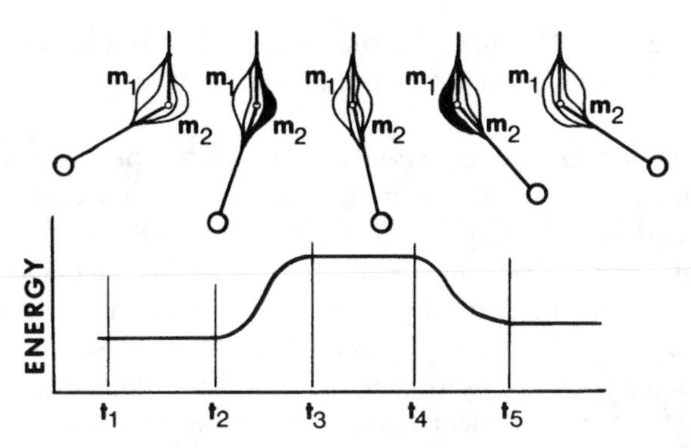

Figure 5.16 Pendulum system with muscles. When positive work is done, the total energy increases; when negative work is done, the total energy decreases.

contracts to slow down the segment. Energy is lost by the segment and is absorbed by m_1. This is negative work being done by the muscle because it is lengthening during its contraction. Thus, at t_5 the segment has a lower total energy than at t_4.

The following major conclusions can be drawn from this example.

1. When muscles do positive work, they increase the total body energy.
2. When muscles do negative work, they decrease the total body energy.
3. During cyclical activity, such as level running, the net energy change per stride equals zero. Thus, the internal positive work done per stride equals the internal negative work done per stride.

5.4 CALCULATION OF INTERNAL AND EXTERNAL WORK

Internal and external work has been calculated in a multitude of ways by different researchers. Some assume that the energy changes in the body center of mass yield the total internal work done by all the muscles. Others look only at the "vertical work" resulting from potential energy increases of the body center of gravity, while others (especially in the exercise physiology area) even ignore internal work completely. It is therefore important to look at all possible sources of muscle activity that have a metabolic cost and keep those readily available on a "checklist" to see how complete any analysis really is. This same list serves a useful purpose in focusing our attention on possible causes of inefficient movement (see Section 5.2).

5.4.1 Internal Work Calculation

The various techniques for calculating the internal work have undergone a general improvement over the years. The vast majority of the research has been done in the area of human gait, and because gait is a complex movement, it will serve well as an example of the dos and don'ts.

5.4.1.1 ***Energy Increases in Segments.*** A number of early researchers attempted a calculation of work based on increases in potential or kinetic energies of the body or of individual segments. Fenn (1929), in his accounting of the flow of energy from metabolic to mechanical, calculated the kinetic and potential energies of each major segment of a sprinter. He then summed the increases in each of these segment energies over the stride period to yield the net mechanical work. Unfortunately, Fenn's calculations ignored two important energy-conserving mechanisms: energy exchanges within segments and passive transfers between segments. Thus, his mechanical work calculations were predictably high: the average power of his sprinters was computed to be 3 horsepower. Conversely, Saunders et al. (1953), Cotes and Meade (1960), and Liberson (1965) calculated the "vertical work" of the trunk as representing the total work done by the body. These calculations ignored the major energy exchange that takes place within the HAT and also the major work done by the lower limbs.

5.4.1.2 ***Center of Mass Approach.*** Cavagna and Margaria proposed in (1966) and in many subsequent papers a technique that is based on the potential and kinetic energies of the body's center of mass. Their data were based on force platform records during walking and running, from which the translational kinetic and potential energies were calculated. Such a model makes the erroneous assumption that the body's center of mass reflects the energy changes in all segments. The body center of mass is a *vector* sum of all segment mass–acceleration products, and, as such, opposite-polarity accelerations will cancel. But energies are scalars, not vectors, and therefore, the reciprocal movements that dominate walking and running will be largely canceled. Thus simultaneous increases and decreases in oppositely moving segments will go unnoticed. Also, Cavagna's technique is tied to force platform data, and nothing is known about the body center of gravity during nonweight-bearing phases of running. Thus, this technique has underestimation errors and limitations that have been documented (Winter, 1979). Also, the center of mass approach does not account for the energy losses due to the simultaneous generation and absorption of energy at different joints.

5.4.1.3 ***Sum of Segment Energies.*** A major improvement on the previous techniques was made by Ralston and Lukin (1969) and Winter et al. (1976). Using displacement transducers and TV imaging techniques, the kinetic and potential energies of the major segments were calculated. A sum of the energy components within each segment recognized the conservation of energy within each segment (see Section 5.3.1) and a second summation across all segments recognized energy transfers between adjacent segments (see Section 5.1.6). The total body work is calculated (Winter, 1979) to be

$$W_b = \sum_{i=1}^{N} |\Delta E_b| \quad \text{J} \tag{5.24}$$

However, this calculation underestimates the simultaneous energy generation and absorption at different joints. Thus, W_b will reflect a low estimate of the positive and negative work done by the human motor system. Williams and Cavanagh (1983) made empirical estimates to correct for these underestimates in running.

5.4.1.4 Joint Power and Work. In Sections 5.1.3 and 5.1.4, techniques for the calculation of the positive and negative work at each joint were presented. Using the time integral of the power curve [Equation (5.6)], we are able to get at the "sources" and "sinks" of all the mechanical energy. Figure 5.17 is an example to show the work phases at the knee during slow running. The power bursts are labeled $K_1, \cdots, K_5$, and the energy generation/absorption resulting from the time integral of each phase is shown (Winter, 1983). In this runner, it is evident that the energies absorbed early in stance (53 J) by the knee extensors and by the knee flexors in later swing (24 J) dominate the profile; only 31 J are generated by the knee extensors in middle and late stance.

It should be noted that this technique automatically calculates any external work that is done. The external power will be reflected in increased joint moments, which, when multiplied by the joint angular velocity, will show an increased power equal to that done externally.

5.4.1.5 Muscle Power and Work. Even with the detailed analysis described in the previous section, we have underestimated the work done by cocontracting muscles. Joint power, as calculated, is the product of the joint moment of force M_j and the angular velocity ω_j. M_j is the net moment resulting from all agonist and antagonist activity, and therefore, cannot account for simultaneous generation by one muscle group and absorption by the antagonist group, or vice versa. For example, if $M_j = 40$ N · m and $\omega_j = 3$ rad/s, the joint power would be calculated to be 120 W. However, if there was a cocontraction, the antagonists might be producing a resisting moment of 10 N · m. Thus, in this case, the agonists would be generating energy at the rate of $50 \times 3 = 150$ W while the antagonists would be absorbing energy at a rate of $10 \times 3 = 30$ W. Thus, the net power and work calculations as described in Section 5.4.1.4 will underestimate both the positive and the negative work done by the muscle groups at each joint. To date, there has been very limited progress to calculate the power and work associated with each muscle's action. The major problem is to partition the contribution of each muscle to the net moment, and this issue has been addressed in Section 4.3.1. However, if the muscle force F_m and the muscle velocity V_m were known, the muscle work W_m would be calculated as

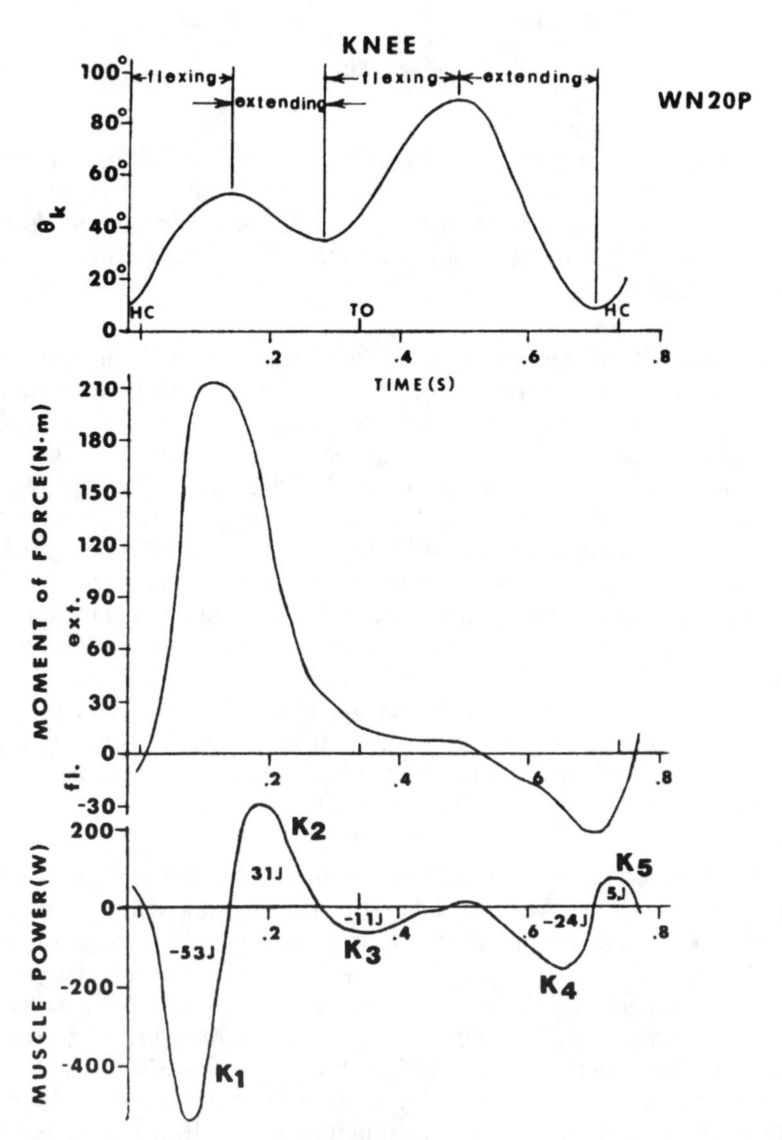

Figure 5.17 Plots of knee angle, moment of force, and power in a slow runner. Five power phases are evident: K_1, energy absorption by knee extensors; K_2, positive work as extensors shorten under tension; K_3, deceleration of backward rotation of leg and foot as thigh drives forward during late stance and early swing; K_4, deceleration of swinging leg and foot by knee flexors prior to heel contact; and K_5, small positive burst to flex the leg slightly and slow down its forward motion to near zero velocity prior to heel contact. (From Winter, 1983. Reproduced by permission of *J. Biomechanics.*)

$$W_m = \int_{t_1}^{t_2} F_m \cdot V_m \, dt \tag{5.25}$$

Morrison (1970) analyzed the power and work in four muscles in normal walking, and some later work (Yack, 1986) analyzed the muscles forces and powers in the three major biarticulate muscle groups during walking.

5.4.1.6 Isometric Work against Gravity. Finally, there is one inefficiency that cannot be resolved at the muscle level: the work requirements to hold a body segment against gravity. Mechanically there is no movement; thus, no mechanical work is being done. Metabolically, however, there is a cost. Such work is not unimportant in many pathologies and in work-related lifting or carrying tasks where loads are held momentarily against gravity or are carried with a forward body lean for extended periods.

5.4.1.7 Summary of Work Calculation Techniques. Table 5.1 summarizes the various approaches described over the past few decades and the different energy components that are not accounted for by each technique.

5.4.2 External Work Calculation

It was noted in Sections 5.4.1.4 and 5.4.1.5 that the work calculations done using Equations (5.6) and (5.25) automatically take into account all work done by the muscles independent of whether that work was internal or external. There is no way to partition the external work except by taking measurements at the interface between human and external load. A cyclist, for example, would require a force transducer on both pedals plus a measure of the velocity of the pedal. Similarly, to analyze a person lifting or lowering a load would need a force transducer between the hands and the load, or an imaging record of the load and the body (from which an inverse solution would calculate the reaction forces and velocity). The external work W_e is calculated as

$$W_e = \int_{t_1}^{t_2} \overline{F}_r \cdot \overline{V}_c \, dt \quad \text{J} \tag{5.26}$$

where $\overline{F}_r$ = reaction force vector, newtons
$\overline{V}_c$ = velocity of contact point, m/s
t_1, t_2 = times of beginning and end of each power phase

5.5 POWER BALANCES AT JOINTS AND WITHIN SEGMENTS

In Section 5.0.2, examples were presented to demonstrate the law of conservation of mechanical energy within a segment. Also, in Section 5.1.3, muscle

TABLE 5.1 Techniques to Calculate Internal Work in Movement

Technique	Work components not accounted for by technique				
Increase PE or KE Fenn (1929) Saunders et al. (1953) Liberson (1965)	Energy exchange within segments and transfers between segments	Simultaneous increases or decreases in reciprocally moving segments	Simultaneous generation and absorption at different joints	Cocontractions	Work against g
Center of mass Cavagna and Margaria (1966)		Simultaneous increases or decreases in reciprocally moving segments	Simultaneous generation and absorption at different joints	Cocontractions	Work against g
Σ Segment energies Winter (1979)			Simultaneous generation and absorption at different joints	Cocontractions	Work against g
Joint power* $\int M_j\omega_j \, dt$ Winter (1983)				Cocontractions	Work against g
Muscle power* $\int F_m V_m dt$ Yack (1986)					Work against g

*Also accounts for external work if it is present.

mechanical power was introduced, and in Section 5.1.6, the concept of passive energy transfers across joints was noted. We can now look at one other aspect of muscle energetics that is necessary before we can effect a complete power balance segment by segment: the fact that active muscles can transfer energy from segment to segment in addition to their normal role of generation and absorption of energy.

5.5.1 Energy Transfer via Muscles

Muscles can function to transfer energy from one segment to the other if the two segments are rotating in the same direction. In Figure 5.18, we have two segments rotating in the same direction but with different angular velocities. The product of $M\omega_2$ is positive (both M and ω_2 have the same polarity), and this means that energy is flowing into segment 2 from the muscles responsible for moment M. The reverse is true as far as segment 1 is concerned, $M\omega_1$ is negative showing that energy is leaving that segment and entering the muscle. If $\omega_1 = \omega_2$ (i.e., an isometric contraction), the same energy rate occurs and a transfer of energy from segment 1 to segment 2 via the isometrically acting muscles. If $\omega_1 > \omega_2$, the muscles are lengthening, and thus, absorption plus a transfer take place, while if $\omega_1 < \omega_2$, the muscles are shortening and a generation as well as a transfer occur. Table 5.2, from Robertson and Winter (1980), summarizes all possible power functions that can occur at a given joint, and if we do not account for these energy transfers through the muscles, we will not be able to account for the total power balance within each seg-

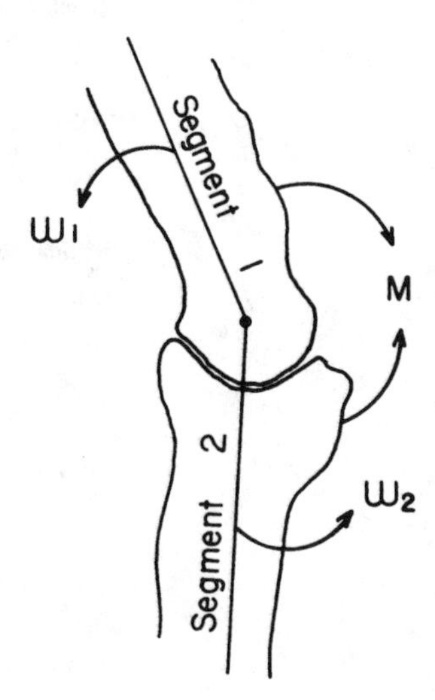

Figure 5.18 Energy transfer between segments occurs when both segments are rotating in the same direction and when there is a net moment of force acting across the joint. See text for detailed discussion.

TABLE 5.2 Power Generation, Transfer, and Absorption Functions

Description of Movement	Type of Contraction	Directions of Segmental Angular Velocities	Muscle Function	Amount, Type, and Direction of Power
Both segments rotating in opposite directions (a) joint angle decreasing	Concentric	 ω_1 M ω_2	Mechanical energy generation	$M\omega_1$ generated to segment 1. $M\omega_2$ generated to segment 2
(b) joint angle increasing	Eccentric	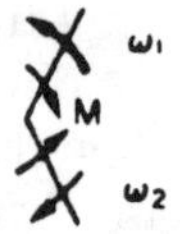 ω_1 M ω_2	Mechanical energy absorption	$M\omega_1$ absorbed from segment 1 $M\omega_2$ absorbed from segment 2
Both segments rotating in same direction (a) joint angle decreasing (e.g. $\omega_1 > \omega_2$)	Concentric	 ω_1 M ω_2	Mechanical energy generation and transfer	$M(\omega_1 - \omega_2)$ generated to segment 1 $M\omega_2$ transferred to segment 1 from 2
(b) joint angle increasing (e.g. $\omega_2 > \omega_1$)	Eccentric	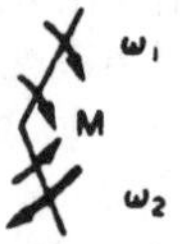 ω_1 M ω_2	Mechanical energy absorption and transfer	$M(\omega_2 - \omega_1)$ absorbed from segment 2. $M\omega_1$ transferred to segment 1 from 2
(c) joint angle constant ($\omega_1 = \omega_2$)	Isometric (dynamic)	 ω_1 M ω_2	Mechanical energy transfer	$M\omega_2$ transferred from segment 2 to 1.
One segment fixed (e.g. segment 1.) (a) joint angle decreasing ($\omega_1 = 0, \omega_2 > 0$)	Concentric	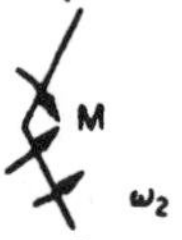 M ω_2	Mechanical energy generation	$M\omega_2$ generated to segment 2.
(b) joint angle increasing ($\omega_1 = 0, \omega_2 > 0$)	Eccentric	 M ω_2	Mechanical energy absorption	$M\omega_2$ absorbed from segment 2.
(c) joint angle constant ($\omega_1 = \omega_2 = 0$)	Isometric (static)	 M	No mechanical energy function	Zero.

From Robertson and Winter (1980). (Reproduced by permission from *J. Biomechanics.*)

ment. Thus, we must modify Equation (5.5) to include the angular velocities of the adjacent segments in order to partition the transfer component. thus ω_j is replaced by $(\omega_1 - \omega_2)$,

$$P_m = M_j(\omega_1 - \omega_2) \quad \text{W} \tag{5.27}$$

Thus, if ω_1 and ω_2 have the same polarity, the rate of transfer will be the lesser of the two power components. Examples are presented in Section 5.5.2 to demonstrate the calculation and to reinforce the sign convention used.

5.5.2 Power Balance Within Segments

Energy can enter or leave a segment at muscles and across joints at the proximal and distal ends. Passive transfer across the joint [Equation (5.9)] and active transfer plus absorption or generation [Equation (5.27)] must be calculated. Consider Figure 5.19*a* as the state of a given segment at any given point in time. The reaction forces and the velocities at the joint centers at the proximal and distal ends are shown plus the moments of force acting at the proximal and distal ends along with the segment angular velocity. The total energy of the segment E_s as calculated by Equation (5.17) must also be known. Figure 5.19*b* is the power balance for that segment, the arrows show-

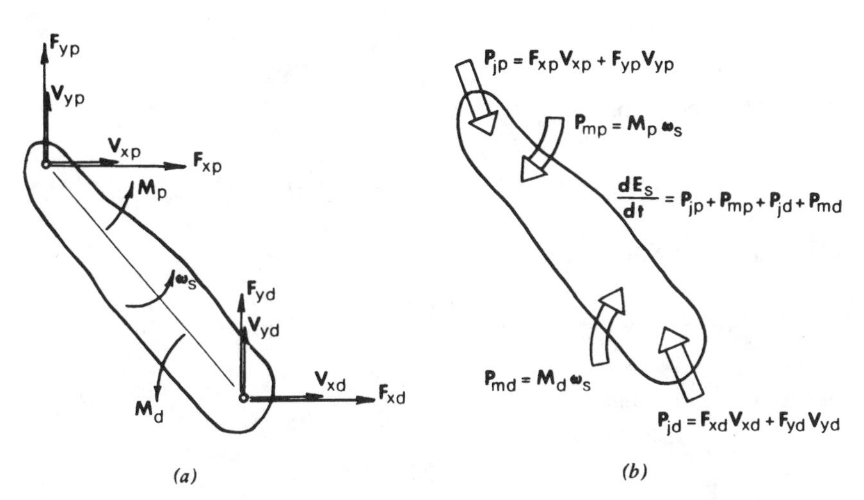

Figure 5.19 (a) Biomechanical variables describing the instantaneous state of a given segment in which passive energy transfers may occur at the proximal and distal joint centers and active transfers through the muscles at the proximal and distal ends. (b) Power balance as calculated using the variables shown in (*a*). The passive power flow at the proximal end P_{jp}, and the distal end P_{jd}, combined with the active (muscle) power at the proximal end P_{mp} and the distal end P_{md}, must equal the rate of change of energy of the segment dE_s/dt.

ing the directions where the powers are positive (energy entering the segment across the joint or through the tendons of the dominant muscles). If the force–velocity or moment–ω product turns out to be negative, this means that energy flow is leaving the segment. According to the law of conservation of energy, the rate of change of energy of the segment should equal the four power terms,

$$\frac{dE_s}{dt} = P_{jp} + P_{mp} + P_{jd} + P_{md} \tag{5.28}$$

An example calculation for two adjacent segments is necessary to demonstrate the use of such power balances and also to demonstrate the importance of passive transfers across joints and across muscles as major mechanisms in the energetics of human movement.

Example 5.5. Carry out a power balance for the leg and thigh segments for frame 5, that is, deduce the dynamics of energy flow for each segment separately and determine the power dynamics of the knee muscles (generation, absorption, transfer?):

Table A.2*a*, hip velocities,

$$V_{xh} = 1.36 \text{ m/s} \qquad V_{yh} = 0.27 \text{ m/s}$$

Table A.2*b*, knee velocities,

$$V_{xk} = 2.61 \text{ m/s} \qquad V_{yk} = 0.37 \text{ m/s}$$

Table A.2*c*, ankle velocities,

$$V_{xa} = 3.02 \text{ m/s} \qquad V_{ya} = 0.07 \text{ m/s}$$

Table A.3*b*, leg angular velocity,

$$\omega_{lg} = 1.24 \text{ rad/s}$$

Table A.3*c*, thigh angular velocity,

$$\omega_{th} = 3.98 \text{ rad/s}$$

Table A.5*a*, leg segment reaction forces and moments,

$$F_{xk} = 15.1 \text{ N}, \quad F_{yk} = 14.6 \text{ N}, \quad F_{xa} = -12.3 \text{ N}, \quad F_{ya} = 5.5 \text{ N},$$
$$M_a = -1.1 \text{ N}\cdot\text{m} \quad M_k = 5.8 \text{ N}\cdot\text{m}$$

Table A.5*b*, thigh segment reaction forces and moments,

$$F_{xk} = -15.1 \text{ N}, \quad F_{yk} = -14.6 \text{ N}, \quad F_{xh} = -9.4 \text{ N}, \quad F_{yk} = 102.8 \text{ N},$$
$$M_k = -5.8 \text{ N}\cdot\text{m}, \quad M_h = 8.5 \text{ N}\cdot\text{m}$$

Table A.6, leg energy,

$$E_{lg} \text{ (frame 6)} = 20.5 \text{ J}, \quad E_{lg} \text{ (frame 4)} = 20.0 \text{ J}$$

Table A.6, thigh energy,

$$E_{th} \text{ (frame 6)} = 47.4 \text{ J}, \quad E_{th} \text{ (frame 4)} = 47.9 \text{ J}$$

1. Leg Power Balance

$$\begin{aligned}
\Sigma \text{ powers} &= F_{xk}V_{xk} + F_{yk}V_{yk} + M_k\omega_{lg} + F_{xa}V_{xa} + F_{ya}V_{ya} + M_a\omega_{lg} \\
&= 15.1 \times 2.61 + 14.6 \times 0.37 + 5.8 \times 1.24 - 12.3 \\
&\quad \times 3.02 + 5.5 \times 0.7 - 1.1 \times 1.24 \\
&= 44.81 + 7.19 - 33.3 - 1.36 \\
&= 17.34 \text{ W}
\end{aligned}$$

$$\frac{\Delta E_{lg}}{\Delta t} = \frac{20.5 - 20.0}{0.0286} = 17.5 \text{ W}$$

$$\text{balance} = 17.5 - 17.34 = 0.16 \text{ W}$$

2. Thigh Power Balance

$$\begin{aligned}
\Sigma \text{ powers} &= F_{xh}V_{xh} + F_{yh}V_{yh} + M_h\omega_{th} + F_{xk}V_{xk} + F_{yk}V_{yk} + M_k\omega_{th} \\
&= -9.4 \times 1.36 + 102.8 \times 0.27 + 8.5 \times 3.98 \\
&\quad -15.1 \times 2.61 - 14.6 \times 0.37 - 5.8 \times 3.98 \\
&= 14.97 + 33.83 - 44.81 - 23.08 \\
&= -19.09 \text{ W}
\end{aligned}$$

$$\frac{\Delta E_{th}}{\Delta t} = \frac{47.4 - 47.9}{0.0286} = -17.5 \text{ W}$$

$$\text{balance} = -17.5 - (-19.09) = 1.59 \text{ W}$$

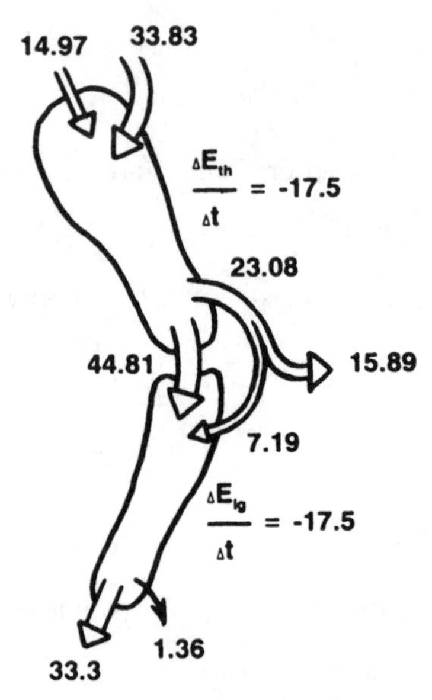

Figure 5.20 Summary of thigh and leg power flows as calculated in Example 5.5. There is power transfer from the thigh to the leg at a rate of 7.19 W through the quadriceps muscles plus passive flow across the knee of 44.81 W.

3. Summary of Power Flows. The power flows are summarized in Figure 5.20 as follows: 23.08 W leaves the thigh into the knee extensors and 7.19 W enters the leg from the same extensors. Thus, the knee extensors are actively transferring 7.19 W from the thigh to the leg and are simultaneously absorbing 15.89 W.

5.6 PROBLEMS BASED ON KINETIC AND KINEMATIC DATA

1. **(a)** Calculate the potential, translational, and rotational kinetic energies of the leg segment for frame 20 using appropriate kinematic data, and check your answer with Table A.6 in Appendix A.
   (b) Repeat Problem 1*a* for the thigh segment for frame 70.
2. **(a)** Plot (every second frame) the three components of energy plus total energy of the leg over the stride period and discuss whether this segment conserves or does not conserve energy over the stride period (frames 28 to 97).
   (b) Repeat Problem 2*a* for the thigh segment.
   (c) Repeat Problem 2*a* for the HAT segment. Using Equation (5.18), calculate the approximate percentage of energy conservation in the HAT segment over the stride period. Compare this percentage with that calculated using the exact Equations (5.19) to (5.22).

3. **(a)** Assuming symmetrical gait, calculate the total energy of the body for frame 28. *Hint:* For a stride period of 68 frames, data for the left side of the body can be estimated using right side data half a stride (34 frames) later.

(b) Scan the total energy of all segments and note the energy changes over the stride period in the lower limb compared with that of the HAT. From your observations deduce whether the movement of the lower limbs or that of HAT makes the major demands on the metabolic system.

4. **(a)** Using segment angular velocity data in Table A.7 in the Appendix plus appropriate data from other tables, calculate the power generation or absorption of the muscles at the following joints. Identify in each case the muscle groups involved that are responsible for the generation or absorption. Check your numerical answers with Table A.7.

(i) Ankle joint for frame 30.
(ii) Ankle joint for frame 50.
(iii) Ankle joint for frame 65.
(iv) Knee joint for frame 35.
(v) Knee joint for frame 40.
(vi) Knee joint for frame 65.
(vii) Knee joint for frame 20.
(viii) Hip joint for frame 50.
(ix) Hip joint for frame 70.
(x) Hip joint for frame 4.

(b) (i) Scan the listings for muscle power in Table A.7 and identify where the major energy generation occurs during walking. When in the gait cycle does this occur and by what muscles?

(ii) Do the knee extensors generate any significant energy during walking? If so, when during the walking cycle?

(iii) What hip muscle group generates energy to assist the swing of the lower limb? When is this energy generated?

5. Using Equation (5.10) (Figure 5.8), calculate the passive rate of energy transfer across the following joints, and check your answers with Table A.7. From what segment to what segment is the energy flowing?

(i) Ankle for frame 20.
(ii) Ankle for frame 33.
(iii) Ankle for frame 65.
(iv) Knee for frame 2.
(v) Knee for frame 20.
(vi) Knee for frame 65.
(vii) Hip for frame 2.
(viii) Hip for frame 20.
(ix) Hip for frame 67.

6. **(a)** Using equations in Figure 5.19*b*, carry out a power balance for the foot segment for frame 20.
   (b) Repeat Problem 6*a* for the leg segment for frame 20.
   (c) Repeat Problem 6*a* for the leg segment for frame 65.
   (d) Repeat Problem 6*a* for the thigh segment for frame 63.
7. Muscles can transfer energy between adjacent segments when they are rotating in the same direction in space. Calculate the power transfer between the following segments and indicate the direction of energy flow. Compare answers with those listed in Table A.7.
   (a) Leg/foot for frame 60.
   (b) Thigh/leg for frame 7.
   (c) Thigh/leg for frame 35.

5.7 REFERENCES

Abbot, B. C., B. Bigland, and J. M. Ritchie. "The Physiological Cost of Negative Work," *J. Physiol.* **117**:380–390, 1952.

Bresler, B. and F. Berry. "Energy Levels during Normal Level Walking," Rep. of Prosthetic Devices Res. Proj., University of California, Berkeley, May 1951.

Cappozzo, A., F. Figura, and M. Marchetti. "The Interplay of Muscular and External Forces in Human Ambulation," *J. Biomech.* **9**:35–43, 1976.

Cavagna, G. A. and R. Margaria. "Mechanics of Walking," *J. Appl. Physiol.* **21**:271–278, 1966.

Cotes, J. and F. Meade. "The Energy Expenditure and Mechanical Energy Demand in Walking," *Ergonomics* **3**:97–119, 1960.

Elftman, H. "Forces and Energy Changes in the Leg during Walking," *Am. J. Physiol.* **125**:339–356, 1939.

Falconer, K. and D. A. Winter. "Quantitative Assessment of Cocontraction at the Ankle Joint in Walking," *Electromyogr. & Clin. Neurophysiol.* **25**:135–148, 1985.

Fenn, W. D. "Frictional and Kinetic Factors in the Work of Sprint Running," *Am. J. Physiol.* **92**:583–611, 1929.

Gaesser, G. A. and G. A. Brooks. "Muscular Efficiency during Steady Rate Exercise: Effects of Speed and Work Rate," *J. Appl. Physiol.* **38**:1132–1139, 1975.

Hill, A. V. "Production and Absorption of Work by Muscle," *Science* **131**:897–903, 1960.

Liberson, W. T. "Biomechanics of Gait: A Method of Study," *Arch. Phys. Med. Rehab.* **46**:37–48, 1965.

Milner-Brown, H., R. Stein, and R. Yemm. "The Contractile Properties of Human Motor Units during Voluntary Isometric Contractions," *J. Physiol.* **228**:285–306, 1973.

Morrison, J. B. "Mechanics of Muscle Function in Locomotion," *J. Biomech.* **3**:431–451, 1970.

Norman, R. W., M. Sharratt, J. Pezzack, and E. Noble, "A Re-examination of Mechanical Efficiency of Horizontal Treadmill Running," *Biomechanics,* Vol. V-B, P. V. Komi, Ed. (University Park Press, Baltimore, MD, 1976), pp. 87–93.

Quanbury, A. O., D. A. Winter, and G. D. Reimer. "Instantaneous Power and Power Flow in Body Segments during Walking," *J. Human Movement Studies* **1**:59–67, 1975.

Ralston, H. J. and L. Lukin. "Energy Levels of Human Body Segments during Level Walking," *Ergonomics* **12**:39–46, 1969.

Robertson, D. G. E. and D. A. Winter. "Mechanical Energy Generation, Absorption, and Transfer amongst Segments during Walking," *J. Biomech.* **13**:845–854, 1980.

Saunders, J. B. D. M., V. T. Inman, and H. D. Eberhart. "The Major Determinants in Normal and Pathological Gait," *J. Bone Jt. Surg.* **35A**:543–558, 1953.

Whipp, B. J. and K. Wasserman. "Efficiency of Muscular Work," *J. Appl. Physiol.* **26**:644–648, 1969.

Williams, K. R., and P. R. Cavanagh. "A Model for the Calculation of Mechanical Power during Distance Running," *J. Biomech.* **16**:115–128, 1983.

Winter, D. A., Biomechanical Model Relating EMG to Changing Isometric Tension. *Dig. 11th Int. Conf. Med. Biol. Eng.,* 1976, pp. 362–363.

Winter, D. A. "Energy Assessments in Pathological Gait," *Physiotherapy Can.* **30**: 183–191, 1978.

Winter, D. A. "A New Definition of Mechanical Work Done in Human Movement," *J. Appl. Physiol.* **46**:79–83, 1979.

Winter, D. A. "Moments of Force and Mechanical Power in Slow Jogging," *J. Biomech.* **16**:91–97, 1983.

Winter, D. A., A. O. Quanbury, and G. D. Reimer. "Analysis of Instantaneous Energy of Normal Gait," *J. Biomech.* **9**:253–257, 1976.

Winter, D. A., and D. G. Robertson. "Joint Torque and Energy Patterns in Normal Gait," *Biol. Cybern.* **29**:137–142, 1978.

Yack, H. J. "The Mechanics of Two-Joint Muscles During Gait," Ph.D. thesis, University of Waterloo, Waterloo, Ont., Canada, 1986.

6

SYNTHESIS OF HUMAN MOVEMENT—FORWARD SOLUTIONS

6.0 INTRODUCTION

The vast majority of kinetic analyses of human movement have been inverse dynamics. As detailed in Chapter 4, this type of analysis took the kinematic measures and combined them with measured external forces (i.e., ground reaction forces) to estimate the internal (joint) reaction forces and moments. We used the outcome measures plus a link-segment model to predict forces that were the cause of the movement. This, of course, is the inverse of what really happens. The real sequence of events begins with a varying neural drive to the muscles, resulting in varying levels of recruitment of agonist and antagonist muscles. The net effect of all muscle forces acting at each joint is the generation of a time-varying moment, which in turn accelerates (or decelerates) the adjacent segments and ultimately causes the displacements that our cameras record. If we model this approach in the computer, we are doing what is called a *forward solution*.

The constraints of forward solution models are considerable when compared with inverse solutions. For example, if we wish to calculate the ankle and knee moments on one limb, we do not need data from anything but the segments concerned (in this case, the foot and the leg). No kinematic or kinetic data are required from the thigh, the opposite limbs, the trunk, the arms, or the head. For a forward solution, we must model the entire body before we start, or in the case where part of the body is fixed in space, all segments that are capable of moving must be modeled. If the link segment that we use is not a valid replication of the anatomical situation, there will likely be major errors in our predictions. The reason why we must model the entire link system is the interlimb coupling of forces. Our inputs could be net muscle moments at each joint plus the initial conditions of positions and velocity. A given joint moment acts on the two adjacent segments and they, in turn, create reaction forces on segments further away from the original cause. For example, in gait the push-off muscles at the ankle (plantarflexors) create a large plantarflexor moment, which results in a rapid ankle plantarflexion. The horizontal and vertical reaction forces at the knee are drastically altered by this plantarflexor moment and they, in turn, act on the thigh to alter its acceleration. In turn, the reaction forces at the hip, contralateral hip, and trunk are also altered. Thus, the acceleration of all body segments is affected by the ankle muscles at this time. If we have errors in any part of our anatomical model, we will have errors in our prediction. If any segment has the wrong mass or a joint has a missing or unrealistic constraint, the entire link system will start to generate displacement errors, and these errors will accumulate with time. Thus, a poor anatomical model, even with valid time histories of moments, will start to accumulate trajectory errors very quickly.

6.0.1 Assumptions and Constraints of Forward Solution Models

1. The link-segment model has the same assumptions as those presented in Section 4.0.1 for the inverse solution.
2. There must be no kinematic constraints whatsoever; the model must be permitted to fall over, jump, or collapse as dictated by the motor inputs.
3. The initial conditions must include the position and velocity of every segment.
4. The only inputs to the model are externally applied forces and internally generated muscle forces or moments.
5. The model must incorporate all important degrees of freedom and constraints. For example, the hip and shoulder joints must have three axes of rotation, but with limitations to the range of movements (due to the passive internal structures such as ligaments), modeled as passive internal forces and moments.
6. External reaction forces must be calculated. For example, the ground reaction forces would be equal to the algebraic summation of the mass–acceleration products of all segments when the feet are on the ground. Partitioning of the reaction forces when two or more points of the body are in contact with external objects is a separate and possibly major problem.

6.0.2 Potential of Forward Solution Simulations

The research and practical potentials of simulations are tremendous, but because of the severe constraints described in the previous section, the potentials unfortunately have not been realized. The kinds of questions that can be posed are prefaced with "What would happen if . . . ." For example, a surgeon might ask the question, "What would happen if I transferred a muscle from one insertion point to another, or what would happen if a muscle tendon was released (lengthened) to reduce specific muscle spasticity?" Or a coach might ask, "Is the movement pattern of my runner optimal, and if not, what changes in motor pattern might improve it?" Or a basic researcher might have a certain theory of the motor control of gait and might wish to test that theory.

However, before any valid answers are forthcoming, the link-segment representation of the anatomy must first be valid. A necessary (but not sufficient) condition that has been tested by researchers is that of internal validity. Such a test requires an inverse solution to calculate the moments at each of the joints in each of the required planes (sagittal, coronal, etc.). Then, by using these motor patterns as inputs along with measured initial conditions, the forward solution should reproduce the originally measured kinematics. If the model does not pass that test, it is fruitless to use the model to answer any functionally related questions. All that will result will be an erroneous set of motor patterns overlaid on an erroneous biomechanical model.

6.1 REVIEW OF FORWARD SOLUTION MODELS

Human locomotion is the movement that has attracted the most attention with researchers. Because of the complexity of the movement and the link-segment model, certain oversimplifications were made or the simulation was confined to short periods of the movements. Townsend and Seireg (1972) modeled the human with massless rigid lower limbs with 1 degree of freedom at each hip (flexion/extension). Hemami (1980) proposed a three-segment three-dimensional model with rigid legs and no feet, and Pandy and Berme (1988) simulated single support only using a five-segment planar model with no feet. Obviously such serious simplifications would not produce valid answers. Even with more complete models, many researchers constrained parts of their models kinematically (Beckett and Chang, 1968; Chao and Rim, 1973; Townsend, 1981) by assuming sinusoidal trajectories of the trunk or pelvic segments. Such constraints violate one of the major requirements of a true simulation. Initial work in our laboratory (Onyshko and Winter, 1980) modeled the body as a seven-segment system (two feet, two legs, two thighs, and an HAT segment), but the model did not satisfy the requirements of internal validity because of certain anatomical constraints (sagittal plane movement only at all three joints plus a rigid foot segment). The model was eventually made to walk, but only by altering the moment patterns. Such results should alert researchers that two wrongs can make a right. An incomplete model will result in a valid movement only if faulty motor patterns are used. More complex models (more segments, more degrees of freedom at each joint) have been introduced (Hemami et al., 1982*b*; Chen et al., 1986), but as yet no internal validity has been attempted.

Simple motions have been modeled reasonably successfully. Phillips et al. (1983) modeled the swinging limbs of a human using the accelerations of the swing hip along with the moments about the hip. Hemami et al. (1982*a*) modeled the sway of the body in the coronal plane with each knee locked. With adductor/abductor actuators at the hip and ankle as input, the stability of the total system was defined.

More recent modeling of three-dimensional (3D) gait has been somewhat more successful. One of the major problems with previous attempts was the modeling of initial contact. These earlier models employed springs to represent the elastic characteristics of the bottom of the feet or shoes, but this resulted in extremely large accelerations of the foot segment and similarly large spikes in the ground reaction forces. This was solved with a viscoelastic model of the foot with an array of parallel springs and dampers under the rigid foot segment (Gilchrist and Winter, 1996). A 3D nine-segment model of walking that used ADAMS software (discussed in Section 6.2) demonstrated some success using the inverse dynamics joint moments as inputs (Gilchrist and Winter, 1997). Nonlinear springs at the knee, ankle, and metatarsal-phalangeal joints constrained those joints to their anatomical range. Linear springs were used at the hips and dampers at all joints to ensure a smooth motion. However, it became apparent that any small errors in the joint moments after about 500 ms resulted in accumulating kinematic errors that ultimately became too large, and the model either became unbalanced or collapsed. The buildup of these errors is an inherent characteristic of any forward solution; the double integration of the segment accelerations caused by the input and reaction moments and forces causes displacement errors that increase over time and that can only be corrected by continuous fine-tuning of the input joint moments. Thus, we are forced to violate some of the constraints of forward solution modeling as listed in Section 6.0.1.

6.2 MATHEMATICAL FORMULATION

For the dynamic analysis of connected segment systems, mathematical models consisting of interconnected mass elements, springs, dampers, and actuators (motion generators) are often used. The motion of such models may be determined by defining the time history of the position of individual segments, or by applications of motor forces, in which case the motion of the segments is determined by the laws of physics.

Until quite recently, the nonlinear nature of the problem has been an impediment to the solution of the general dynamic model. With the advancement in digital computers, researchers are in a position to extricate themselves from the problem of nonlinearity and utilize the computer in solving models that consist of many elements. In the following pages, a systematic method for writing the equations of motion for a general model configuration is explained. The method is suitable for hand derivation of the equations of motion of simple to moderate model configurations. For complex models, the method can be adapted easily to write computer programs using symbolic computer languages such as LISP, PROLOG, and MAPLE and to generate the equations of motion. It is also possible to use computer languages such as C, FORTRAN, or BASIC in writing self-formulating computer programs general enough to accept model description as an input and provide model response as an output. In fact, Mechanical Dynamics, Inc., of Ann Arbor, Mich., have developed a computere program that automates the dynamic simulation process. Their product is marketed under the name ADAMS (Automatic dynamic analysis of mechanical systems; Chace, 1984).

Formulating the equations of motion may be done in several ways. The first and most direct, but possibly the least efficient way, is to apply Newton's laws of dynamics to each segment in the model. Although the reaction forces and torques are obtained as byproducts of the solution, the method is cumbersome and does not lend itself easily to a general dynamic simulation program. However, if some concepts of graph theory are incorporated into Newton's laws of motion, the result is a methodical procedure that can be used for writing self-formulating dynamic simulation programs. Three computer programs that utilize these ideas have been developed at the University of Waterloo: VECNET (vector network), a three-dimensional package for particles (Andrews and Kesavan, 1975); PLANET (plane network) for planar mechanisms; and ADVNET (advanced network) for 3D systems (Andrews, 1977; Singhal and Kesavan, 1983).

The second method to formulate the equations of motion is to utilize Lagrangian dynamics (Wells, 1967). Lagrange's equations require the concepts of virtual displacement and employ system energy and work as functions of the generalized coordinates to obtain a set of second-order differential equations of motion. To a large extent, the method reduces the entire field of dynamics to a single procedure involving the same basic steps, regardless of the number of segments considered, the type of coordinates employed, the number of constraints on the model, and whether or not the constraints are in motion. Alternative methods have also been used in this research, such as methods of virtual work extended using D'Alembert principles as in DYMAC (Paul, 1978).

Which method is more suitable? Each method has its advantages and disadvantages. However, the Lagrangian method is characterized by simplicity and is applicable in any suitable coordinates. The task of this chapter is to encode a procedure based on Lagrange's dynamics and suitable for computer implementation using symbolic manipulation language. In doing so, system elements are described by lists that are linked together. Each list has its name and index number and a stack of parameters associated with it. The first element in a given list is usually an integer that establishes a link between the list and other relevant lists in the system. The equations of motion are then obtained by systematic manipulation of these lists. Before we go any further, a review of Lagrange's method is necessary.

6.2.1 Lagrange's Equations of Motion

The following lines are a brief description of Lagrange's equations of motion of the second type. The derivation of these equations can be found in any advanced classical dynamics textbook, such as Greenwood (1977) or Wells (1967).

6.2.2 The Generalized Coordinates and Degrees of Freedom

One of the first things to be determined in a given system is the number of coordinates that represent the model. It is important to realize that any set of time-dependent parameters that give an unambiguous representation of the system configuration will serve as system coordinates. These parameters are known as generalized coordinates, and for the sake of generality, they are denoted by $[\mathbf{q}]^t$,

$$[\mathbf{q}]^t = [q_1, q_2, \ldots, q_n] \tag{6.1}$$

The symbol $[\,]^t$ is used to indicate the transpose of an array or a matrix. For example, consider the particle p in the plane xy, as shown in Figure 6.1. The point p may be located by several pairs of quantities, such as (x, y), (r, θ), $(A, \sin\theta)$, $(r, \sin\theta)$, and so on. There is, in fact, no limit to the number of different coordinate systems that can be defined, which could be used to indicate the position. Furthermore, the relations between these pairs can be found easily. Using the first and last pairs, and letting $q_1 = r$ and $q_2 = \sin\theta$, it can be proven that $x = q_1[1 - (q_2)^2]^{0.5}$ and $y = q_1 q_2$.

A spatial segment can have as many as six degrees of freedom (DOF) (i.e., possible coordinates), while a point mass has a maximum of three DOF. A system degrees of freedom is defined as the sum of the degrees of freedom of its elements. The degrees of freedom of a system consisting of segments, S, particles, P, and constraints, C, is given by

$$\text{DOF} = 6S + 3P - C \tag{6.2}$$

The constraints in a system are described by equations of constraint $[\mathbf{\Phi}] = [0]$. They may be caused by joints (i.e., a pin joint forces two points on adjacent segments to have the same trajectory) or by an externally imposed

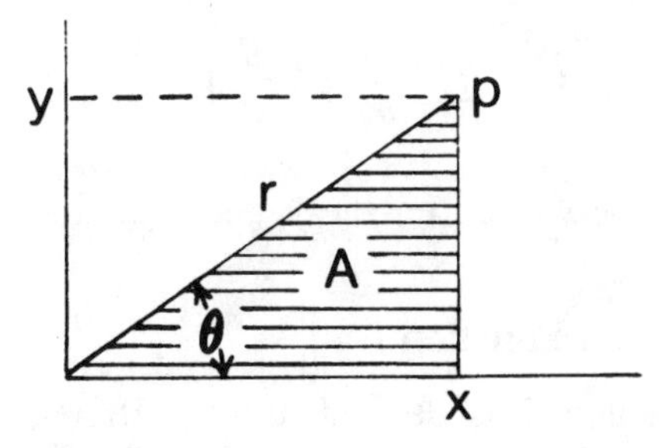

Figure 6.1 Point p whose position can be specified as a function of several other variables: r, x, y, θ, and A.

motion pattern. The constraints enter the equations of motion in the form of constraint forces $[\boldsymbol{\lambda}]^t$ rather than in geometric terms. There exists a constraint force associated with each constraint equation and analogous to a reaction force. The constraint forces $[\boldsymbol{\lambda}]^t$ are known as Lagrange multipliers. In the previous example, the particle has two DOF since it is restricted to planar motion. Hence, the constraint equation $\boldsymbol{\Phi}$ is in the form $z = 0$. The constraint force associated with this constraint equation is the force in the z axis that keeps the particle in the xy plane of motion. A totally constrained system (DOF $= 0$) can be solved kinematically. It is impossible to solve an overconstrained system (DOF < 0) without removing the redundant constraints.

The independent generalized coordinates are those coordinates that can be varied independently without violating the constraints, and must equal the degrees of freedom of the model. The use of the independent generalized coordinates allows the analysis of most models to be made without solving for the forces of constraint. Any additional coordinates in the model are known as superfluous or dependent coordinates. The relations between the independent and the superfluous coordinates are in fact constraint equations $[\boldsymbol{\Phi}] = [0]$. If the dependent coordinates can be eliminated, then the system is called *holonomic*. Nonholonomic systems always require more coordinates for their description than there are degrees of freedom. By definition, the first and second derivatives of a generalized coordinate q_i with respect to time are called the *generalized velocity* $\dot{q}_i$ and the *generalized acceleration* $\ddot{q}_i$, respectively. The relation between the position vector $\mathbf{r}_i$ of a point i in the system and the generalized coordinates $[\mathbf{q}]^t$ are called the *transformation equations*. It is assumed that these equations are in the form

$$
\begin{aligned}
x_i &= f_{xi}(q_1, q_2, q_3, \ldots, q_n, t) \\
y_i &= f_{yi}(q_1, q_2, q_3, \ldots, q_n, t) \\
z_i &= f_{zi}(q_1, q_2, q_3, \ldots, q_n, t)
\end{aligned}
\tag{6.3}
$$

where some or all of the generalized coordinates may be present. The velocity component in the x direction is obtained by taking the time derivative of the x_i equation,

$$
\dot{x}_i = \sum_{j=1}^{n} \left(\frac{\partial x_i}{\partial q_j}\right)\left(\frac{\partial q_j}{\partial t}\right) + \frac{\partial x_i}{\partial t} \tag{6.4}
$$

Similar relations can be written for the $\dot{y}_i$ and $\dot{z}_i$ components.

6.2.3 The Lagrangian Function L

The Lagrangian function L is defined as the difference between the total kinetic energy KE and the total potential energy PE in the system,

$$L = \text{KE} - \text{PE} \tag{6.5}$$

The kinetic energy for a segment is defined as the worrk done on the segment to increase its velocity from rest to some value **v**, where **v** is measured relative to a global (inertial) reference system. The existence of an inertial reference system is a fundamental postulate of classical dynamics. Potential energy exists if the system is under the influence of conservative forces. Hence, segment potential energy is defined as the energy possessed by virtue of a segment (or particle) position in a gravity field relative to a selected datum level (usually ground level) in the system. In case of a spring, potential energy is the energy stored in the spring due to its elastic deformation.

6.2.4 Generalized Forces [Q]

A nonconservative force $\mathbf{F}_j$ acting on a segment can be resolved into components corresponding to each generalized coordinate (q_i, $i = 1, \ldots, n$) in the system. This is also true for constraint forces. A generalized force Q_i is the component of the forces that do work when q_i is varied and all other generalized coordinates are kept constant. In more useful terms, if f forces are acting on the system, then

$$Q_i = \sum_{j=1}^{f} \lambda_j \left(\frac{F_{xj} \partial R_{xj}}{\partial q_i} + \frac{F_{yj} \partial R_{yj}}{\partial q_i} + \frac{F_{zj} \partial R_{zj}}{\partial q_i} \right) \tag{6.6}$$

where $\mathbf{R}_j$ is the position vector of the force $\mathbf{F}_j$. Moments that are generated by these forces or externally applied moments are greatly affected by the choice of the angular system. More on this topic later. In the case of m constraint equations,

$$Q_i = \sum_{j=1}^{m} \lambda_j \frac{\partial \Phi_j}{\partial q_i} \tag{6.7}$$

Equations (6.6) and (6.7) are added together before they are used.

6.2.5 Lagrange's Equations

One of the principal forms of the Lagrange equations for a system with n generalized coordinates and m constraint equations is

$$\frac{\partial(\partial L / \partial \dot{q}_i)}{\partial t} - \frac{\partial L}{\partial q_i} = Q_i, \qquad (i = 1, \ldots, n) \tag{6.8}$$

$$[\mathbf{\Phi}] = [0] \tag{6.9}$$

As we shall demonstrate, these equations consist of n second-order nonlinear differential equations and m constraint equations. The set has $n + m$ unknowns in the form of $[\mathbf{q}]^t$ and $[\boldsymbol{\lambda}]^t$.

6.2.6 Points and Reference Systems

A moving point pt in a given Cartesian reference system (RS) has a maximum of three degrees of freedom (DOF). If a point is constrained to a certain motion, its DOF are reduced accordingly. For a given RS, in addition to the translational DOF of its original point, it has a maximum of three rotational DOF. In this context, points can represent the origin of an RS, a segment's center of mass, a point where an external force is applied, a joint center, a muscle insertion, and other points of interest. An RS that is conveniently chosen represents either a segment's local reference system (LRS) or the global reference system (GRS).

A moving reference frame LRS is defined if its point of origin and its orientation relative to an already defined GRS are given. On the other hand, a moving point is defined if its RS is given and its local coordinates are known. By using the notation of linked lists and the indices i, j, k, a point pt(i) moving relative to an RS_j is represented by the list

$$\mathrm{pt}(i) = [j, x_i, y_i, z_i] \tag{6.10}$$

In a similar way, the moving LRS(j) with origin at pt(k) and given orientation (usually relative to the zero reference system GRS) is represented by the list

$$\mathrm{LRS}(j) = [k, \theta_{1j}, \theta_{2j}, \theta_{3j}, 0] \tag{6.11}$$

For a two-dimensional (2D) system, Equations (6.10) and (6.11) are reduced to the forms

$$\mathrm{pt}(i) = [j, x_i, y_i] \qquad \text{and} \qquad \mathrm{LRS}(j) = [k, \theta_j]$$

Starting at some convenient stationary point pt(0) in the system (point zero), the GRS is constructed. It is given the index zero. Other points in the domain of the GRS can be specified by Equation (6.10), where the index j is set to zero. Some of these points are origins of all or some of the LRSs in the model. Each point should have its unique index number. The model's LRSs are specified by Equation (6.11); they are also given their own unique index numbers. The process is repeated until all the points and LRSs in the model are represented. Thus, the model configuration is reduced to a vector network in the form of two groups of linked lists.

As an illustrative example, consider a block of mass m_2 slides on another block of mass m_1, which, in turn, slides on a horizontal surface, as shown in

Figure 6.2. In this example, the blocks are treated as particles. The system has two DOF, and hence two independent generalized coordinates q_1, q_2 are selected as shown. The system is described by one LRS and two pt as follows:

$$\begin{aligned} \text{LRS}(1) &= [1, \theta] \\ \text{pt}(1) &= [0, q_1, 0] \\ \text{pt}(2) &= [1, d_1 - q_2, 0] \end{aligned} \tag{6.12a}$$

where x and y of Equation (6.10) are replaced by the appropriate transformations. It should be noted that the q are treated as implicit time-dependent variables, while other symbols are considered constants. To calculate the force of the constraint that keeps the two blocks together at a distance d_2 between the center of mass m_2 and the sliding surface of m_1, the mass m_2 is given an additional degree of freedom using the generalized coordinate q_3. A constraint is then added to restore the system to its original state as follows:

$$\begin{aligned} \text{pt}(2) &= [1, d_1 - q_2, q_3] \\ 0 &= q_3 - d_2 \end{aligned} \tag{6.12b}$$

Once these lists are established, with the aid of the first element in each list, the displacement and velocity vectors of any point in the system can be derived easily. The system is completely specified when the mass and the center of mass of the two blocks are given. This is done using the lists

$$\begin{aligned} \text{seg}(1) &= [1, m_1] \\ \text{seg}(2) &= [2, m_2] \end{aligned} \tag{6.12c}$$

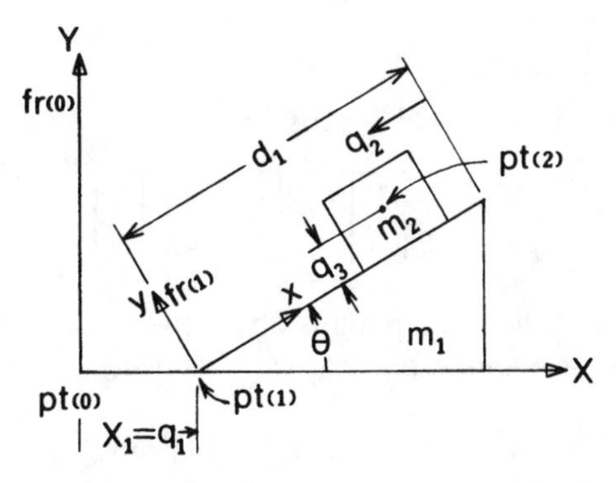

Figure 6.2 Sliding mass example where mass m_1 moves horizontally and mass m_2 moves along the top of the sloped surface of m_1.

The term *segment* seg is used to describe a group of lists that represent any mass element in the system, including particles and rigid segments. More will be said about this example (Figure 6.2) later, with the specific lists or equations numbered (6.12*d*) to (6.12*g*).

6.2.7 Displacement and Velocity Vectors

In the following discussion, the notations $\mathbf{r}$, $\mathbf{v}$, x, y, z, . . . are used for LRS while $\mathbf{R}$, $\mathbf{V}$, X, Y, Z, . . . are used for GRS. The letters c and s are used to denote cosine and sine, respectively. The derived displacement and velocity vectors can be saved in two groups of lists, disp and velo, where an entry (i) is coded as follows:

$$\text{disp}(i) = [x_i, y_i, z_i] \tag{6.13}$$

$$\text{velo}(i) = [\dot{x}_i, \dot{y}_i, \dot{z}_i] \tag{6.14}$$

Equations (6.13) and (6.14) represent the displacement and the velocity, respectively, of pt(i). Saving the derived components rather than deriving them when they are needed will save considerable time, especially in the case of complex systems. The angular velocity vectors of LRSs can also be saved in a group, omga. An entry (j) in this group represents components of the angular velocity of LRS(j):

$$\text{omga}(j) = [\omega_{jx}, \omega_{jy}, \omega_{jz}] \tag{6.15}$$

6.2.7.1 Two-Dimensional Systems. Let pt(a) be a moving point in the LRS(i) (Figure 6.3). Let $\mathbf{R}_a$ be the absolute position vector for the same point transformed to the directions of the GRS. From vector algebra,

$$\mathbf{R}_a = \mathbf{R}_i + [\phi]\mathbf{r}_{ia} \tag{6.16}$$

where ϕ is the angle between x_i and X.

The same equation in expanded form is

$$\begin{bmatrix} X \\ Y \end{bmatrix}_a = \begin{bmatrix} X \\ Y \end{bmatrix}_i + \begin{bmatrix} c & -s \\ s & c \end{bmatrix} \begin{bmatrix} x \\ y \end{bmatrix}_{ia} \tag{6.16$'$}$$

The velocity vector of the same point is obtained by taking the time derivative of Equation (6.16). Remembering that $\omega_z = d\theta_z/dt$,

$$\mathbf{V}_a = \mathbf{V}_i + [\phi]\mathbf{v}_{ia} + [\phi][\tilde{\boldsymbol{\omega}}]\mathbf{r}_{ia} \tag{6.17}$$

where $[\tilde{\boldsymbol{\omega}}]$ is a 2×2 skew symmetric angular velocity matrix generated from the angular velocity vector,

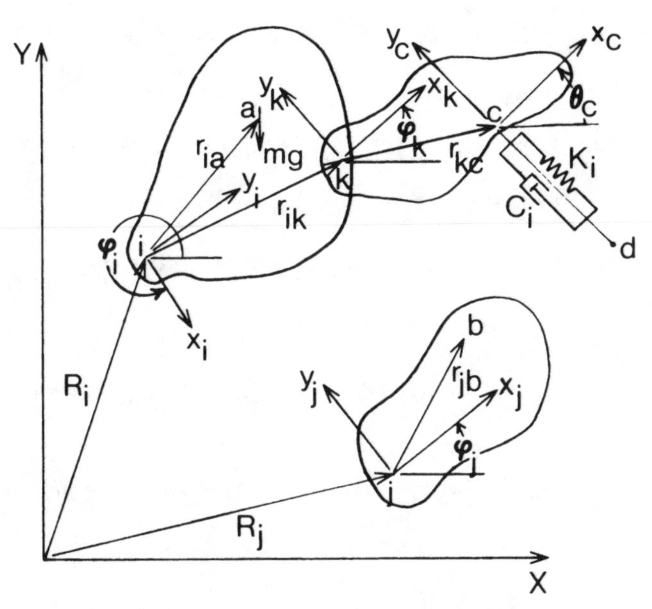

Figure 6.3 Link-segment diagram showing planar modeling of three segments and a spring damper. Three local references systems are shown. See text for details.

$$[\tilde{\boldsymbol{\omega}}] = \begin{bmatrix} 0 & -\omega_z \\ \omega_z & 0 \end{bmatrix} \tag{6.18}$$

Now let us return to the sliding block example (Figure 6.2). Setting $\theta = 45°$, then $c = \cos(45) = \sin(45) = 0.707$, and $\omega = 0$. The displacement and velocity lists are

$$\begin{aligned} \text{DISP}(1) &= [q_1, 0] \\ \text{VELO}(1) &= [\dot{q}_1, 0] \end{aligned} \tag{6.12d}$$

Since the origin of LRS(1) is pt(1), then, from Equation (6.16),

$$\begin{aligned} \text{DISP}(2) &= \text{DISP}(1) + [\phi][d_1 - q_2\ q_3]^t \\ &= [q_1 + 0.707(d_1 - q_2) - 0.707q_3, 0.707(d_1 - q_2) + 0.707q_3] \\ \text{VELO}(2) &= [\dot{q}_1 - 0.707(\dot{q}_2 + \dot{q}_3), 0.707(\dot{q}_3 - \dot{q}_2)] \end{aligned} \tag{6.12e}$$

From lists (6.12*c*), (6.12*d*), and (6.12*e*), system energy and, hence the Lagrangian [Equation (6.5)], are

$$\text{KE} = \tfrac{1}{2}m_1(\dot{q}_1)^2 + \tfrac{1}{2}m_2\{[q_1 - 0.707(\dot{q}_2 + \dot{q}_3)]^2 + [0.707(\dot{q}_3 - \dot{q}_2)]^2\}$$

$$\text{PE} = 0.707m_2g(d_1 + q_3 - q_2)$$

$$L = \tfrac{1}{2}(m_1 + m_2)(\dot{q}_1)^2 + \tfrac{1}{2}m_2[(\dot{q}_2)^2 + (\dot{q}_3)^2 - 1.414(\dot{q}_1)(\dot{q}_2 + \dot{q}_3)] - 0.707m_2g(d_1 + q_3 - q_2) \quad (6.12f)$$

From Equations (6.7) and (6.12*b*), the constraint forces [**Q**] are

$$Q_1 = 0, \quad Q_2 = 0, \quad Q_3 = \lambda_3 \quad (6.12g)$$

From Equations (6.8), (6.9), (6.12*f*), and (6.12*g*), the equations of motion are for q_1,

$$0 = (m_1 + m_2)\ddot{q}_1 - (0.707\, m_2)\ddot{q}_2 - (0.707\, m_2)\ddot{q}_3$$

for q_2,

$$0 = -(0.707\, m_2)\ddot{q}_1 + (\mathrm{m}_2)\ddot{q}_2 - (0.707\, m_2 g)$$

and for q_3,

$$\lambda_3 = -(0.707\, m_2)\ddot{q}_1 + (m_2)\ddot{q}_2 + (0.707\, m_2 g)$$

$$0 = q_3 - d_2$$

These four equations are in four variables (q_1, q_2, q_3, λ_3). If the constraint force is not required, then it can be seen that the first two equations less the q_3 term are the equations needed.

For computer implementation using symbolic computer language, it is a simple task to encode a general program that accepts system variables and parameters in the form of linked lists as an input and gives the equations of motions as an output. If the task is hand derivation of the equations, tables have to be created, each containing one group of relevant lists. Some of these tables are used for intermediate derivations. The task is then as simple as filling in the blanks in these tables.

6.2.7.2 Three-Dimensional Systems. For 3D systems, Equations (6.16) and (6.17) are rewritten for a moving pt(j) in the domain of an LRS with an origin at pt(i), as shown in Figure 6.4,

$$\mathbf{R}_j = \mathbf{R}_i + [\phi]\,\mathbf{r}_{ij}$$

$$\mathbf{V}_j = \mathbf{V}_i + [\phi]\,\mathbf{v}_{ij} + [\phi][\tilde{\omega}]\,\mathbf{r}_{ij}$$

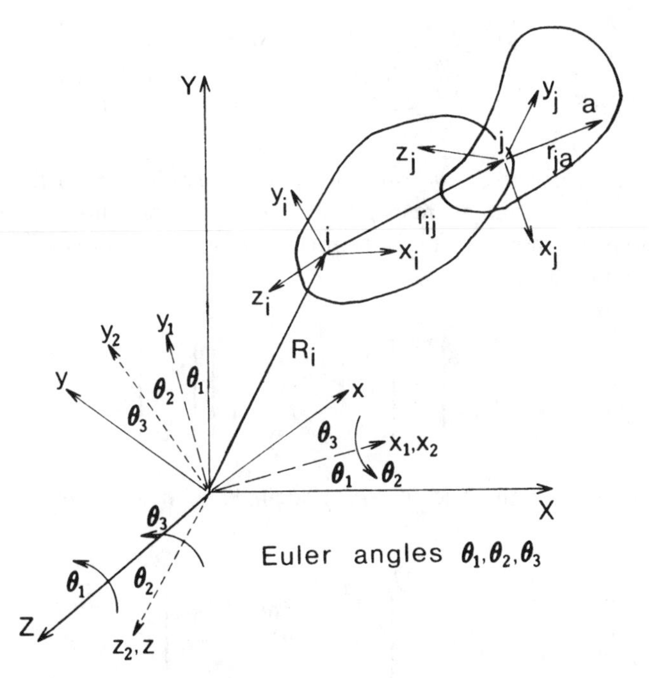

Figure 6.4 Three-dimensional link-segment system showing the three Euler angles θ_1, θ_2, and θ_3 and the two LRS for the *i*th and *j*th segments.

where $[\phi]$ is a 3×3 transformation matrix known as the direction cosines matrix (DCM) and $[\tilde{\omega}]$ is a 3×3 angular velocity matrix. The elements of both matrices are greatly dependent on the angular system used. Among the common parameters used to describe the angular orientation of a segment in space are Euler angles. For Euler angles that follow the order of rotation *zxz*, the angular orientation of an LRS (reference *xyz* in Figure 6.4) is represented as the result of a sequence of three rotations. The first rotation θ_1 is carried out about the *Z* axis. It results in the auxiliary reference (x_1, y_1, z_1). The second rotation through the angle θ_2 about the x_1 axis produces a second reference (x_2, y_2, z_2). The third rotation through the angle θ_3 about the z_2 axis gives the final orientation. Let c_1 and s_1 be $\cos(\theta_1)$ and $\sin(\theta_1)$, respectively. It can be seen that the transformation matrices associated with these rotations are

$$[\phi_1] = \begin{bmatrix} c_1 & s_1 & 0 \\ -s_1 & c_1 & 0 \\ 0 & 0 & 1 \end{bmatrix}, \quad [\phi_2] = \begin{bmatrix} 1 & 0 & 0 \\ 0 & c_2 & s_2 \\ 0 & -s_2 & c_2 \end{bmatrix}, \quad [\phi_3] = \begin{bmatrix} c_3 & s_3 & 0 \\ -s_3 & c_3 & 0 \\ 0 & 0 & 1 \end{bmatrix}$$

The DCM is then obtained by combining the effect of the three rotations in the same order, $[\phi] = [\phi_3][\phi_2][\phi_1]$. In expanded form,

$$[\phi] = \begin{bmatrix} c_1c_3 - s_1c_2s_3 & -c_3s_1 + c_1c_2c_3 & s_2s_3 \\ s_1s_3 + c_1c_2s_3 & -s_1s_3 + c_1c_2c_3 & c_3c_2 \\ s_2s_3 & -c_1s_2 & c_2 \end{bmatrix} \tag{6.19}$$

The first time derivatives (ω_1, ω_2, ω_3) of Euler angles are in fact vectors in the directions Z, x_1, and z_2, respectively. These components expressed in the xyz directions are more useful for segment energy calculation. They are given by the transformation

$$\begin{bmatrix} \omega_x \\ \omega_y \\ \omega_z \end{bmatrix} = \begin{bmatrix} s_2s_3 & c_3 & 0 \\ s_2c_3 & -s_3 & 0 \\ c_3 & 0 & 1 \end{bmatrix} \begin{bmatrix} \omega_1 \\ \omega_2 \\ \omega_3 \end{bmatrix} \tag{6.20}$$

Finally, the skew-symmetric matrix associated with $\boldsymbol{\omega} = [\omega_x, \omega_y, \omega_z]^t$ is in the form

$$[\tilde{\boldsymbol{\omega}}] = \begin{bmatrix} 0 & -\omega_z & \omega_y \\ \omega_z & 0 & -\omega_x \\ -\omega_y & \omega_x & 0 \end{bmatrix} \tag{6.21}$$

It should be noted that the externally applied torques τ (or $\mathbf{T}$) are always expressed as components in the directions xyz. The relations between these torque components and the generalized torques required by Lagrange equations are given by

$$\begin{bmatrix} Q_{\theta 1} \\ Q_{\theta 2} \\ Q_{\theta 3} \end{bmatrix} = \begin{bmatrix} s_2s_2 & s_2c_3 & c_3 \\ c_3 & -s_3 & 0 \\ 0 & 0 & 1 \end{bmatrix} \begin{bmatrix} \tau_x \\ \tau_y \\ \tau_z \end{bmatrix} \tag{6.22}$$

6.3 SYSTEM ENERGY

At any given time, the system energy may consist of the energy of segments due to their motion and position in the system, the energy stored in springs due to their elastic deformation, and the dissipation energy due to friction in the system. The first and second types of energy are included in the Lagrangian L of the system. The third type can be treated as an external force applied to the system at the proper points rather than energy, and hence it is covered under external forces. In the case of dampers, it is possible to write an energy expression that may be included in Lagrangian equations. The three types of energy are covered in more detail below.

6.3.1 Segment Energy

As was previously discussed, a segment is replaced by an arbitrary LRS. Let pt(j) be the origin of the segment LRS and pt(c) its center of mass. The segment has a mass M and inertia tensor [**J**]. The formula for calculating the kinetic energy of the segment has the form (Wittenburg, 1977):

$$\mathbf{KE} = \tfrac{1}{2}M[\boldsymbol{v}_j]^t[\boldsymbol{v}_j] + M[\boldsymbol{v}_j]^t[\tilde{\boldsymbol{\omega}}][\mathbf{r}_{jc}] + \tfrac{1}{2}[\boldsymbol{\omega}]^t[\mathbf{J}][\boldsymbol{\omega}] \tag{6.23}$$

A closer look at the first and third terms in this formula reveals the well-known kinetic energy formula $\frac{1}{2}(mv^2 + I\omega^2)$. The second term reflects the selection of the LRS origin at a point rather than the center of mass of the segment. When pt(c) and pt(j) coincide, the second term vanishes. The inertia tensor [**J**] is a 3×3 matrix containing the mass moments of inertia (I_{xx}, I_{yy}, I_{zz}) and the mass products of inertia (I_{xy}, I_{xz}, I_{yz}) of the segment relative to its LRS. The inertia tensor has the general form

$$[\mathbf{J}] = \begin{bmatrix} I_{xx} & -I_{xy} & -I_{xz} \\ -I_{yx} & I_{yy} & -I_{yz} \\ -I_{zx} & -I_{zy} & I_{zz} \end{bmatrix} \tag{6.24}$$

It is possible to select LRS in such a way that the products of inertia vanish and the tensor [**J**] becomes a diagonal matrix. The axes of the selected LRS are then called the *principal axes of the segment.*

If the Z axis of the GRS is selected in the direction of the gravity field and its origin is taken as the zero level of the system, then the potential energy of a segment j is defined as

$$\mathrm{PE}_j = MgZ_c \tag{6.25}$$

where g is the gravity constant and Z_c is the z component of $\mathbf{R}_c$. It should be noted that the first element of the pt(c) list is the segment LRS index [Equation (6.11)]. Hence the pt(c) list contains the necessary information about segment kinematics. It follows that a segment i is completely specified when its mass, inertia, and the index of its center point are given. This information can be stacked in the seg(i) list as follows:

$$\mathrm{seg}(i) = [c, M, I_{xx}, I_{yy}, I_{zz}, I_{xy}, I_{xz}, I_{yz}] \tag{6.26}$$

There are many variations of this list. When segment LRS is a principal system, the last three elements are not required. A segment in a 2D system is completely specified by the first three elements in the list, while a particle is specified by two elements only. It should be noted that a massless segment (i.e., a rigid link that joins two points) need not be specified. Hence, its LRS replaces it completely.

6.3.2 Spring Potential Energy and Dissipative Energy

In case of a linear spring damper with spring constant k_l, damping coefficient c_l, free length l_s, and end terminals at pt(e) and pt(f) as shown in Figure 6.3, the potential energy is calculated by

$$\mathrm{PE}_s = \tfrac{1}{2}k_l(l - l_s)^2 \tag{6.27}$$

where l is the length of $\mathbf{r}_{ef}$. A linear spring j is thus defined by its end points and properties in the form of the list

$$\mathrm{spl}(j) = [e, f, k_1, c_1, l_s] \tag{6.28}$$

For a torsional spring with spring constant k_t and end terminals attached to two adjacent references LRS(g) and LRS(h), respectively, the potential energy is calculated by $\mathrm{PE}_s = \tfrac{1}{2}k_t(\theta - \theta_s)^2$, where θ is the angle between the two LRS about a common axis of rotation and θ_s is the angle when the spring is unloaded. The torsional spring list spt(j) is then

$$\mathrm{spt}(j) = [g, h, k_t, c_t, \theta_s] \tag{6.29}$$

The dissipative energy (DE) associated with dampers is obtained by the aid of Rayleigh's dissipation function $p = cv^{n+1}/(n + 1)$ where $n = 1$ for viscous friction and v is the relative velocity between damper end points. For a system of d dampers, the DE is defined by

$$\mathrm{DE} = \sum_{j=1}^{d} \tfrac{1}{2}c_j(v_j)^2 \tag{6.30}$$

The generalized force Q_i due to the dampers in the system is given by

$$Q_i = -\frac{\partial \mathrm{DE}}{\partial \dot{q}_i} \tag{6.31}$$

Either the left-hand side or the right-hand side of this equation is added to the Lagrange equations.

6.4 EXTERNAL FORCES AND TORQUES

As part of the model description, an external force j acting at a point a is given by a force list frc that contains force components and the point of application,

$$\text{frc}(j) = [a, F_x, F_y, F_z] \tag{6.32}$$

Bidirectional forces (force motors and actuators) are treated as two forces equal in magnitude and opposite in direction, acting on two segments. Equation (6.32) is then modified to include the other end point. Similarly, an external torque or torque motor i acting between frames j and k is given by the list

$$\text{trq}(i) = [j, k, \tau_x, \tau_y, \tau_z] \tag{6.33}$$

where (τ_x, τ_y, τ_z) are the components of the externally applied torque. The contribution of the externally applied forces and torques to the generalized forces of the system is obtained by

$$Q_i = \frac{\partial W}{\partial q_i} \tag{6.34}$$

where W is the total work done by the external forces and torques [similar to Equation (6.6)].

6.5 DESIGNATION OF JOINTS

Under Lagrangian dynamics, there are two ways of treating joints in a given model due to the fact that action and reaction at joints do not perform work. If joint reaction forces are not required, then a joint can be viewed as a common point between two adjacent segments. To obtain joint forces, a joint is set apart into two points, each on the appropriate segment LRS. An equation, known as *loop closure* (or *constraint*) *equation* Φ, is then written to state the relation between these points. The reaction forces are those forces λ that maintain the constraint. In the case of pin joints or ball and socket joints, the relation is that the vector between the two points is zero. Other joint types are beyond the limitations of this introductory topic.

6.6 ILLUSTRATIVE EXAMPLE

A standing human is modeled as an inverted pendulum. The three elements shown in Figure 6.5*a* represent the leg, thigh, and trunk segments. The seg-

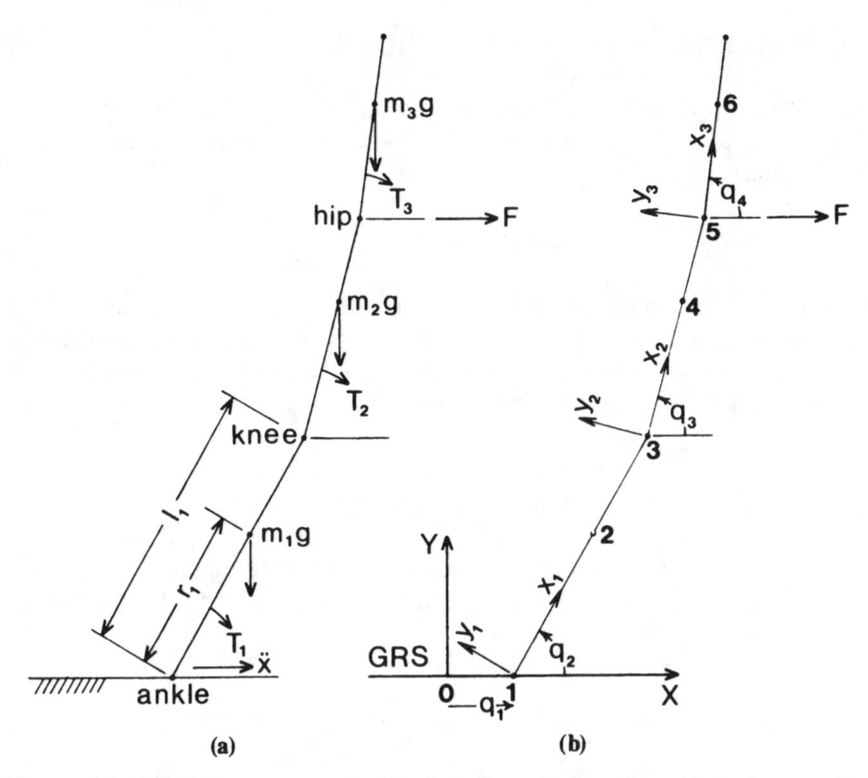

Figure 6.5 (*a*) Three-segment model of the leg, thigh, and trunk during standing balance. The feet (ankle) can be perturbed with a horizontal acceleration $\ddot{x}$ and the hip with a horizontal force F. The system responds with three joint torques, T_1, T_2, and T_3. (*b*) Link-segment model of (*a*) showing the global reference system and three local reference systems along with output displacement variables q_1, q_2, q_3, and q_4.

ment lengths are represented by l. The locations of the segment centers of mass are represented by r, all measured from the distal end. A horizontal disturbance force F is applied at the hip. Horizontal feet disturbances were simulated as a fixed acceleration x of the ground contact point. The joint torques T_1, T_2, and T_3 represent the ankle, knee, and hip torques, respectively.

The system has four variables, $q_1(= x)$, $q_2(= \theta_1)$, $q_3(= \theta_2)$, and $q_4(= \theta_3)$. Therefore, four differential equations are expected. As a first step, the model is replaced by the appropriate references LRS and points pt. Model parameters are described relative to these points and references. Selecting LRS such that their origins are segments cg simplifies the formulation of the equations, but for the sake of generality let the LRS origins be at the distal ends of the segments. Also, the variables (θ_2, θ_3, θ_4) can be selected to represent joint angles, but we rather let these variables be segment angles. It is also assumed that the reaction forces at the joints are not needed. Based on these choices, the model is completely specified using three LRSs, six points, three segments, three torques, and one force, as shown in Figure 6.5*b*. They are stated in the following lists:

$$\text{LRS}(1) = [1, q_2], \quad \text{LRS}(2) = [3, q_3], \quad \text{LRS}(3) = [5, q_4]$$

$$\text{pt}(1) = [0, q_1, 0], \quad \text{pt}(2) = [1, r_1, 0], \quad \text{pt}(3) = [1, l_1, 0],$$

$$\text{pt}(4) = [2, r_2, 0] \quad \text{pt}(5) = [2, l_2, 0] \quad \text{pt}(6) = [3, r_3, 0]$$

$$\text{seg}(1) = [2, m_1, I_1], \quad \text{seg}(2) = [4, m_2, I_2], \quad \text{seg}(3) = [6, m_3, I_3]$$

$$\text{trq}(1) = [0, 1, T_1], \quad \text{trq}(2) = [1, 2, T_2], \quad \text{trq}(3) = [2, 3, T_3],$$

$$\text{frc}(1) = [5, F, 0]$$

The steps that follow are to create displacement and velocity lists that contain all points and use the results to obtain system Lagrangian and the generalized forces that are necessary for the Lagrange equations. The following is a pseudo code for the derivation procedures:

For $p = 1, \ldots, 6$, derive displacement and velocity expressions for point (p) [Equations (6.16), (6.17)]
For $s = 1, \ldots, 3$, calculate KE and PE for seg(s) and update the Lagrangian [Equations (6.5), (6.23), (6.25)]
For $t = 1, \ldots, 3$, find the work done by torque trq(t) and update system work W [Equation (6.34)]
For $f = 1$, find the work done by the force frc(f) and update W [Equation (6.34)]
For $i = 1, \ldots, 4$, apply Lagrangian equations for coordinate q_i [Equations (6.8), (6.9)]

The intermediate results for the relevant parts of these steps are as follows.

1. Displacement vectors,

$$\text{disp}(1) = [q_1, 0]$$

$$\text{disp}(2) = [q_1 + r_1 \cos(q_2), r_1 \sin(q_2)]$$

$$\text{disp}(3) = [q_1 + l_1 \cos(q_2), l_1 \sin(q_2)]$$

$$\text{disp}(4) = [q_1 + l_1 \cos(q_2) + r_2 \cos(q_3), l_1 \sin(q_2) + r_2 \sin(q_3)]$$

$$\text{disp}(5) = [q_1 + l_1 \cos(q_2) + l_2 \cos(q_3), l_1 \sin(q_2) + l_2 \sin(q_3)]$$

$$\text{disp}(6) = [q_1 + l_1 \cos(q_2) + l_2 \cos(q_3) + r_3 \cos(q_4), l_1 \sin(q_2) + l_2 \sin(q_3) + r_3 \sin(q_4)]$$

2. Velocity vectors for selected points,

$$\text{velo}(2) = [\dot{q}_1 - r_1\dot{q}_2 \sin(q_2), r_1\dot{q}_2 \cos(q_2)]$$

$$\text{velo}(4) = [\dot{q}_1 - l_1\dot{q}_2 \sin(q_2) - r_2\dot{q}_3 \sin(q_3),$$
$$l_1\dot{q}_2 \cos(q_2) + r_2\dot{q}_3 \cos(q_3)]$$

$$\text{velo}(6) = [\dot{q}_1 - l_1\dot{q}_2 \sin(q_2) - l_2\dot{q}_3 \sin(q_3) - r_3\dot{q}_4 \sin(q_4),$$
$$l_1\dot{q}_2 \cos(q_2) + l_2\dot{q}_3 \cos(q_3) + r_3\dot{q}_4 \cos(q_4)]$$

3. Lagrangian $L = \text{KE} - \text{PE}$

$$L = \tfrac{1}{2}m_1\{[\dot{q}_1 - r_1\dot{q}_2 \sin(q_2)]^2 + [r_1\dot{q}_2 \cos(q_2)]^2\} + \tfrac{1}{2}I_1(\dot{q}_2)^2$$
$$- m_1g[r_1 \sin(q_2)] + \tfrac{1}{2}m_2\{[\dot{q}_1 - l_1\dot{q}_2 \sin(q_2) - r_2\dot{q}_3 \sin(q_3)]^2$$
$$+ [l_1\dot{q}_2 \cos(q_2) + r_2\dot{q}_3 \cos(q_3)]^2\} + \tfrac{1}{2}I_2(\dot{q}_3)^2 - m_2g[l_1 \sin(q_2)$$
$$+ r_2 \sin(q_3)] + \tfrac{1}{2}m_3\{[\dot{q}_1 - l_1\dot{q}_2 \sin(q_2) - l_2\dot{q}_3 \sin(q_3)$$
$$- r_3\dot{q}_4 \sin(q_4)]^2 + [l_1\dot{q}_2 \cos(q_2) + l_2\dot{q}_3 \cos(q_3)$$
$$+ r_3\dot{q}_4 \cos(q_4)]^2\} + \tfrac{1}{2}I_3(\dot{q}_4)^2 - m_3g[l_1 \sin(q_2) + l_2 \sin(q_3)$$
$$+ r_3 \sin(q_4)]$$

The Lagrangian L is simplified by expanding the previous expression and collecting terms,

$$L = \tfrac{1}{2}(\dot{q}_1)^2(m_1 + m_2 + m_3) + \tfrac{1}{2}(\dot{q}_2)^2[m_1(r_1)^2 + I_1 + m_2(l_1)^2 + m_3(l_1)^2]$$
$$+ \tfrac{1}{2}(\dot{q}_3)^2[m_2(r_2)^2 + I_2 + m_3(l_2)^2] + \tfrac{1}{2}(\dot{q}_4)^2[m_3(r_3)^2 + I_3]$$
$$- \dot{q}_1\dot{q}_2 \sin(q_2)(m_1r_1 + m_2l_1 + m_3l_1) - \dot{q}_1\dot{q}_3 \sin(q_3)$$
$$\cdot (m_2r_2 + m_3l_2) - \dot{q}_1\dot{q}_4(m_3r_3) + \dot{q}_2\dot{q}_3l_1 \cos(q_3 - q_2)$$
$$\cdot (m_2r_2 + m_3l_2) + \dot{q}_2\dot{q}_4l_1 \cos(q_4 - q_2)(m_3r_3)$$
$$+ \dot{q}_3\dot{q}_4l_2 \cos(q_4 - q_3)(m_3r_3) - m_1gr_1 \sin(q_2) - m_2g[l_1 \sin(q_2)$$
$$+ r_2 \sin(q_3)] - m_3g[l_1 \sin(q_2) + l_2 \sin(q_3) + r_3 \sin(q_4)]$$

4. External work W,

$$W = T_1(q_2 - 0) + T_2(q_3 - q_2) + T_3(q_4 - q_3) + F(q_1 + l_1 \cos(q_2)$$
$$+ l_2 \cos(q_3))$$

5. Finally the equations of motion are for the generalized coordinates q_1, q_2, q_3, q_4, respectively,

(a) $F = \ddot{q}_1(m_1 + m_2 + m_3)$

$- [(\dot{q}_2)^2 \cos(q_2) + \ddot{q}_2 \sin(q_2)]$

$\cdot (m_1 r_1 + m_2 l_2 + m_3 l_1)$

$- [(\dot{q}_3)^2 \cos(q_3) + \ddot{q}_3 \sin(q_3)]$

$\cdot (m_2 r_2 + m_3 l_2) - \ddot{q}_4 m_3 r_3$

(b) $T_1 - T_2 - F l_1 \sin(q_2) = -\ddot{q}_1 \sin(q_2)(m_1 r_1 + m_2 l_1 + m_3 l_1)$

$+\ddot{q}_2[m_1(r_1)^2 + I_1 + m_2(l_1)^2 + m_3(l_1)^2]$

$+ [\ddot{q}_3 \cos(q_3 - q_2) + (\dot{q}_3)^2$

$\cdot \sin(q_3 - q_2)](m_2 r_2 l_1 + m_3 l_1 l_2)$

$+ [\ddot{q}_4 \cos(q_4 - q_2) + (\dot{q}_4)^2$

$\cdot \sin(q_4 - q_2)](m_3 r_3 l_1)$

$+ g(m_1 r_1 + m_2 l_1 + m_3 l_1) \cos(q_2)$

(c) $T_2 - T_3 - F l_2 \sin(q_3) = -\ddot{q}_1 \sin(q_3)(m_2 r_2 + m_3 l_2)$

$+ [\ddot{q}_2 \cos(q_3 - q_2) - (\ddot{q}_2)^2$

$\cdot \sin(q_3 - q_2)](m_2 r_2 l_1 + m_3 l_1 l_2)$

$+ \ddot{q}_3[m_2(r_2)^2 + m_3(l_2)^2 + I_2]$

$+ [\ddot{q}_4 \cos(q_4 - q_3) - (\dot{q}_4)^2$

$\cdot \sin(q_4 - q_3)](m_3 r_3 l_2)$

$+ g(m_2 r_2 + m_3 l_2) \cos(q_3)$

(d) $T_3 = -\ddot{q}_1 m_3 r_3 \sin(q_4) + [\ddot{q}_2 \cos(q_4 - q_2)$

$+ (\dot{q}_2)^2 \sin(q_4 - q_2)](m_3 r_3 l_1) + [\ddot{q}_3 \cos(q_4 - q_3) + (\dot{q}_3)^2$

$\cdot \sin(q_4 - q_3)](m_3 r_3 l_2) + \ddot{q}_4[m_3(r_3)^2 + I_3] + m_3 g r_3 \cos(q_4)$

These four equations are the minimum number of equations that describe the model. They are highly nonlinear, tightly coupled, lengthy, and prone to errors. By breaking up the model at joints and introducing quasi DOF in the form of additional generalized coordinates and constraint equations that counteract these superfluous coordinates, a larger set of equations are obtained that are shorter, manageable, and contain more information about the model.

The tight coupling between these four equations demonstrates the potential errors that were mentioned in Section 6.0.2. An error in the mass of any segment, for example, will result in errors in all four coordinates. Mass m_3, the mass of the HAT, appears in all four equations and will affect the calculation of q_1, q_2, q_3, and q_4. The errors will also accumulate over time. Thus, it is essential before the model is used to answer a research question that all the anthropometrics and internal joint constraints be almost perfect such that an internal validation is achieved.

6.7 CONCLUSIONS

In this chapter, a procedure for generating the equations of motion for a given model was presented. The method replaces model configuration and gathers parameters by groups of linked lists. The first element of each list links the list to other relevant lists in the model. Once these lists are written correctly, deriving the equations of motion follows the same systematic procedure. This makes the method more suitable for computer implementation. As for hand derivation, the only variations from a simple model to a complicated one is the amount of work involved.

6.8 REFERENCES

Andrews, G. C. "A General Restatement of the Laws of Dynamics Based on Graph Theory," in *Problem Analysis in Science and Engineering,* F. H. Branin, Jr., Ed. (Academic Press, New York, 1977).

Andrews, G. C. and H. K. Kesavan. "The Vector Network Model: A New Approach to Vector Dynamics," *J. Mechanisms and Mach. Theory* **10**:57–75, 1975.

Beckett, R. and K. Chang. "On the Evaluation of the Kinematics of Gait by Minimum Energy," *J. Biomech.* **1**:147–159, 1968.

Chace, M. A. *Methods and Experience in Computer Aided Design of Large Displacement Mechanical Systems,* NATO ASI Series, vol. F9. (Springer, Berlin, 1984).

Chao, E. Y. and K. Rim. "Applications of Optimization Principles in Determining the Applied Moments in Human Leg Joints During Gait," *J. Biomech.* **6**:497–510, 1973.

Chen, B., M. J. Hines, and H. Hemami. "Dynamic Modeling for Implementation of a Right Turn in Bipedal Walking," *J. Biomech.* **19**:195–206, 1986.

Gilchrist, L. A. and D. A. Winter. "A Two-Part, Viscoelastic Foot Model for Use in Gait Simulations," *J. Biomech.* **29**:795–798, 1996.

Gilchrist, L. A. and D. A. Winter. "A Multisegment Computer Simulation of Normal Human Gait," *IEEE Trans. Rehab. Eng.* **5**:290–299, 1997.

Greenwood, D. T. *Classical Dynamics.* (Prentice-Hall, Englewood Cliff, NJ, 1977).

Hemami, H. "A Feedback On-Off Model of Biped Dynamics," *IEEE Trans. on Systems, Man and Dynamics* **SMC-10**:376–383, 1980.

Hemami, H., M. J. Hines, R. E. Goddard, and B. Friedman. "Biped Sway in the Frontal Plane with Locked Knees," *IEEE Trans. System. Man. Cybern.* **SMC-12**:577–582, 1982*a*.

Hemami, H., Y. F. Zhang, and M. J. Hines. "Initiation of Walk and Tiptoe of a Planar Nine-Link Biped," *Math. Biosci.* **61**:163–189, 1982*b*.

Onyshko, S. and D. A. Winter. "A Mathematical Model for the Dynamics of Human Locomotion," *J. Biomech.* **13**:361–368, 1980.

Pandy, M. G. and N. Berme. "Synthesis of Human Walking: A Planar Model of Single Support," *J. Biomech.* **21**:1053–1060, 1988.

Paul, B. "DYMAC: A Computer Program to Simulate Machinery Dynamics," *J. Agric. Eng.* **59**:15–16, 1978.

Phillips, S. J., E. M. Roberts, and T. C. Huang. "Quantification of Intersegmental Reactions During Rapid Swing Motion," *J. Biomech.* **16**:411–417, 1983.

Singhal, K. and H. K. Kesavan. "Dynamic Analysis of Mechanisms via Vector-Network Model," *J. Mechanisms Mach. Theory* **18**:175–180, 1983.

Townsend, M. A. "Dynamics and Coordination of Torso Motions in Human Locomotion," *J. Biomech.* **14**:727–738, 1981.

Townsend, M. A. and A. Seireg. "The Synthesis of Bipedal Locomotion," *J. Biomech.* **5**:71–83, 1972.

Wells, D. A. *Lagrangian Dynamics with a Treatment of Euler's Equations of Motion.* (McGraw-Hill, New York, 1967).

Wittenburg, J. *Dynamics of Systems of Rigid Bodies.* (Teubner, Stuttgart, West Germany, 1977).

7

THREE-DIMENSIONAL KINEMATICS AND KINETICS

7.0 INTRODUCTION

Over the past fifteen years, there have been major commercial developments in three-dimensional (3D) hardware and software. Chapter 2 has included descriptions of some of the 3D imaging systems that have been introduced. The majority of the systems are television-based, with multiple-camera arrangements requiring passive reflective markers, while the remainder use active infrared emitting diodes (IRED) markers and infrared sensors. Regardless of the system used, the output of the data collection stage is a file of x, y, z coordinates of each of the markers at each sample point in time. These coordinates are in the Global Reference System (GRS) that is fixed in the laboratory or data collection space. The purpose of this chapter is to go through the steps where these coordinate data are transformed into the anatomical axes of the body segments so that a kinetic analysis can be done in a similar manner, as has been detailed for two-dimensional (2D) analyses in Chapters 4 and 5.

7.1 AXES SYSTEMS

There are several axes reference systems that must be introduced in addition to the GRS already introduced above. The markers that are placed on each segment provide a marker axis system that is a local reference system (LRS) for each individual segment. A second LRS is the axis system that defines the principal axis of each segment. Because skeletal landmarks are used to define these axes, this system is referred to as the anatomical axis system.

7.1.1 Global Reference System

For purposes of convenience, we will be consistent in our axis directions for the GRS: X is the forward/backward direction, Y is the vertical (gravitational) axis, and Z is the left/right (medial/lateral) axis. Thus, the XZ plane is the horizontal plane and by definition is orthogonal to the vertical axis. The directions of these GRS axes are the same as those of the axes in the force plate. To ensure that this is so, a spatial calibration system (a rigid cubic frame or a rigid 3D mechanical axis) is instrumented with markers and is placed on one of the force plates and aligned along the X and Z axes of the force platform. The position of each of the markers relative to the origin of the force plate is known and fed into the computer. If more than one force platform is used, the origin of each additional platform is recorded by an X and Z offset from the primary platform. An additional offset in the Y direction would be necessary if the additional platform were at a different height from the first (as would be necessary for a biomechanical analysis of stairway or ramp walking). Many laboratories have a fixed arrangement of cameras so there is no need to recalibrate the GRS every day; such is the case in clinical gait laboratories. Such a system is illustrated in a previous chapter; see Figure 2.12. However, in many research situations, the cameras are rearranged to best capture the new movement, and therefore require a new calibration of the GRS. Once the calibration is complete, the cameras cannot be moved, and care must be taken to ensure they are not accidentally displaced.

7.1.2 Local Reference Systems and Rotation of Axes

Students are referred back to several sections in Chapter 6 and are asked to review Sections 6.2.6 through 6.2.7.2. These sections cover reference systems, displacement, and velocity vectors for 2D and 3D systems. The symbols introduced in those sections are continued in this chapter.

Within each segment, the anatomical axis system is set with its origin at the center of mass (COM) of the segment and its principal y-axis usually along the long axis of the segment, or in the case of segments like the pelvis, along a line defined by skeletal landmarks such as PSIS and ASIS. The other local axis system is constructed on the segment using a set of surface markers. A total of two transformations are necessary to get from the GRS to the marker axis system and from the marker to the anatomical axis system. Figure 7.1 shows how one of those rotations is done. The axis system x, y, z needs to be rotated into the system denoted by x''', y''', z'''. Many sequences of rotation are possible; here we use the common Cardan sequence x–y–z. This means we rotate about the x axis first, about the new y axis second, and about the new z axis last. The first rotation is θ_1 about the x axis to get x', y', z'. Because we have rotated about the x axis, x will not be changed and $x' = x$, while the y axis changes to y' and the z axis to z'. The second rotation is θ_2 about the new y' axis to get x'', y'', z''. Because this rotation has been about

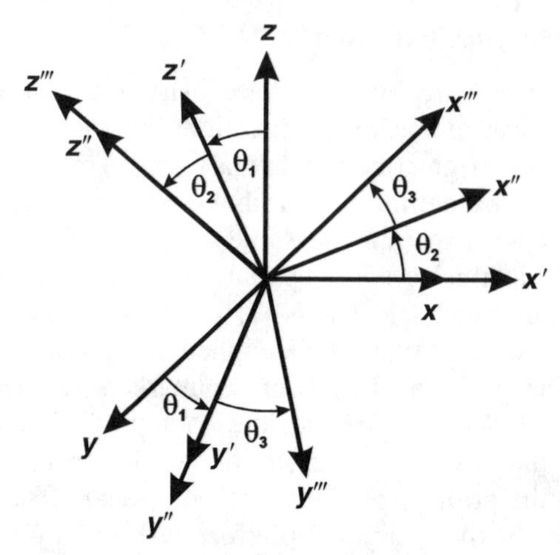

Figure 7.1 Cardan sequence of three rotations about the x, y, z axes. The first rotation, θ_1, about the x axis to get x', y', z'; the second rotation, θ_2, about the new y' axis to get x'', y'', z'', and the final rotation, θ_3, about the new z'' axis to get the desired x''', y''', z'''.

the y' axis, $y'' = y'$. The final rotation is θ_3 about the new z'' axis to get the desired x''', y''', z'''. Assuming that we have a point with coordinates x_0, y_0, z_0 in the original x, y, z axis system, that same point will have coordinates x_1, y_1, z_1 in the x', y', z' axis system. Based on the rotation θ_1:

$$x_1 = x_0$$

$$y_1 = y_0 \cos\theta_1 + z_0 \sin\theta_1$$

$$z_1 = -y_0 \sin\theta_1 + z_0 \cos\theta_1$$

Using the shorthand notation $c_1 = \cos\theta_1$ and $s_1 = \sin\theta_1$, in matrix notation this may now be written as:

$$\begin{bmatrix} x_1 \\ y_1 \\ z_1 \end{bmatrix} = \begin{bmatrix} 1 & 0 & 0 \\ 0 & c_1 & s_1 \\ 0 & -s_1 & c_1 \end{bmatrix} \begin{bmatrix} x_0 \\ y_0 \\ z_0 \end{bmatrix} = [\Phi_1] \begin{bmatrix} x_0 \\ y_0 \\ z_0 \end{bmatrix} \tag{7.1}$$

After the second rotation θ_2 about y', this point will have coordinates x_2, y_2, z_2 in the x'', y'', z'' axis system.

$$\begin{bmatrix} x_2 \\ y_2 \\ z_2 \end{bmatrix} = \begin{bmatrix} c_1 & 0 & -s_1 \\ 0 & 1 & 0 \\ s_2 & 0 & c_2 \end{bmatrix} \begin{bmatrix} x_1 \\ y_1 \\ z_1 \end{bmatrix} = [\Phi_2] \begin{bmatrix} x_1 \\ y_1 \\ z_1 \end{bmatrix} \tag{7.2}$$

Finally, the third rotation θ_3 about z'' yields the coordinates x_3, y_3, z_3 in the x''', y''', z''' axis system.

$$\begin{bmatrix} x_3 \\ y_3 \\ z_3 \end{bmatrix} = \begin{bmatrix} c_3 & s_3 & 0 \\ -s_3 & c_3 & 0 \\ 0 & 0 & 1 \end{bmatrix} \begin{bmatrix} x_2 \\ y_2 \\ z_2 \end{bmatrix} = [\Phi_3] \begin{bmatrix} x_2 \\ y_2 \\ z_2 \end{bmatrix} \tag{7.3}$$

Combining Equations (7.1), (7.2), and (7.3), we get:

$$\begin{bmatrix} x_3 \\ y_3 \\ z_3 \end{bmatrix} = [\Phi_3][\Phi_2][\Phi_1] \begin{bmatrix} x_0 \\ y_0 \\ z_0 \end{bmatrix} \tag{7.4}$$

Note that the matrix multiplication as shown in Equation (7.4) is not commutative, which means that the order of the transformations must be such that $[\Phi_1]$ is done first, $[\Phi_2]$ second, and $[\Phi_3]$ last. In other words, $[\Phi_1]\,[\Phi_2] \neq [\Phi_2]\,[\Phi_1]$. An expansion of Equation (7.4) yields

$$\begin{bmatrix} x_3 \\ y_3 \\ z_3 \end{bmatrix} = \begin{bmatrix} c_2c_3 & s_3c_1 + s_1s_2c_3 & s_1s_3 - c_1s_2c_3 \\ -c_2s_3 & c_1c_3 - s_1s_2s_3 & s_1c_3 + c_1s_2s_3 \\ s_2 & -s_1c_2 & c_1c_2 \end{bmatrix} \begin{bmatrix} x_0 \\ y_0 \\ z_0 \end{bmatrix} \tag{7.5}$$

7.1.3 Other Possible Rotation Sequences

In theory, there are twelve possible correct rotation sequences; all were introduced by the Swiss mathematician, Leonhard Euler (1707–1783). The list below gives all possible valid rotation sequences. The example explained above is generally referred to as the Cardan system, which is commonly used in biomechanics, while the z–x–z rotation sequence, generally referred to as the Euler system, is commonly used in mechanical engineering.

x–y'–x''	x–y'–z'' (Cardan)	x–z'–x''	x–z'–y''
y–x'–y''	y–x'–z''	y–z'–x''	y–z'–y''
z–x'–y''	z–x'–z'' (Euler)	z–y'–x''	z–y'–z''

7.2 MARKER AND ANATOMICAL AXES SYSTEMS

The following description outlines the steps that are necessary to transform the x, y, z marker coordinates from the GRS to the anatomical axes of the segments of the person whose movement is being analyzed. Figure 7.2 presents the axis systems involved for a given segment whose COM is at c and whose axes x–y–z are as shown. The GRS has axes X–Y–Z, and they are fixed for any given camera arrangement. The second axis system, x_m–y_m–z_m, is the marker axis system for each segment, and this can vary from laboratory to laboratory. Even within a given laboratory, each experiment could have a different arrangement of markers. For a correct 3D analysis, there must be at least three independent markers per body segment, and there must not be

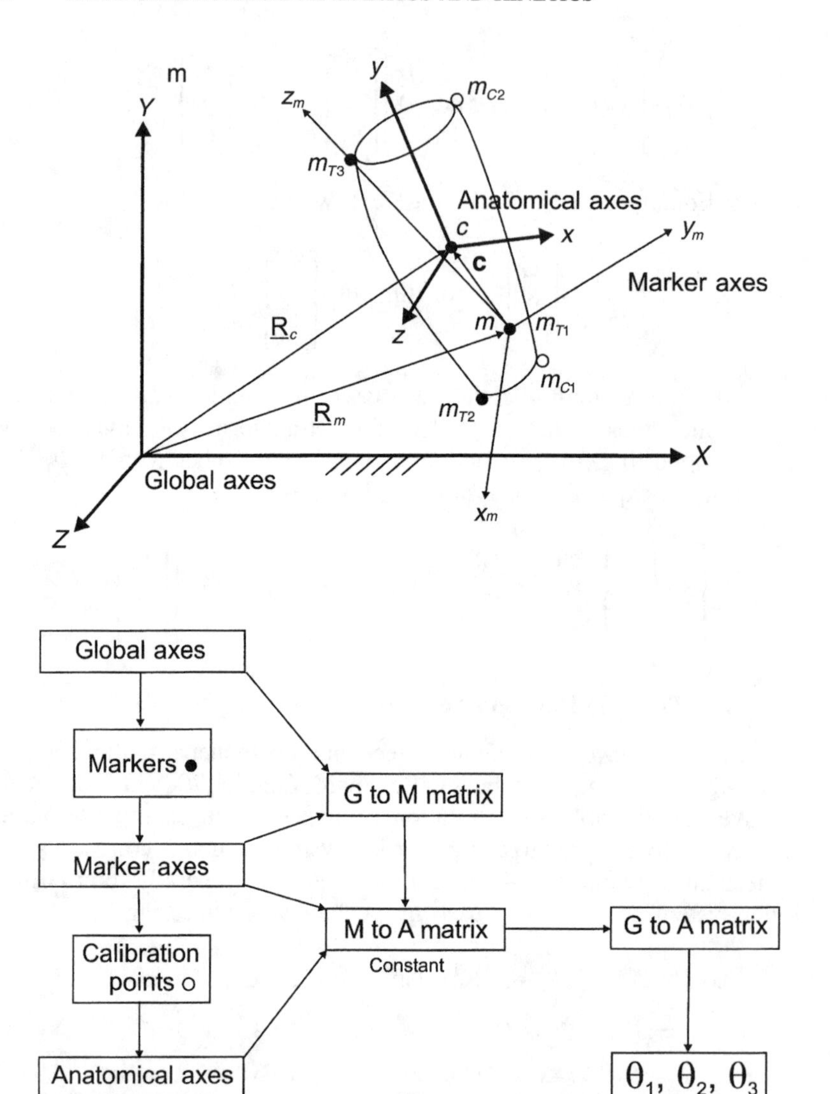

Figure 7.2 An anatomical segment showing the GRS, the marker axes, and anatomical axes. Three tracking markers, m_{T1}, m_{T2}, and m_{T3}, in conjunction with two calibration markers, m_{C1} and m_{C2}, define the constant [M to A] matrix. The three tracking markers in the GRS define the [G to M] matrix. The product of the variable [G to M] and the fixed [M to A] matrices gives the variable [G to A] matrix from which a new θ_1, θ_2, and θ_3 are defined for each frame.

common markers between adjacent segments. The markers on each segment must not be collinear, which means they must not be in a straight line. They must form a plane in 3D space; as shown in Figure 7.2, the three tracking markers m_{T1}, m_{T2}, and m_{T3} define the tracking marker plane. This plane is assumed to contain the x_m and z_m axes such that all three markers are in the $+ve$ x_m and the $+ve$ z_m quadrant. One point on this marker plane is arbitrarily chosen as the origin of the marker axes system; here m_{T1} is chosen and is labeled m. The line from m_{T1} to m_{T3} defines the $+ve$ z_m axis; y_m is normal to the tracking plane, and x_m is orthogonal to the plane defined by y_m–z_m to form a right-hand system.

The anatomical calibration process is to find the relation between the marker axes, x_m–y_m–z_m, and the anatomical axes, x–y–z. This process requires the subject to assume a well-defined position; usually the anatomical position is used. At this time, extra calibration markers may be placed temporarily on the segment to define well-known anatomical points from which the segment's anatomical axes can be defined. For example, for the leg segment the three tracking markers could be placed on the head of the fibula (m_{T3}), the lateral malleolus (m_{T2}), and at the midpoint on the anterior surface of the tibia (m_{T1}). During calibration, temporary markers, m_{C1} and m_{C2}, could be placed on the medial malleolus and the medial epicondyle of the tibia, respectively. With the subject standing still for about a second, the coordinates of the three tracking and the two calibration markers are recorded and averaged over the calibration time. The long axis of the leg segment (y axis) would then be defined as the line joining the midpoint between the lateral and medial malleolii (m_{T2} and m_{C1}) and the midpoint between the head of the fibula and the medial epicondyle of the tibia (m_{T3} and m_{C2}). These midpoints are the ankle and knee joints, respectively. The leg y axis and the line from m_{C1} to m_{T2} define a plane normal to which is the leg x axis. The direction of the leg z axis would be defined as a line normal to the leg x–y plane such that the leg x–y–z is a dextral system. The anatomical axes of the leg are now defined relative to the three tracking markers. The location of the center of mass of the leg would be a known distance along the y axis of the leg from the ankle joint; thus, the vector, $\mathbf{c}$, from m, the origin of the tracking marker axis system, is also known. The two calibration markers are now removed and are no longer needed because the orientation of the axis system of the three tracking markers is now known and assumed to be fixed relative to the newly defined anatomical axes.

In clinical gait laboratories, it is impossible for many patients, such as cerebral palsy or stroke patients, to assume the anatomical position for even a short period of time. Thus, the clinical gait team have developed a consistent marker arrangement combined with a number of specific anthropometric measures. These include such measures as ankle and knee diameters which, when combined with generic X-ray anthropometric measures, allow the team to input an algorithm so that offset displacements from the tracking markers to the joint centers are known. Patients are then asked to assume a static standing position with a number of temporary calibration markers similar to what has been described above. In effect, the major single difference in the clinical laboratory is that the patient is calibrated in a comfortable standing position rather than the anatomical position. For the complete detailed steps in arriving at the joint kinetics in an operational clinical laboratory, the reader is referred to Davis et al. (1991) and Õunpuu et al. (1996).

In Figure 7.2 we see two matrix rotations. [G to M] is a 3 × 3 rotation matrix that rotates from the GRS to the tracking marker axes, x_m–y_m–z_m. This is a time-varying matrix because the tracking marker axes will be continuously changing relative to the GRS. [M to A] is a 3 × 3 matrix that rotates from the tracking marker axes to the anatomical axes. This matrix is assumed to be constant and results from the calibration protocol. The combination of these two rotation matrices gives us the [G to A] rotation matrix which, when solved for a selected angle sequence, yields the three time-varying rotation angles, θ_1, θ_2, θ_3. With this final matrix, we can get the orientation of the anatomical axes directly from the tracking marker coordinates that are collected in the GRS.

However, from Figure 7.2, we are not yet finished. We also have to find a translational transformation to track the 3D coordinates of the segment COM, **c**, over time. The location of **c** is defined by the vector, $\mathbf{R}_c$, which is a vector sum of $\mathbf{R}_m$ + **c**. The vector $\mathbf{R}_m$ is the GRS coordinates of the tracking marker, m_{T1}, while **c** is a constant vector joining m to c, as defined earlier.

7.2.1 Example of a Kinematic Data Set

7.2.1.1 Calibration—Calculation of* [*Marker to Anatomical*] *Matrix. Let us now look at an example of numerical data to see how the various transformations are calculated. The leg segment in Figure 7.2 will be used as an example. Recall that there are three tracking markers on this segment plus two calibration markers whose coordinates were digitized during the calibration period when the subject stood steady in the anatomical position. In Figure 7.2, the subject would be facing the $+ve$ X direction and the left leg was being analyzed. Table 7.1 gives the X, Y, and Z coordinates in the GRS, averaged over one second.

In this calibration position, the ankle = $(m_{T2} + m_{C1})/2$, $X_a = 2.815$, $Y_a = 10.16$, $Z_a = 22.685$.

TABLE 7.1 Marker Coordinates During Standing Calibration

Marker	Location	X (cm)	Y (cm)	Z (cm)
m_{T1}	Midleg	9.39	30.02	21.90
m_{T2}	Lateral malleolus	2.92	10.10	18.85
m_{T3}	Fibular head	5.05	41.90	15.41
m_{C1}	Medial malleolus	2.71	10.22	26.52
m_{C2}	Medial condyle	8.29	41.88	26.52

The coordinates of the knee $= (m_{T3} + m_{C2})/2$, $X_k = 6.67$, $Y_k = 41.89$, $Z_k = 20.965$.

The COM of the leg $= 0.567 \times$ knee $+ 0.433 \times$ ankle, $X_c = 5.001$, $Y_c = 28.151$, $Z_c = 21.710$.

Now we have to locate the anatomical x, y, and z-axes. Let the line joining the ankle to the knee be the y axis, and the line joining the lateral malleolus to the medial malleolus be an interim z axis (because it is not exactly normal to the y axis, but will be corrected later). These two axes now form a plane, and the x axis, by definition, is normal to the yz plane and therefore is the "cross product" of y and z or $x_{an} = (y_{an} \times z_{an})$. Using the subscript "*an*" to indicate anatomical axes;

$$z_{an} = (m_{C1} - m_{T2}): \quad x_z = -0.21, \quad y_z = 0.21, \quad z_z = 7.67$$

$$y_{an} = (\text{knee-ankle}): \quad x_y = 3.855, \quad y_y = 31.73, \quad z_y = -1.72$$

$$x_{an} = (y_{an} \times z_{an}): \quad x_x = 243.575, \quad y_x = -29.207, \quad z_x = 7.126$$

One final correction must be done to our anatomical model. z_{an} is the line joining the medial to lateral malleolii and is approximately 90° from the long axis y_{an}. To ensure that all three anatomical axes are at right angles to each other, z_{an} must be corrected: $z_{an} = (x_{an} \times y_{an})$: $x_z = -176.872$, $y_z = 446.42$, $z_z = 7841.27$. Note that none of these vectors are unity length, and they are normally reported as a unit vector. For example, the length of x_{an} is 245.423 cm; thus, dividing each coordinate by the length of x_{an} yields a unity vector for x_{an}: $x_x = 0.9925$, $y_x = -0.1190$, $z_x = 0.290$. Similarly for y_{an} as a unit vector, $x_y = 0.1204$, $y_y = 0.9913$, $z_y = -0.0537$. Similarly as a unit vector, the corrected z_{an} is: $x_z = -0.0225$, $y_z = 0.0568$, $z_z = 0.9981$. We can now construct the leg anatomical-to-global matrix [LA to G] and it is:

$$\begin{bmatrix} 0.9925 & 0.1204 & -0.0225 \\ -0.1190 & 0.9913 & 0.0568 \\ 0.0290 & -0.0537 & 0.9981 \end{bmatrix}$$

Note that the diagonal values are almost = 1, indicating that the subject stood in the calibration position with his three anatomical axes almost perfectly lined up with the global axes. A more useful transformation matrix is the leg global-to-anatomical [LG to A], which is the transpose of [LA to G]:

$$\begin{bmatrix} 0.9925 & -0.1190 & 0.0290 \\ 0.1204 & 0.9913 & -0.0537 \\ -0.0225 & 0.0568 & 0.9981 \end{bmatrix}$$

These anatomical axes have their origins at the ankle joint. However, from our inverse dynamics it is more convenient that the origin of the anatomical axes (see Figure 7.2) be located at the COM of the leg segment with coordinates 0, 0, 0. We must now establish the local coordinates of the ankle and knee joints and the three tracking markers relative to this new origin at the COM.

From the COM the ankle vector = (global ankle – global COM),

$$x_{al} = X_a - X_c = 2.815 - 5.001 = -2.186,\ y_{al} = Y_a - Y_c = 10.16 - 28.151$$
$$= -17.991,\ z_{al} = Z_a - Z_c = 22.685 - 21.710 = 0.975.$$

The anatomical ankle vector is the product of [LG to A][ankle vector]:

$$\begin{bmatrix} 0.9925 & -0.1190 & 0.0290 \\ 0.1204 & 0.9913 & -0.0537 \\ -0.0225 & 0.0568 & 0.9981 \end{bmatrix} \begin{bmatrix} -2.186 \\ -17.991 \\ 0.975 \end{bmatrix} = \begin{bmatrix} \approx 0 \\ -18.15 \\ \approx 0 \end{bmatrix}$$

This anatomical ankle vector lies along the line joining the ankle to knee and the ankle is 18.15 cm distal of the COM. The x and z components of this ankle vector are theoretically 0 but were calculated ≈0 because of the limited number of decimal points in our numbers. If we were to repeat this procedure for the anatomical knee vector and for the anatomical vectors of the three tracking markers m_{T1}, m_{T2} and m_{T3}, we would calculate the following:

Anatomical knee vector	Anatomical m_{T1} vector	Anatomical m_{T2} vector	Anatomical m_{T3} vector
$\begin{bmatrix} \approx 0 \\ 13.86 \\ \approx 0 \end{bmatrix}$	$\begin{bmatrix} 4.139 \\ 2.371 \\ 0.197 \end{bmatrix}$	$\begin{bmatrix} 0.000 \\ -17.991 \\ -3.833 \end{bmatrix}$	$\begin{bmatrix} -1.770 \\ 13.974 \\ -5.508 \end{bmatrix}$

We are now ready to calculate the constant marker-to-anatomical matrix ([M to A] in Figure 7.2). The three tracking markers form a plane in the GRS, and we can now define our maker axes in that plane. m_{T2} is chosen as the origin of the marker plane and the line joining m_{T2} to m_{T3} is chosen to be the z axis, labeled z_m. The line joining m_{T2} to m_{T1} is a vector labeled **a** (an interim vector to allow us to calculate y_m and x_m). y_m is normal to the plane defined by z_m and A and x_m, is normal to the plane defined by y_m and z_m.

$$z_m = \text{local } m_{T3} - \text{local } m_{T2} : [-1.770, 31.965, -1.675]$$
$$A \text{ vector} = \text{local } m_{T1} - \text{local } m_{T2} : [4.139, 20.362, 4.030]$$
$$y_m = (z_m \times A \text{ vector}): [162.925, 0.200, -168.344]$$
$$x_m = (y_m \times z_m): [5380.78, 570.87, 5208.25]$$

The normalized axis for this leg anatomical-to-marker matrix [LA to M] is:

$$\begin{bmatrix} 0.7164 & 0.0760 & 0.6935 \\ 0.6954 & 0.0008 & -0.7186 \\ -0.0552 & 0.9971 & -0.0522 \end{bmatrix}$$

The fixed leg marker-to-anatomical matrix [LM to A] is the transpose of [LA to M]:

$$\begin{bmatrix} 0.7164 & 0.6954 & -0.0552 \\ 0.0760 & 0.0008 & 0.9971 \\ 0.6935 & -0.7186 & -0.0522 \end{bmatrix}$$

7.2.1.2 Tracking Markers—Calculation of [Global to Marker] Matrix. We are now ready to calculate the [G to M] matrix in Figure 7.2. Table 7.2 lists representative GRS coordinates for the leg segment for three successive frames of walking taken during the swing phase. The procedure to calculate this [G to M] matrix is exactly the same as the latter part of the calculation of the [M to A] matrix. Consider the coordinates for frame 6.

$$\mathbf{z}_m = (m_{T3} - m_{T2})\text{: } [X_z = 24.34,\ Y_z = 19.99,\ Z_z = -3.64]$$

$$\mathbf{a}\text{ vector} = (m_{T1} - m_{T2}) : [X_a = 18.86,\ Y_a = 9.15,\ Z_z = 4.09]$$

$$y_m = (z_m \times \mathbf{a})\text{: } [115.065,\ -168.201,\ -154.30]$$

$$x_m = (y_m \times z_m)\text{: } [3696.709,\ -3336.825,\ 6394.162]$$

The normalized axis for this leg [G to M] matrix for frame 6 is

$$\begin{bmatrix} 0.4561 & -0.4117 & 0.7889 \\ 0.4501 & -0.6580 & -0.6037 \\ 0.7677 & 0.6305 & -0.1148 \end{bmatrix}$$

7.2.1.3 Calculation of [Global to Anatomical] Matrix. From Figure 7.2, the final step is to calculate the [G to A] matrix that is the product of the fixed [M to A] matrix and the variable [G to M] matrix; for frame 6 this product is

TABLE 7.2 Tracking Markers During Walking

	m_{T1}			m_{T2}			m_{T3}		
Frame	X	Y	Z	X	Y	Z	X	Y	Z
5	20.65	33.87	35.95	1.30	25.74	32.14	26.52	44.43	28.10
6	25.46	34.47	35.95	6.60	25.32	31.86	30.94	45.31	28.22
7	30.18	34.97	35.94	11.98	24.64	31.60	35.08	46.10	28.36

$$\begin{bmatrix} 0.7164 & 0.6954 & -0.0552 \\ 0.0760 & 0.0008 & -0.9971 \\ 0.6935 & 0.7186 & -0.0522 \end{bmatrix} \begin{bmatrix} 0.4561 & -0.4117 & 0.7887 \\ 0.4501 & -.6580 & 0.6037 \\ 0.7677 & 0.6305 & -0.1148 \end{bmatrix}$$
$$= \begin{bmatrix} 0.5974 & -0.7873 & 0.1515 \\ 0.8000 & 0.5969 & -0.0550 \\ -0.0472 & 0.1544 & 0.9868 \end{bmatrix}$$

From Equation (7.5), this [G to A] matrix is equal to

$$\begin{bmatrix} c_2c_3 & s_3c_1 + s_1s_2c_3 & s_1s_3 - c_1s_2c_3 \\ -c_2s_3 & c_1c_3 - s_1s_2s_3 & s_1c_3 + c_1s_2s_3 \\ s_2 & -s_1c_2 & c_1c_2 \end{bmatrix}$$

We now solve this matrix to get θ_1, θ_2, and θ_3. Equating the three terms in the bottom row: $s_2 = -0.0472$, $-s_1c_2 = 0.1544$, $c_1c_2 = 0.9868$. $\therefore$ $\theta_2 = -2.71°$ or 177.29°; assuming $\theta_2 = -2.71°$, $c_2 = 0.99888$ or -0.99888, $c_1c_2 = 0.9868$, $\therefore c_1 = 0.9868/0.99888 = 0.9879$, and $\theta_1 = 8.92°$ or $-8.92°$. $\therefore s_1 = 0.1550$ or -0.1550. We now validate that $\theta_1 = -8.92°$ because $-s_1c_2 \approx 0.1544$. We now use the first two terms in the first column to calculate and validate θ_3: $c_2c_3 = 0.5974$, $-c_2s_3 = 0.8000$. $c_3 = 0.5974/0.99888 = 0.5981$, $\therefore \theta_3 = 53.27°$ or $-53.27°$, and $s_3 = 0.8014$ or -0.8014. The only valid solution is $\theta_3 = -53.27°$ because $-c_2s_3 \approx 0.8000$. To summarize the results of these three rotations (see Figure 7.1), to bring the global axes in line with the anatomical axes requires an initial rotation about the global X axis of $-8.92°$. This will create new Y' and Z' axes, and will be followed by a rotation of $-2.71°$ about the Y' axis. This rotation creates new X'' and Z'' axes. The final rotation is the largest (because we are analyzing the leg segment during swing), and it is $-53.27°$, which creates the final X''', Y''', and Z''' axes. These final axes are the anatomical x–y–z axes shown in Figure 7.2.

Finally, to get the COM of the segment, we must calculate **c** in GRS coordinates. We have **c** in the leg anatomical reference and it is = − [Anatomical m_{T2} vector] = [0.0000, 17.991, 3.833]. In the GRS,

$$\mathbf{c} = [\text{A to G}]\ [0.0000, 17.991, 3.833]$$
$$= \begin{bmatrix} 0.5974 & 0.8000 & -0.0472 \\ -0.7873 & 0.5969 & 0.1544 \\ 0.1515 & -0.0550 & 0.9868 \end{bmatrix} \begin{bmatrix} 0.000 \\ 17.991 \\ 3.833 \end{bmatrix} = \begin{bmatrix} 14.212 \\ 11.346 \\ 2.793 \end{bmatrix}$$

From Figure 7.2, the global vector $\mathbf{R}_c = \mathbf{R}_m + \mathbf{c} = [20.812, 36.646, 34.653]$.

As an exercise students can repeat these calculations for frames 5 and 7 with the answers:

Frame 5: $\theta_1 = -8.97°$, $\theta_2 = -1.31°$, $\theta_3 = -56.02°$,
$\mathbf{R}_c = [16.120, 36.325, 34.697]$

Frame 7: $\theta_1 = -8.56°$, $\theta_2 = -4.08°$, $\theta_3 = -49.89°$,
$\mathbf{R}_c = [25.429, 36.818, 34.623]$

7.3 DETERMINATION OF SEGMENT ANGULAR VELOCITIES AND ACCELERATIONS

Recall from Section 7.1.2 and Figure 7.2 that we had to determine three time-varying rotation angles, θ_1, θ_2, and θ_3, prior to transforming from the GRS to the anatomical axes. The first time-derivative of these transformation angles yields the components of the segment angular velocities:

$$\boldsymbol{\omega} = d\theta_1/dt \cdot \mathbf{e}_x + d\theta_2/dt \cdot \mathbf{e}_{y'} + d\theta_3/dt \cdot \mathbf{e}_{z''} \tag{7.6a}$$

where $\mathbf{e}_x$, $\mathbf{e}_{y'}$ and $\mathbf{e}_{z''}$ denote the unit vectors of the three rotation axes x, y', and z'' shown in Figure 7.1. Consider an angular velocity, $\boldsymbol{\omega}'$, about axis x; here $\boldsymbol{\omega}' = d\theta_1/dt \cdot \mathbf{e}_x$ and there is no rotation of θ_2 or θ_3. This angular velocity can be expressed as

$$\boldsymbol{\omega}' = \begin{bmatrix} \dot{\theta}_1 \\ 0 \\ 0 \end{bmatrix}$$

The second angular velocity, $\boldsymbol{\omega}'' = d\theta_2/dt \cdot \mathbf{e}_{y'}$, plus the component of $\boldsymbol{\omega}'$ that is transformed by $[\Phi_2]$ in Equation (7.2), can be expressed as

$$\boldsymbol{\omega}'' = \begin{bmatrix} 0 \\ \dot{\theta}_2 \\ 0 \end{bmatrix} + \begin{bmatrix} c_2 & 0 & -s_2 \\ 0 & 1 & 0 \\ s_2 & 0 & c_2 \end{bmatrix} \begin{bmatrix} \dot{\theta}_1 \\ 0 \\ 0 \end{bmatrix} = \begin{bmatrix} 0 \\ \dot{\theta}_2 \\ 0 \end{bmatrix} + \begin{bmatrix} c_2\dot{\theta}_1 \\ 0 \\ s_2\dot{\theta}_1 \end{bmatrix} = \begin{bmatrix} c_2\dot{\theta}_1 \\ \dot{\theta}_2 \\ s_2\dot{\theta}_1 \end{bmatrix}$$

Similarly, the third angular velocity, $\boldsymbol{\omega}''' = d\theta_3/dt \cdot \mathbf{e}_{z''}$, plus the contribution from $\boldsymbol{\omega}''$ that is transformed by $[\Phi_3]$ in Equation (7.3), gives us

$$\boldsymbol{\omega}''' = \begin{bmatrix} 0 \\ 0 \\ \dot{\theta}_3 \end{bmatrix} + \begin{bmatrix} c_3 & s_3 & 0 \\ -s_3 & c_3 & 0 \\ 0 & 0 & 1 \end{bmatrix} \begin{bmatrix} c_2\dot{\theta}_1 \\ \dot{\theta}_2 \\ s_2\dot{\theta}_1 \end{bmatrix} = \begin{bmatrix} 0 \\ 0 \\ \dot{\theta}_3 \end{bmatrix} + \begin{bmatrix} c_3c_2\dot{\theta}_1 + s_3\dot{\theta}_2 \\ -s_3c_2\dot{\theta}_1 + c_3\dot{\theta}_2 \\ s_2\dot{\theta}_1 \end{bmatrix}$$

$$= \begin{bmatrix} c_3c_2\dot{\theta}_1 + s_3\dot{\theta}_2 \\ -s_3c_2\dot{\theta}_1 + c_3\dot{\theta}_2 \\ s_2\dot{\theta}_1 + \dot{\theta}_3 \end{bmatrix}$$

Decomposing $\boldsymbol{\omega}'''$ into its three components along the three anatomical axes:

$$\boldsymbol{\omega} = \begin{bmatrix} \omega_x \\ \omega_y \\ \omega_z \end{bmatrix} = \begin{bmatrix} c_2c_3 & s_3 & 0 \\ -c_2s_3 & c_3 & 0 \\ s_2 & 0 & 1 \end{bmatrix} \begin{bmatrix} \dot{\theta}_1 \\ \dot{\theta}_2 \\ \dot{\theta}_3 \end{bmatrix} \tag{7.6b}$$

We can now calculate the three segment angular velocities, ω_x, ω_y, and ω_z, that are necessary to solve the 3D inverse dynamics equations developed in the next section. Recall that the time varying θ_1, θ_2, and θ_3 are calculated from Equation (7.5) and the time derivatives of these angles are individually calculated using the same finite difference technique used in two dimensions; see Equation (2.21). The three segment angular accelerations, α_x, α_y, and α_z, can now be calculated using either of the finite difference Equations (2.22) or (2.23). We now have all the kinematic variables necessary for our 3D kinetic analyses.

7.4 KINETIC ANALYSIS OF REACTION FORCES AND MOMENTS

Having developed the transformation matrices from global to anatomical and from anatomical to global, we are now in a position to begin calculating the reaction forces and moments at each of the joints. Because the ground reaction forces are measured in the GRS and the moments of inertia are known in the anatomical axes, these previously determined transformation matrices are used extensively in the kinetic calculations. All joint reaction forces are initially calculated in the GRS and all joint moments are calculated in the anatomical axes.

7.4.1 Newtonian Three-Dimensional Equations of Motion for a Segment

All reaction forces are calculated in the GRS and, because the gravitational forces and the segment COM accelerations are readily available in the GRS, it is convenient to calculate all segment joint reaction forces in the GRS. Students are referred to the 2D link-segment equations and free-body diagram equations in Section 4.1. Figure 7.3 is now presented to demonstrate the steps required to calculate kinetics for this segment. The only addition is the third dimension, z. We are given the three distal reaction forces either as measures from a force plate or from the analysis of the adjacent distal segment. It should be noted that the reaction forces and moments at the distal end are in the reverse direction from those at the proximal end, the same convention as was used in Section 4.1.

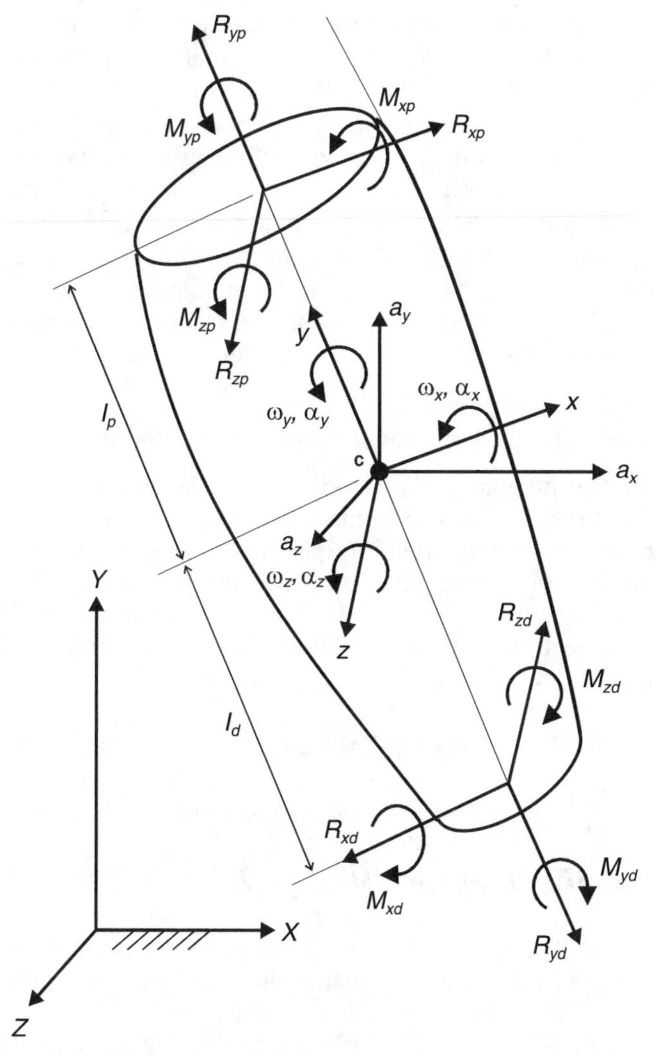

Figure 7.3 3D free-body diagram for solution of the inverse dynamics equations. Known are the distal reaction forces and moments, the COM linear accelerations, and the segment's angular velocities and accelerations. Using the kinetic Equations (7.7) and (7.8), we calculate the proximal reaction forces and moments.

Step 1: Calculate the reaction forces at the proximal end of the segment in the GRS

$$\Sigma F_X = ma_X \qquad \text{or} \qquad R_{XP} - R_{XD} = ma_X \tag{7.7a}$$

$$\Sigma F_Y = ma_Y \qquad \text{or} \qquad R_{YP} - R_{YD} - mg = ma_Y \tag{7.7b}$$

$$\Sigma F_Z = ma_Z \qquad \text{or} \qquad R_{ZP} - R_{ZD} = ma_Z \tag{7.7c}$$

where a_X, a_Y, a_Z are the segment COM accelerations in the X, Y, Z GRS directions and R_{XP}, R_{XD}, R_{YP}, R_{YD}, R_{ZP}, R_{ZD} are the proximal and distal reaction forces in the X, Y, and Z axes.

Step 2: Transform both proximal and distal reaction forces into the anatomical axes using the [G to A] matrix transformation based on θ_1, θ_2, and θ_3 [see Equation (7.5)]. We will now have the proximal and distal reaction forces in the anatomical axes x, y, z: R_{xp}, R_{xd}, R_{yp}, R_{yd}, R_{zp}, R_{zd}.

Step 3: Transform the distal moments from those previously calculated in the GRS using the [G to A] matrix to the anatomical axes, as before, based on θ_1, θ_2, and θ_3: M_{xd}, M_{yd}, M_{zd}. We now have all the variables necessary to calculate the proximal moments in the anatomical axes.

7.4.2 Euler's Three-Dimensional Equations of Motion for a Segment

The equations of motion for the 3D kinetic analyses are the Euler equations. Considerable simplification can be made in the rotational equations of motion if these equations are written with respect to the principal (anatomical) axes of the segment with their origin at the COM of the segment. Thus, the x–y–z axes of the segment in Figure 7.3 satisfies those conditions. The angular velocity of the segment in its coordinate system is ω. The rotational equations of motion are

$$I_x\alpha_x + (I_z - I_y)\,\omega_y\omega_z = \Sigma M_x = R_{zd}l_d + R_{zp}l_p + M_{xp} - M_{xd} \quad (7.8a)$$

$$I_y\alpha_y + (I_x - I_z)\,\omega_x\omega_z = \Sigma M_y = M_{yp} - M_{yd} \quad (7.8b)$$

$$I_z\alpha_z + (I_y - I_x)\,\omega_x\omega_y = \Sigma M_z = -R_{xd}l_d - R_{xp}l_p + M_{zp} - M_{zd} \quad (7.8c)$$

where I_x, I_y, I_z = moments of inertia about x–y–z axes
ω_x, ω_y, and ω_z = components of angular velocity $\boldsymbol{\omega}$ about x–y–z axes
α_x, α_y, α_z = components of angular acceleration about x–y–z axes
M_{xd}, M_{yd}, M_{zd} = previously transformed distal moments (not shown in Figure 7.3) about x–y–z axes
R_{xd}, R_{xp}, R_{yd}, R_{yp}, R_{zd}, R_{zp} = previously transformed joint reaction forces about x–y–z axes
l_p, l_d = distances from COM to proximal and distal joints

The unknowns in these three equations are the three moments (M_{xp}, M_{yp} and M_{zp}) about the proximal x, y, z axes. Note that Equations (7.8) are in the same form as the 2D Equation (4.3) with the additional term $(I_1 - I_2)\omega_1\omega_2$ to account for the interaction of the angular velocities in the other two axes. Also note that the moments about the y axis (long axis of the segment) does not involve the proximal and distal reaction forces because these forces have zero moment arms about that axis.

7.4.3 Example of a Kinetic Data Set

The set of kinematic and kinetic data shown in Tables 7.3 and 7.4 is taken from stance phase of walking, and we will confine our analysis to the leg segment (see Figure 7.3). The following anthropometric measures apply: $m = 3.22$ kg, $I_x = 0.0138$ kg · m², $I_y = 0.0024$ kg · m², $I_z = 0.0138$ kg · m², $l_d = 13.86$ cm, $l_p = 18.15$ cm, frame rate = 60 Hz.

From the kinematic and kinetic measures presented in Tables 7.3 and 7.4, we complete the analysis of frame 6: $a_X = 7.029$ m/s², $a_Y = 1.45$ m/s², $a_Z = -.348$ m/s². Refer to the three steps in Section 7.4.1.

Step 1

$$R_{XP} - R_{XD} = ma_X, \qquad R_{XP} = -119.04 + 3.22 \times 7.029 = -96.41 \text{ N}$$

$$R_{YP} - R_{YD} - mg = ma_Y, \qquad R_{YP} = -791.44 + 3.22 \times 9.814 + 3.22 \times 1.45$$

$$= -755.17 \text{ N}$$

$$R_{ZP} - R_{ZD} = ma_Z, \qquad R_{ZP} = -12.03 + 3.22 \times (-.348) = -13.15 \text{ N}$$

Step 2

$$\theta_1 = -0.1478 \text{ rad} = -8.468°, \qquad \theta_2 = -0.025 \text{ rad} = -1.432°,$$

$$\theta_3 = -0.4605 \text{ rad} = -26.385°$$

$$\cos\theta_1 = 0.9891, \qquad \sin\theta_1 = -.1473$$

$$\cos\theta_2 = 0.9997, \qquad \sin\theta_2 = -0.025$$

$$\cos\theta_3 = 0.8958, \qquad \sin\theta_3 = -0.4444$$

TABLE 7.3 Ankle Reaction Forces (N) and Moments (N · m) in GRS

Frame	R_{XD}	R_{YD}	R_{ZD}	M_{XD}	M_{YD}	M_{ZD}
4	−86.55	−766.65	−13.81	10.16	6.74	−97.55
5	−102.71	−790.27	−12.12	11.94	4.65	−102.50
6	−119.04	−791.44	−12.03	14.29	2.11	−103.85
7	−134.33	−763.38	−10.26	16.13	−0.49	−101.32
8	−146.37	−704.00	−5.26	16.64	−2.88	−94.59

TABLE 7.4 Leg Angular Displacements (rad) and Velocities (rad/s)

Frame	θ_1	θ_2	θ_3	$\dot{\theta}_1$	$\dot{\theta}_2$	$\dot{\theta}_3$
4	−0.14246	−0.02646	−0.37248	−0.1752	0.2459	−2.358
5	−0.14512	−0.02362	−0.41396	−0.1607	0.0434	−2.641
6	−0.14781	−0.02501	−0.46053	−0.0911	−0.1048	−2.909
7	−0.14815	−0.02705	−0.51094	−0.1139	−0.0923	−3.209
8	−0.15161	−0.02809	−0.56749	−0.1109	0.2175	−3.472

Substituting these values in the [G to A] matrix (Equation (7.5)], we get

$$\begin{bmatrix} c_2c_3 & s_3c_1 + s_1s_2c_3 & s_1s_3 - c_1s_2c_3 \\ -c_2s_3 & c_1c_3 - s_1s_2s_3 & s_1c_3 + c_1s_2s_3 \\ s_2 & -s_1c_2 & c_1c_2 \end{bmatrix} = \begin{bmatrix} 0.8955 & -0.4363 & 0.0875 \\ 0.4443 & 0.8877 & -0.121 \\ -0.025 & 0.1473 & 0.9888 \end{bmatrix}$$

We can now transform the reaction forces from the global to the anatomical axes system:

$$\begin{bmatrix} R_{xd} \\ R_{yd} \\ R_{zd} \end{bmatrix} = \begin{bmatrix} 0.8955 & -0.4363 & 0.0875 \\ 0.4443 & 0.8877 & -0.121 \\ -0.025 & 0.1473 & 0.9888 \end{bmatrix} \begin{bmatrix} R_{XD} \\ R_{YD} \\ R_{ZD} \end{bmatrix}$$

$$= \begin{bmatrix} 0.8955 & -0.4363 & 0.0875 \\ 0.4443 & 0.8877 & -0.121 \\ -0.025 & 0.1473 & 0.9888 \end{bmatrix} \begin{bmatrix} -119.04 \\ -791.44 \\ -12.03 \end{bmatrix} = \begin{bmatrix} 237.65 \\ -754.00 \\ -125.50 \end{bmatrix}$$

$$\begin{bmatrix} R_{xp} \\ R_{yp} \\ R_{zp} \end{bmatrix} = \begin{bmatrix} 0.8955 & -0.4363 & 0.0875 \\ 0.4443 & 0.8877 & -0.121 \\ -0.025 & 0.1473 & 0.9888 \end{bmatrix} \begin{bmatrix} R_{XP} \\ R_{YP} \\ R_{ZP} \end{bmatrix}$$

$$= \begin{bmatrix} 0.8955 & -0.4363 & 0.0875 \\ 0.4443 & 0.8877 & -0.121 \\ -0.025 & 0.1473 & 0.9888 \end{bmatrix} \begin{bmatrix} -96.41 \\ -755.17 \\ -13.15 \end{bmatrix} = \begin{bmatrix} 241.99 \\ -711.61 \\ -121.83 \end{bmatrix}$$

Step 3: In a similar manner, we transform the ankle moments from the global to the anatomical axes system:

$$\begin{bmatrix} M_{xd} \\ M_{yd} \\ M_{zd} \end{bmatrix} = \begin{bmatrix} 0.8955 & -0.4363 & 0.0875 \\ 0.4443 & 0.8877 & -0.121 \\ -0.025 & 0.1473 & 0.9888 \end{bmatrix} \begin{bmatrix} M_{XD} \\ M_{YD} \\ M_{ZD} \end{bmatrix}$$

$$= \begin{bmatrix} 0.8955 & -0.4363 & 0.0875 \\ 0.4443 & 0.8877 & -0.121 \\ -0.025 & 0.1473 & 0.9888 \end{bmatrix} \begin{bmatrix} 14.29 \\ 2.11 \\ -103.85 \end{bmatrix} = \begin{bmatrix} 2.79 \\ 20.79 \\ -102.73 \end{bmatrix}$$

Using Equation (7.6*b*), we calculate the angular velocities and accelerations required for the solution of Euler's kinetic Equations (7.8). As previously calculated for frame 6, $c_2 = 0.9997$, $c_3 = 0.8958$, $s_2 = -0.025$, $s_3 = -0.4444$.

$$\begin{bmatrix}\omega_x\\ \omega_y\\ \omega_z\end{bmatrix} = \begin{bmatrix} c_2c_3 & s_3 & 0\\ -c_2s_3 & c_3 & 0\\ s_2 & 0 & 1\end{bmatrix}\begin{bmatrix}\dot{\theta}_1\\ \dot{\theta}_2\\ \dot{\theta}_3\end{bmatrix} = \begin{bmatrix} 0.8955 & -0.4444 & 0\\ 0.4443 & 0.8958 & 0\\ -0.025 & 0 & 1\end{bmatrix}\begin{bmatrix}-0.0911\\ -0.1048\\ -2.909\end{bmatrix}$$

$$= \begin{bmatrix}-0.0350\\ -.1344\\ -2.907\end{bmatrix}$$

Similarly for frame 5,

$$\begin{bmatrix}\omega_x\\ \omega_y\\ \omega_z\end{bmatrix} = \begin{bmatrix}-0.1646\\ -0.0249\\ -2.637\end{bmatrix}$$

and for frame 7,

$$\begin{bmatrix}\omega_x\\ \omega_y\\ \omega_z\end{bmatrix} = \begin{bmatrix}-0.0542\\ -0.13621\\ -3.206\end{bmatrix}$$

We now calculate the angular accelerations of the segment where Δt is the sampling period.

$$\alpha_x \text{ (fr. 6)} = [\omega_x(\text{fr. 7}) - \omega_x(\text{fr. 5})]/2\Delta t = [-0.0542 - (-0.1646)]/0.03333$$
$$= 3.312 \text{ r/s}^2$$

$$\alpha_y \text{ (fr. 6)} = [\omega_y(\text{fr. 7}) - \omega_y(\text{fr. 5})]/2\Delta t = [-0.1362 - (-0.0249)]/0.03333$$
$$= -3.337 \text{ r/s}^2$$

$$\alpha_z \text{ (fr. 6)} = [\omega_z(\text{fr. 7}) - \omega_z(\text{fr. 5})]/2\Delta t = [-3.206 - (-2.637)]/0.03333$$
$$= -17.07 \text{ r/s}^2$$

We are now ready to solve Euler's Equations (7.8) for the proximal moments:

$$I_x\alpha_x + (I_z - I_y)\omega_y\omega_z = \Sigma M_x = R_{zd}l_d + R_{zp}l_p + M_{xp} - M_{xd}$$

$$M_{xp} = 2.79 + 121.83 \times 0.1815 + 125.5 \times 0.1386 + 0.0138 \times 3.312$$
$$+ (0.0138 - 0.0024) \times 0.1344 \times 2.907 = 42.35 \text{ N} \cdot \text{m}$$

$$I_y\alpha_y + (I_x - I_z)\,\omega_x\omega_z = \Sigma M_y = M_{yp} - M_{yd}$$

$$M_{yp} = 20.79 - 0.0024 \times 3.337 + (0.0138 - 0.0138) \times 0.035 \times 2.907$$
$$= 20.78 \text{ N} \cdot \text{m}$$

$$I_z\alpha_z + (I_y - I_x)\,\omega_x\omega_y = \Sigma M_z = -R_{xd}l_d - R_{xp}l_p + M_{zp} - M_{zd}$$

$$M_{zp} = -102.73 + 241.99 \times 0.1815 + 237.65 \times 0.1386 - 0.0138$$
$$\times 17.07 + (0.0024 - 0.0138) \times 0.035 \times 0.1343 = -26.11 \text{ N} \cdot \text{m}$$

The interpretation of these moments for the left knee is as follows as the subject bears weight during single support. M_{xp} is $+ve$, thus is a counterclockwise moment, hence an abductor moment acting at the knee to counter the large gravitational load of the upper body acting downwards and medial of the support limb. M_{yp} is the axial moment acting along the long axis of the leg and reflects the action of the left hip internal rotators actively rotating the pelvis, upper body, and right limb in a forward direction to gain extra step length. M_{zp} is $-ve$, indicating a clockwise (flexor) knee moment in the sagittal plane that would assist in starting the knee to flex late in stance shortly before toe-off.

The next stage of the kinetic analysis is to transform these knee moments, M_{xp}, M_{yp}, and M_{zp} (which are in the leg anatomical axis system), to the global system so that the inverse dynamics analysis can continue for the thigh segment. This transformation is accomplished by the [A to G] matrix for the leg and yields a new set of distal moments, M_{XD}, M_{YD}, and M_{ZD}, for the thigh analysis.

As an exercise, students may wish to repeat the above calculations for frame 5 or 7. For frame 5: $a_X = 5.89 \text{ m/s}^2$, $a_Y = 1.30 \text{ m/s}^2$, $a_Z = -1.66 \text{ m/s}^2$; for frame 7: $a_X = 9.60 \text{ m/s}^2$, $a_Y = 1.94 \text{ m/s}^2$, $a_Z = -.020 \text{ m/s}^2$. For frame 5: $M_{xp} = 40.76 \text{ N} \cdot \text{m}$, $M_{yp} = 21.61 \text{ N} \cdot \text{m}$, $M_{zp} = -31.23 \text{ N} \cdot \text{m}$; for frame 7: $M_{xp} = 41.85 \text{ N} \cdot \text{m}$, $M_{yp} = 19.20 \text{ N} \cdot \text{m}$, $M_{zp} = -19.80 \text{ N} \cdot \text{m}$.

7.4.4 Joint Mechanical Powers

The joint mechanical power generated or absorbed at the distal and proximal joints, P_d and P_p, can now be calculated using Equation (5.5) for each of three moment components and their respective angular velocities:

$$P_d = M_{xd}\omega_{xd} + M_{yd}\omega_{yd} + M_{zd}\omega_{zd} \tag{7.9a}$$

$$P_p = M_{xp}\omega_{xp} + M_{yp}\omega_{yp} + M_{zp}\omega_{zp} \tag{7.9b}$$

where the moments are previously defined and ω_{xd}, ω_{yd}, ω_{zd}, ω_{xp}, ω_{yp}, ω_{zp} are the joint angular velocities at the distal and proximal ends (not shown on Figure 7.3). The angular velocities are in rad/s, the moments are in N · m, and the powers are in W. If these products are $+ve$ the muscle group concerned is generating energy, and if they are $-ve$, the muscle group is absorbing energy.

7.4.5 Sample Moment and Power Curves

Figure 7.4 presents a set of intersubject averaged 3D joint moments at the ankle, knee, and hip during one walking stride (Eng and Winter, 1995). Heel contact (HC) is at 0% stride and toe-off (TO) is at 60%. Figure 7.5 shows the 3D power curves for these intersubject averages (Eng and Winter, 1995). The curves are normalized relative to body mass; the moments are reported in $\text{N} \cdot \text{m/kg}$ and the powers in W/kg. Convention plots extensor moments in the sagittal plane as positive; in the transverse plane, external rotation moments are positive, and in the frontal plane, evertor moments are positive. A detailed explanation of the specific function of each of these moments is beyond the scope of this text; however, a few comments will be made on the larger or functionally more important moments.

1. The largest moment during walking is seen at the ankle in the sagittal plane. Immediately at heel contact (HC), there is a small dorsiflexor moment to lower the foot to the ground; this is followed by a large increase in plantarflexor moment reaching a peak at about 50% of stride to cause the ankle to rapidly plantarflex and achieve an upward and forward "push off" of the lower limb as the subject starts swing at toe-off (TO) (see A2-S power generation burst in Figure 7.5).

2. The knee extensors are active at 8–25% of stride to control knee flexion as the limb accepts weight (K1-S absorption phase), then the moment reverses to a flexor pattern as a by-product of the gastrocnemic's contribution to the increasing ankle plantarflexor moment. Then, just before and after TO, a small knee extensor moment acts to limit the amount of knee flexion in late stance and early swing (K3-S absorption phase). The final burst of flexor activity just before HC is to decelerate the swinging leg prior to HC (K4-S absorption phase).

3. The hip pattern is characterized by an extensor moment for the first half of stance followed by a flexor for the latter half. During the first half, the extensors stabilize the posture of the trunk by preventing it from flexing forward under the influence of a large posterior reaction force at the hip; this also assists the knee extensors in preventing collapse of the knee joint and in addition contributes to forward propulsion by what has been described as a "push-from-behind" (H1-S generation burst). The flexor moment during the second half of stance serves two functions: first, to stabilize the trunk posture by preventing it from flexing backwards under the influence of the forward reaction force at the hip; second, in the last phase of stance and early swing (50–75% of stride), to achieve a "pull-off" of the thigh into swing (H3-S generation burst).

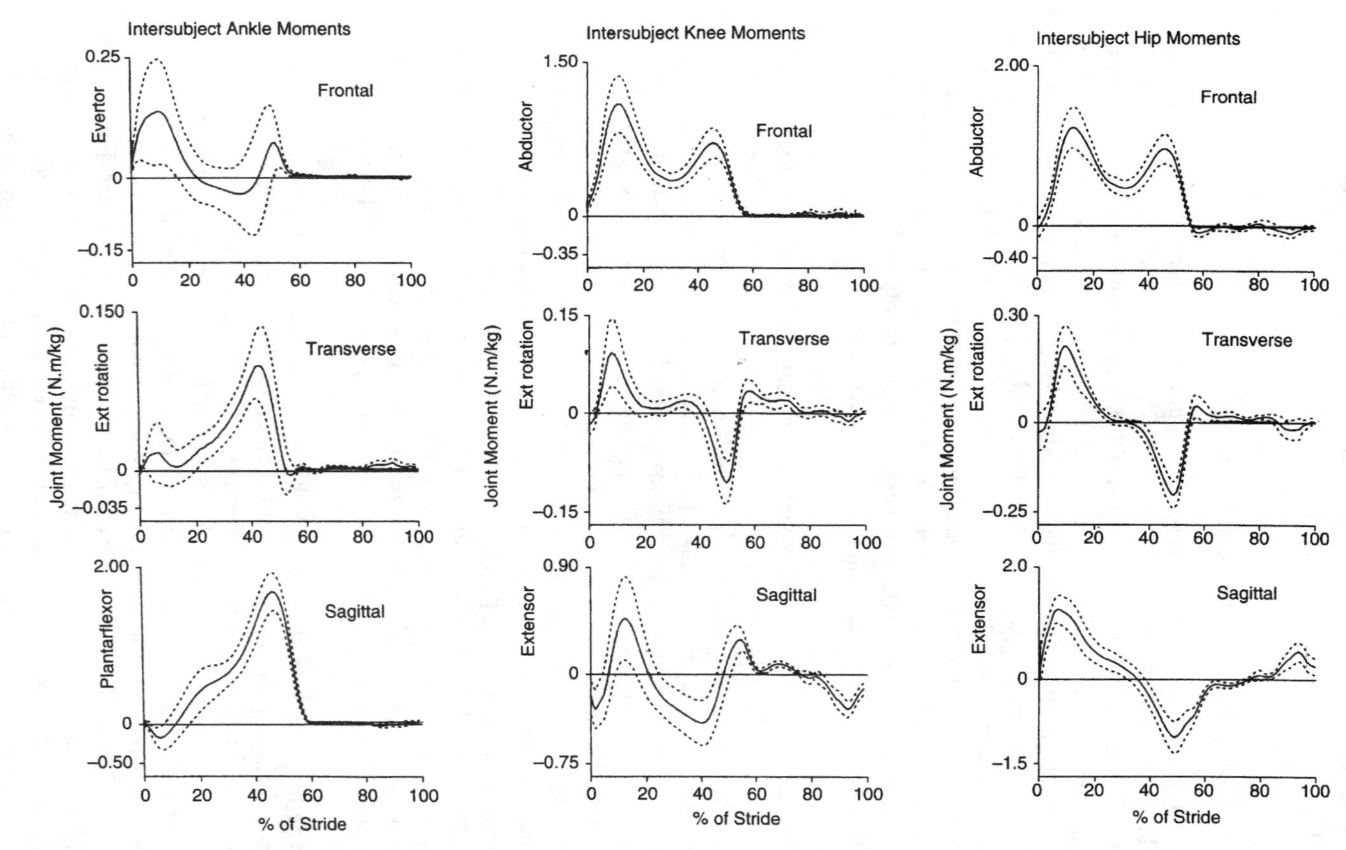

Figure 7.4 Typical 3D moments at the ankle, knee, and hip during one stride of walking (heel contact at 0 and 100%). Profiles are intersubject averages; solid line is the average curve with one standard deviation plotted as a dotted line. For interpretation of the more important profiles, see text.

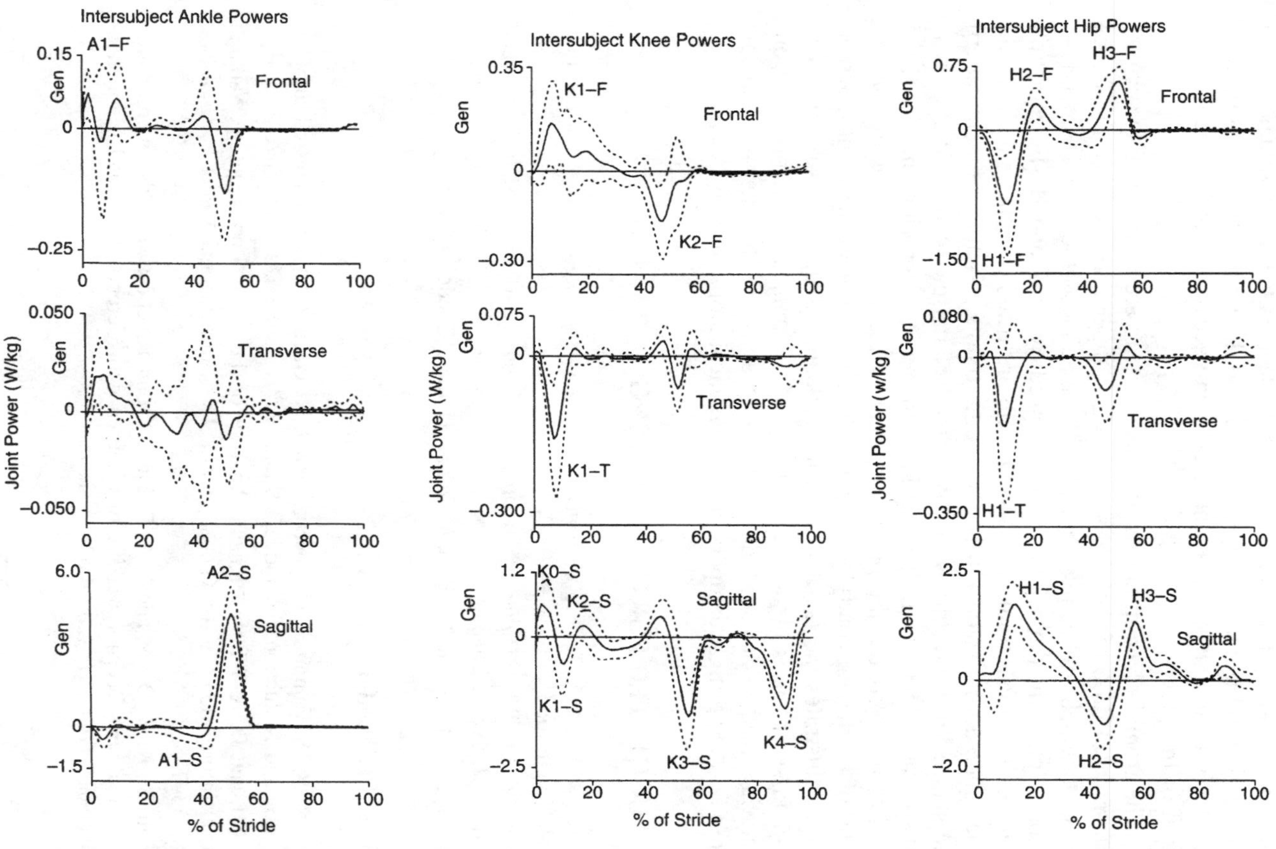

Figure 7.5 Mechanical power generation and absorption at the ankle, knee, and hip for the same averaged trials as in Figure 7.4. Generation power is $+ve$ and absorption power is $-ve$. For interpretation of selected profiles, see text.

4. The major transverse activity is at the hip. During the first half of stance, the external rotators of the stance limb act to decelerate the horizontal rotation of the pelvis (and trunk) over the stance limb (H1-T absorption phase); then, during the second half, the internal rotators are active to stabilize the forward rotation of the pelvis and swing limb.

5. The frontal plane moments at the stance hip are a strong abductor pattern to prevent drop of the pelvis (and entire upper body) against the forces of gravity, which are acting about 10 cm medial of the stance hip (H1-F absorption phase as the pelvis drops, followed by H2-F and H3-F generation phases as the pelvis, trunk, and swing limb are lifted to help achieve a safe toe clearance during swing). It is interesting to note a similar moment pattern at the stance knee, but this is not as a result of muscle activity. Rather, it is the response of the weight-bearing knee to the large gravitational load; the knee tries to invert but the passive loading of the medial condyles and unloading of the lateral condyles creates an internal abductor moment. Such an example illustrates the fact that internal skeletal and ligament structures can aid (or in some cases hinder) the activity of the muscles. Rehabilitation engineers involved in prothesis design must be aware of the internal moments created by the springs, dampers, and mechanical stops in their design.

7.5 SUGGESTED FURTHER READING

D'Sousa, A. F. and V. J. Garg. *Advanced Dynamics Modeling and Analysis* (Prentice-Hall, Englewood Cliffs, NJ, 1984).

Greenwood, D. T. *Principles of Dynamics,* 2nd *edition*. (Prentice-Hall, Englewood Cliffs, NJ, 1988), Chapter 7.

Zatsiorsky, V. M. *Kinetics of Human Motion* (Human Kinetics, Champaign, IL, 2002).

7.6 REFERENCES

Davis, R. B., S. Õunpuu, D. Tyburski, and J. R. Gage. "A Gait Analysis Data Collection and Reduction Technique," *Human Movement Sci.* **10**:575–587, 1991.

Eng, J. J. and D. A. Winter. "Kinetic Analysis of the Lower Limbs During Walking: What Information Can Be Gained from a Three-dimensional Model," *J. Biomech.* **28**:753–758, 1995.

Õunpuu, S., R. B. Davis, and P. A. DeLuca. "Joint Kinetics: Methods, Interpretation and Treatment Decision-Making in Children with Cerebral Palsy and Myelomeningocele," *Gait & Posture* **4**:62–78, 1996.

8

MUSCLE MECHANICS

8.0 INTRODUCTION

The most intriguing and challenging area of study in biomechanics is probably the muscle itself. It is the "living" part of the system. The neural control, metabolism, and biomechanical characteristics of muscle are subjects of continuing research. The purpose of this chapter is to report the state of knowledge with regard to the biophysical characteristics of individual motor units, connective tissue, and the total muscle itself. The characteristics of the individual units are described in detail, and it is shown how these characteristics influence the biomechanical function of the overall muscle.

8.0.1 The Motor Unit

The smallest subunit that can be controlled is called a *motor unit* because it is innervated separately by a motor axon. Neurologically the motor unit consists of a synaptic junction in the ventral root of the spinal cord, a motor axon, and a motor end plate in the muscle fibers. Under the control of the motor unit are as few as three muscle fibers or as many as 2000, depending on the fineness of the control required (Feinstein et al., 1955). Muscles of the fingers, face, and eyes have a small number of shorter fibers in a motor unit, while the large muscles of the leg have a large number of long fibers in their motor units. A muscle fiber is about 100 μm in diameter and consists of fibrils about 1 μm in diameter. Fibrils in turn consist of filaments about 100 Å in diameter. Electron micrographs of fibrils show the basic mechanical structure of the interacting actin and myosin filaments. In the schematic diagram of Figure 8.1, the darker and wider myosin protein bands are interlaced

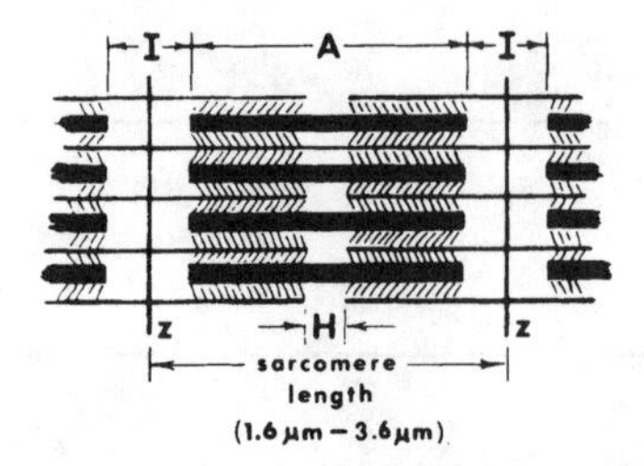

Figure 8.1 Basic structure of the muscle contractile element showing the *Z* lines and sarcomere length. Wider dark myosin filament interacts across cross bridges (cross-hatched lines) with the narrower actin filament. Darker and lighter bands (*A, H,* and *I*) are shown.

with the lighter and smaller actin protein bands. The space between them consists of a cross-bridge structure, and it is here that the tension is created and the shortening or lengthening takes place. The term *contractile element* is used to describe the part of the muscle that generates the tension, and it is this part that shortens and lengthens as positive or negative work is done. The basic length of the myofibril is the distance between the *Z* lines and is called the *sarcomere length.* It can vary from 1.5 μm at full shortening to 2.5 μm at resting length to about 4.0 μm at full lengthening.

The structure of the muscle is such that many filaments are in parallel and many sarcomere elements are in series to make up a single contractile element. Consider a motor unit of a cross-sectional area of 0.1 cm^2 and a resting length of 10 cm. The number of sarcomere contractile elements in series would be 10 cm/2.5 μm = 40,000 and the number of filaments (each with an area of 10^{-8} cm^2) in parallel would be $0.1/10^{-8} = 10^7$. Thus, the number of contractile elements of sarcomere length packed into this motor unit would be 4×10^{11}.

The active contractile elements are contained within another fibrous structure of connective tissue called *fascia.* These tissue sheaths enclose the muscles, separating them into layers and groups, and ultimately connecting them to the tendons at either end. The mechanical characteristics of connective tissue are important in the overall biomechanics of the muscle. Some of the connective tissue is in series with the contractile element, while some is in parallel. The effect of this connective tissue has been modeled as springs and viscous dampers, and is discussed in detail in Section 8.3.

8.0.2 Recruitment of Motor Units

Each muscle has a finite number of motor units, each of which is controlled by a separate nerve ending. Excitation of each unit is an all-or-nothing event. The electrical indication is a motor unit action potential; the mechanical result is a twitch of tension. An increase in tension can therefore be accomplished in two ways: by an increase in the stimulation rate for that motor unit or by the excitation (recruitment) of an additional motor unit. Figure 8.2 shows the electromyogram (EMG) of a needle electrode in a muscle as the tension was gradually increased. The upper tracing shows one motor unit firing, the middle trace two motor units, and the lower trace three units. Initially muscle tension increases as the firing rate of the first motor unit increases. Each unit has a maximum firing rate, and this rate is reached well after the next unit is recruited (Erim et al., 1996). When the tension is reduced, the reverse process occurs. The firing rates of all recruited units decrease until the last-recruited unit drops out at a rate usually less than when it was originally recruited. As the tension further decreases, the remainder of the units drop out in the reverse order from that in which they were recruited. The firing rate of each unit increases monotonically with force; however the firing rates increase nonlinearly as force increases from 0 to 100% maximum voluntary contraction (MVC) (Erim et al., 1996). Mathematical models describing recruitment have been reported by Wani and Guha (1975), Milner-Brown and Stein (1975), Fuglevand et al. (1993), and Erim et al. (1996).

8.0.3 Size Principle

Considerable research and controversy has taken place over the past two decades over how the motor units are recruited. Which ones are recruited first? Are they always recruited in the same order? It is now generally accepted

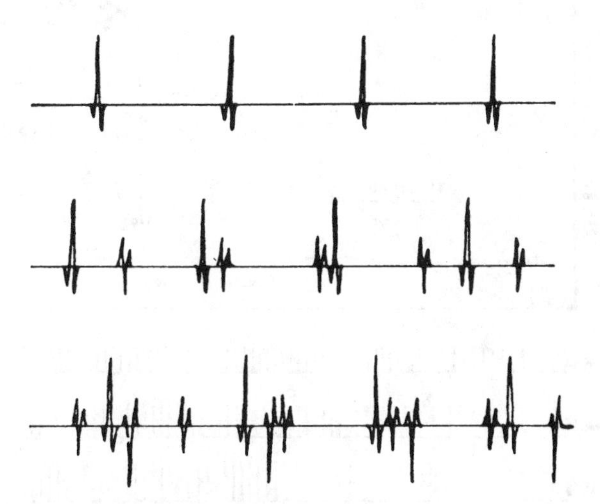

Figure 8.2 EMG from an indwelling electrode in a muscle as it begins to develop tension. The smallest motor unit is first recruited, and as its rate increases, a second, then a third motor unit are recruited. Each motor action potential has a characteristic shape at a given electrode, which depends on the size of the motor unit and the distance from the electrode to the fibers of the motor unit (see Section 9.1.3).

that they are recruited according to the *size principle* (Henneman, 1974*a*), which states that the size of the newly recruited motor unit increases with the tension level at which it is recruited. This means that the smallest unit is recruited first and the largest unit last. In this manner, low tension movements can be achieved in finely graded steps. Conversely, those movements requiring high forces but not needing fine control are accomplished by recruiting the larger motor units. Figure 8.3 depicts a hypothetical tension curve resulting from successive recruitment of several motor units. The smallest motor unit (MU 1) is recruited first, usually at an initial frequency ranging from about 5 to 13 Hz. Tension increases as MU 1 fires more rapidly, until a certain tension is reached at which MU 2 is recruited. Here, MU 2 starts firing at its initial low rate and further tension is achieved by the increased firing of both MU 1 and 2. At a certain tension, MU 1 will reach its maximum firing rate (15–60 Hz) and will therefore be generating its maximum tension. This process of increasing tension, reaching new thresholds, and recruiting another larger motor unit continues until maximum voluntary contraction is reached (not shown in Figure 8.3). At that point, all motor units will be firing at their maximum frequencies. Tension is reduced by the reverse process: successive reduction of firing rates and dropping out of the larger units first (DeLuca et

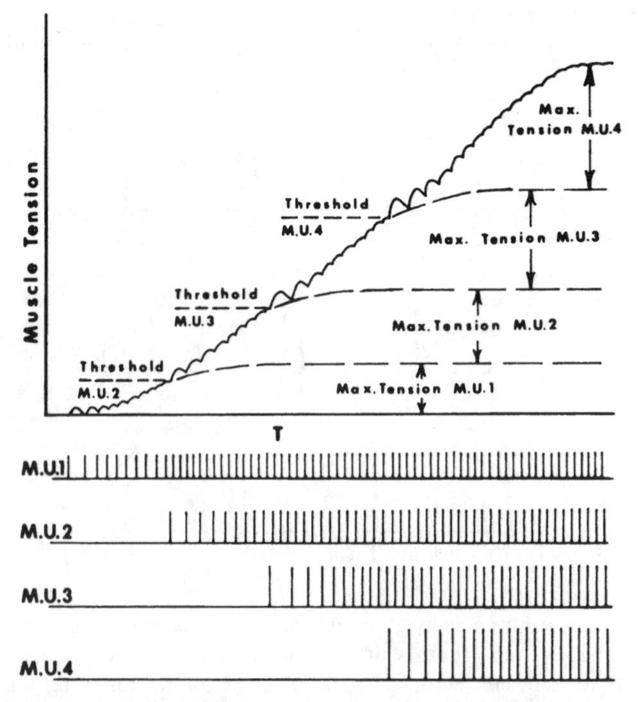

Figure 8.3 Size principle of recruitment of motor units. Smaller motor units are recruited first; successively larger units begin firing at increasing tension levels. In all cases, the newly recruited unit fires at a base frequency, then increases to a maximum.

al., 1982). During rapid ballistic movements, the firing rates of individual motor units have been estimated to reach 120 Hz (Desmedt and Godaux, 1978).

The motor unit action potential increases with the size of the motor unit with which it is associated (Milner-Brown and Stein, 1975). The reason for this appears to be twofold. The larger the motor unit, the larger the motoneuron that innervates it, and the greater the depolarization potentials seen at the motor end plate. Second, the greater the mass of the motor unit, the greater the voltage changes seen at a given electrode. However, it is never possible from a given recording site to predict the size of a motor unit because its action potential also decreases with the distance between the recording site and the electrode. Thus, a large motor unit located a distance from the electrode may produce a smaller action potential than that produced by a small motor unit directly beneath the electrode.

8.0.4 Types of Motor Units—Fast- and Slow-Twitch Classification

There have been many criteria and varying terminologies associated with the types of motor units present in any muscle (Henneman, 1974*b*). Biochemists have used metabolic or staining measures to categorize the fiber types. Biomechanics researchers have used force (twitch) measures (Milner-Brown et al., 1973*b*), and electrophysiologists have used EMG indicators (Wormolts and Engel, 1973; Milner-Brown and Stein, 1975). The smaller slow-twitch motor units have been called *tonic units*. Histochemically they are the smaller units (type I), and metabolically they have fibers rich in mitochondria, are highly capillarized, and therefore have a high capacity for aerobic metabolism. Mechanically they produce twitches with a low peak tension and a long time to peak (60–120 ms). The larger fast-twitch motor units are called *phasic units* (type II). They have less mitochondria, are poorly capillarized, and therefore rely on anaerobic metabolism. The twitches have larger peak tensions in a shorter time (10–50 ms). Figure 8.4 is a typical histochemical stain of fibers of a muscle that contains both slow- and fast-twitch fibers. An ATPase-type stain was used here, so slow-twitch fibers appear dark and fast-twitch fibers appear light. If an indwelling microelectrode were present in the area of these fibers, the muscle action potential from these darker slow-twitch fibers would be smaller than that from the lighter stained fast-twitch fibers.

In spite of the histochemical classification described, there is growing biophysical evidence (Milner-Brown et al., 1973*b*) that the motor units controlled by any motoneuron pool form a continuous spectrum of sizes and excitabilities.

8.0.5 The Muscle Twitch

Thus far, we have said very little about the smallest increment of tension, that of the individual twitch itself. As was described in Section 8.0.4, each motor unit has its unique time course of tension. Although there are individual dif-

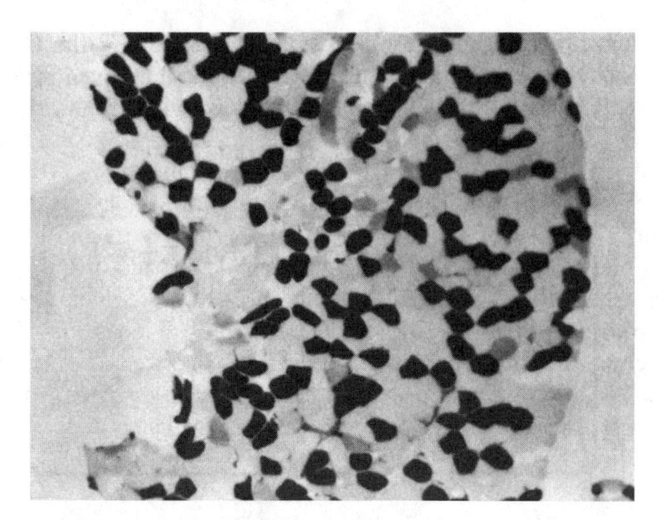

Figure 8.4 Histochemical stain showing dark slow-twitch fibers and light fast-twitch fibers. A myofibrillar ATPase stain, pH 4.3, was used to stain the vastus lateralis of a female volleyball player. (Reproduced by permission of Professor J. A. Thomson, University of Waterloo, Waterloo, Ont., Canada.)

ferences in each newly recruited motor unit, they all have the same characteristic shape. The time-course curve follows quite closely that of the impulse response of a critically damped second-order system (Milner-Brown et al., 1973*a*). The electrical stimulus of a motor unit, as indicated by this action potential, is of short duration and can be considered an impulse. The mechanical response to this impulse is the much longer duration twitch. The general expression for a second-order critically damped impulse response is

$$F(t) = F_0 \frac{t}{T} e^{-t/T} \tag{8.1}$$

For the curve plotted in Figure 8.5, the twitch time T is the time for the tension to reach a maximum, and F_0 is a constant for that given motor unit. T is the contraction time and is larger for the slow-twitch fibers than for the fast-twitch motor units, while F_0 increases for the larger fast-twitch units. Muscles tested by Buchthal and Schmalbruch (1970) using submaximal stimulations showed a wide range of contraction times. Muscles of the upper limbs generally had short T values compared with the leg muscles. Typical mean values of T and their range were:

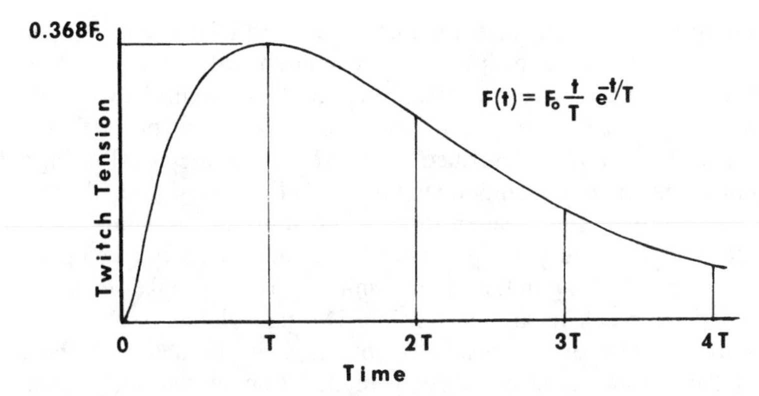

Figure 8.5 Time course of a muscle twitch, modeled as the impulse response of a second-order critically damped system. Contraction time T varies with each motor unit type, from about 20 to 100 ms. Effective tension lasts until about $4T$.

Triceps brachii	44.5 ms (16–68 ms)
Biceps brachii	52.0 ms (16–85 ms)
Tibialis anterior	58.0 ms (38–80 ms)
Soleus	74.0 ms (52–100 ms)
Medial gastrocnemius	79.0 ms (40–110 ms)

These same researchers found that T increased in all muscles as they were cooled; for example, the biceps brachialis had a contraction time that increased from 54 ms at 37°C to 124 ms at 23°C. It is evident that the cause of the delayed buildup of tension is the slower metabolic rates and increased muscle viscosity seen at lower temperatures.

Many other researchers have tested other muscles using different techniques. Milner-Brown et al. (1973*b*) looked at the first dorsal interossei during voluntary contractions and measured T to average 55 ms (± 12 ms), while Desmedt and Godaux (1978) calculated a mean of 60 ms on the same muscle but used supramaximal stimulations. These latter researchers, using the same technique, reported T for the tibialis anterior as 84 ms and for the soleus as 100 ms. Bellemare et al. (1983), also using supramaximal stimulations, reported soleus $T = 116$ ms (± 9 ms) and biceps $T = 66$ ms (± 9 ms). Thus, it appears that the contraction time varies considerably depending on the muscle, the person, and the experimental technique employed.

8.0.6 Shape of Graded Contractions

The shape of a voluntary tension curve depends to a certain extent on the shape of the individual muscle twitches. For example, if we elicit a maximum contraction in a certain muscle, the rate of increase of tension depends on the individual motor units and how they are recruited. Even if all motor units were turned on at the same instant and fired at their maximum rate, maximum tension could never be achieved in less than the average contraction time for that muscle. However, as described in Figure 8.2 and depicted in Figure 8.3, the recruitment of motor units does not take place all at once during a voluntary contraction. The smaller slow-twitch units are recruited first in accordance with the size principle, with the largest units not being recruited until tension has built up in the smaller units. Thus, it can take several hundred milliseconds to reach maximum tension. During voluntary relaxation of the same muscle, the drop in tension is governed by the shape of the trailing edge of the twitch curve. This delay in the drop of tension is even more pronounced than the rise. This delay, combined with the delay in dropping out of the motor units themselves according to the size principle, means that a muscle takes longer to turn off than to turn on. Typical turn-on times could be 200 ms and turn-off times 300 ms, as shown in Figure 8.6, where maximum rapid turn-on and turn-off are plotted showing the tension curve and the associated EMG.

8.1 FORCE–LENGTH CHARACTERISTICS OF MUSCLES

As indicated in Section 8.0.1, the muscle consists of an active element, called the *contractile element,* and passive connective tissue. The net force–length

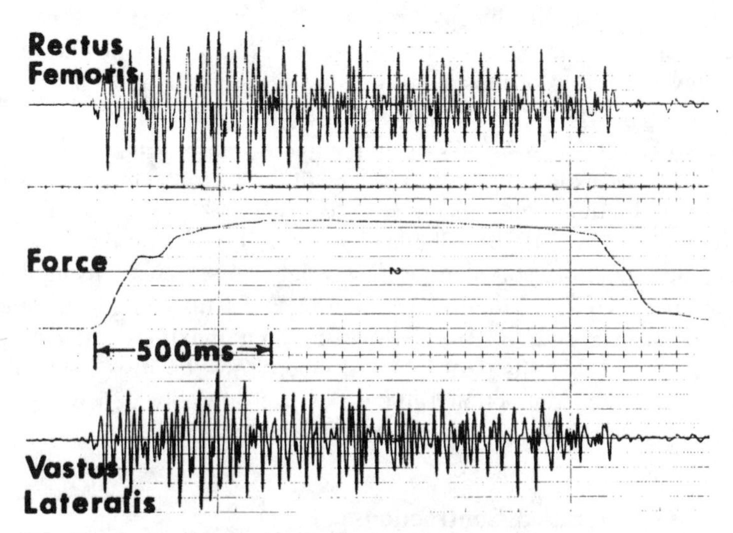

Figure 8.6 Tension buildup and decrease during a rapid maximum voluntary contraction and relaxation. The time to peak tension can be 200 ms or longer, mainly because of the recruitment according to the size principle and because of delay between each motor unit action potential and twitch tension. Note the presence of tension for about 150 ms after the cessation of EMG activity.

characteristics of a muscle are a combination of the force–length characteristics of both active and passive elements.

8.1.1 Force–Length Curve of the Contractile Element

The key to the shape of the force–length curve is the changes of the structure of the myofibril at the sarcomere level (Gordon et al., 1966). Figure 8.7 is a schematic representation of the myosin and actin cross-bridge arrangement. At resting length, about 2.5 μm, there are a maximum number of cross bridges between the filaments, and therefore a maximum tension is possible. As the muscle lengthens, the filaments are pulled apart, the number of cross bridges reduce, and tension decreases. At full length, about 4.0 μm, there are no cross bridges and the tension reduces to zero. As the muscle shortens to less than resting length, there is an overlapping of the cross bridges and an interference takes place. This results in a reduction of tension that continues until a full overlap occurs, at about 1.5 μm. The tension doesn't drop to zero, but is drastically reduced by these interfering elements.

8.1.2 Influence of Parallel Connective Tissue

The connective tissue that surrounds the contractile element influences the force–length curve. It is called the *parallel elastic component,* and it acts much like an elastic band. When the muscle is at resting length or less, the parallel elastic component is in a slack state with no tension. As the muscle lengthens, the parallel element is no longer loose, so tension begins to build up, slowly at first and then more rapidly. Unlike most springs, which have a

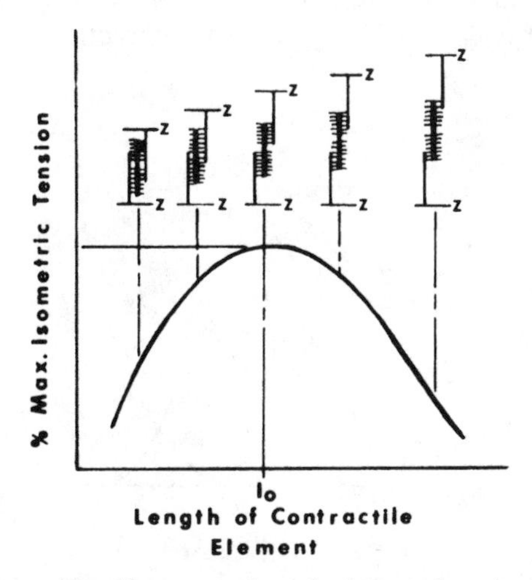

Figure 8.7 Tension produced by a muscle as it changes length about its resting length l_0. Drop of tension on either side of maximum can be explained by interactions of cross-bridge attachments in the contractile elements.

linear force–length relationship, the parallel element is quite nonlinear. In Figure 8.8 we see the force–length curve of this element F_p combined with that of the overall contractile component F_c. If we sum the forces from both elements, we see the overall muscle force–length characteristic F_t. The force–length curve typically presented is usually for a maximum contraction. The passive force F_p of the parallel element is always present, but the amount of active tension in the contractile element at any given length is under voluntary control. Thus, the overall force–length characteristics are a function of the percentage of excitation, as seen in Figure 8.9.

The student can demonstrate the drop of tension at either end of the force–length curve by two simple experiments. The hamstrings, as a two-joint muscle, can be made to shorten as follows: the person stands on one leg, leaning backward with the swing hip fully extended, then contracts the hamstrings to flex the leg. He will feel the tension decrease drastically when the hamstring muscles shorten before the knee is completely flexed. The converse situation can be realized if the person attempts to extend the hip joint while the knee is fully extended.

8.1.3 Series Elastic Tissue

All connective tissue in series with the contractile component, including the tendon, is called the *series elastic element*. Under isometric contractions, it will lengthen slightly as tension increases. However, during dynamic situations the series elastic element, in conjunction with viscous components, does influence the time course of the muscle tension.

During isometric contractions, the series elastic component is under tension and therefore is stretched a finite amount. Because the overall length of the muscle is kept constant, the stretching of the series elastic element can only occur if there is an equal shortening of the contractile element itself. This is

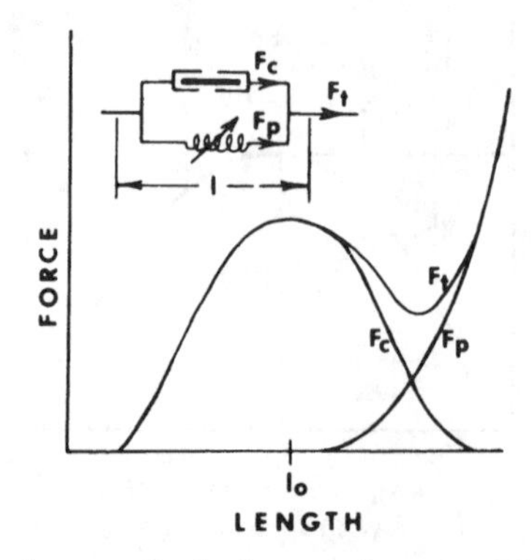

Figure 8.8 Contractile element producing maximum tension F_c along with the tension from F_p from the parallel elastic element. Tendon tension is $F_t = F_c + F_p$.

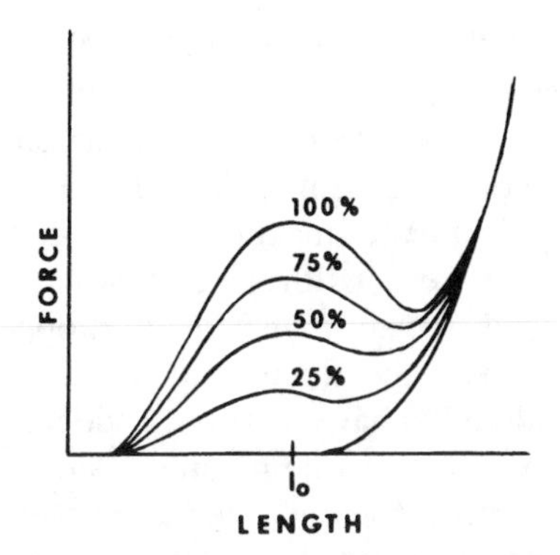

Figure 8.9 Tendon tension resulting from various levels of muscle activation. Parallel elastic element generates tension independent of the activation of the contractile element.

described as internal shortening. Figure 8.10 illustrates this point at several contraction levels. Although the external muscle length L is kept constant, the increased tension from the contractile element causes the series elastic element to lengthen by the same amount as the contractile element shortens internally. The amount of internal shortening from rest to maximum tension is only a few percent of the resting length in most muscles (van Ingen Schenau, 1984), but has been shown to be as high as 7% in others (Bahler, 1967). It is widely inferred that these series elastic elements store large amounts of energy as muscles are stretched prior to an explosive shortening in athletic movements. However, van Ingen Schenau (1984) has shown that the elastic capacity of these series elements is far too small to explain im-

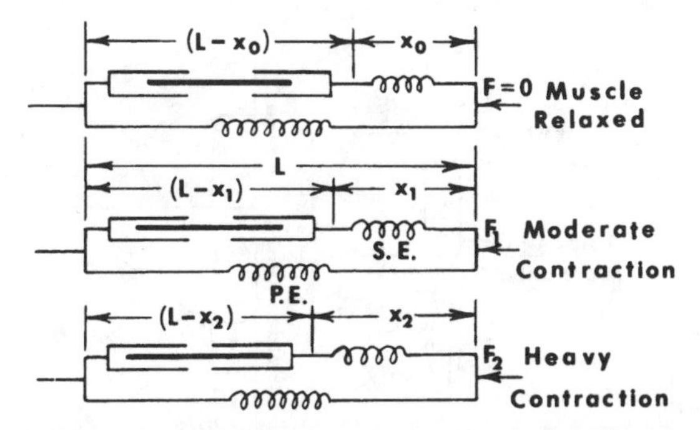

Figure 8.10 Introduction of the series elastic (S.E.) element. During isometric contractions, the tendon tension reflects a lengthening of the series element and an internal shortening of the contractile element. During most human movement, the presence of the series elastic element is not too significant, but during high-performance movements, such as jumping, it is responsible for storage of energy as a muscle lengthens immediately prior to rapid shortening.

proved performances resulting from prestretch. He argues that some other mechanism at the cross bridges must be responsible.

Experiments to determine the force–length characteristics of the series element can only be done on isolated muscle and will require dynamic changes of force or length. A typical experimental setup is shown in Figure 8.11. The muscle is stimulated to a certain level of tension while held at a certain isometric length. The load (tension) is suddenly dropped to zero by releasing one end of the muscle. With the force suddenly removed, the series elastic element, which is considered to have no mass, suddenly shortens to its relaxed length. This sudden shortening can be recorded, and the experiment repeated for another force level until a force-shortening curve can be plotted. During the rapid shortening period (about 2 ms), it is assumed that the contractile element has not had time to change length, even though it is still generating tension. Two other precautions must be observed when conducting this experiment. First, the isometric length prior to the release must be such that there is no tension in the parallel elastic element. This will be so if the muscle is at resting length or shorter. Second, the system that records the sudden shortening of the series elastic element must have negligible mass and viscosity. Because of the high acceleration and velocity associated with the rapid shortening, the resisting force of the attached recording transducer should be negligible. If such conditions are not possible, a correction should be made to account for the mass or viscosity of the transducer.

8.1.4 In Vivo Force–Length Measures

Length–tension relationships of striated muscle have been well researched in mammalian and amphibian species, but in humans the studies have been lim-

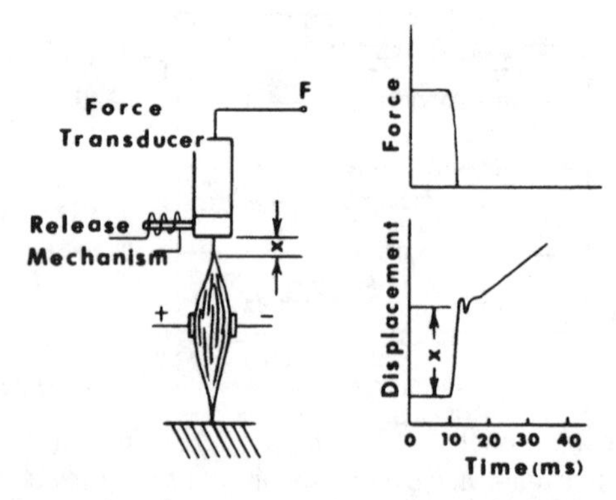

Figure 8.11 Experimental arrangement to determine the spring constant of the series elastic element. Muscle is stimulated; after the tension builds up in the muscle, the release mechanism activates, allowing an almost instantaneous shortening *x*, while the force change is measured on a force transducer.

ited to changes in joint moments during maximum, voluntary contraction (MVC) as the joint angle is altered. Two problems arise in such studies. First, it is usually impossible to generate an MVC for a single agonist without activating the remaining agonists. Thus, the joint moment is the sum of the moments generated by several agonists, each of which is likely to be operating at a different point in its force–length curve. Also, the moment generated by any muscle is a product of its force and its moment arm length, both of which change as the joint angle changes. Thus, any change in joint moment with joint angle cannot be attributed uniquely to the length–tension characteristics of the muscle.

One in vivo study that did yield some meaningful results was reported by Sale et al. (1982), who quantified the plantarflexor moment changes as the ankle angle was positioned over a range of 50° (30° plantarflexion to 20° dorsiflexion). The contribution of the gastrocnemius was reduced by having the knee flexed to 90°, thus causing it to be slack. Thus, the contribution of the soleus was considered to be dominant. Also, the range of movement of the ankle was considered to have limited influence on the moment arm length of the soleus. Thus the moment–ankle angle curve was considered to reflect the force–length characteristics of the soleus muscle. The results of the MVC, peak twitch tensions, and a 10-Hz tetanic stimulation were almost the same: the peak tension was seen at 15° dorsiflexion and dropped off linearly to near zero as the ankle plantarflexed to 30°. Such a curve agreed qualitatively with a muscle whose resting length was at 15° dorsiflexion and whose cross bridges reached a full overlap of 30° plantarflexion.

8.2 FORCE–VELOCITY CHARACTERISTICS

The previous section was concerned primarily with isometric contractions, and most physiological experiments are conducted under such conditions. However, movement cannot be accomplished without a change in muscle length. Alternate shortening and lengthening occurs regularly during any given movement, so it is important to see the effect of muscle velocity on muscle tension.

8.2.1 Concentric Contractions

The tension in a muscle decreases as it shortens under load. The characteristic curve that describes this effect is called a *force–velocity curve* and is shown in Figure 8.12. The usual curve is plotted for a maximum (100%) contraction. However, this condition is rarely seen except in athletic events, and then only for short bursts of time. In Figure 8.12 the curves for 75%, 50%, and 25% contractions are shown as well. Isometric contractions lie along the zero-velocity axis of this graph and should be considered as nothing more than a special case within the whole range of possible velocities. It should be noted

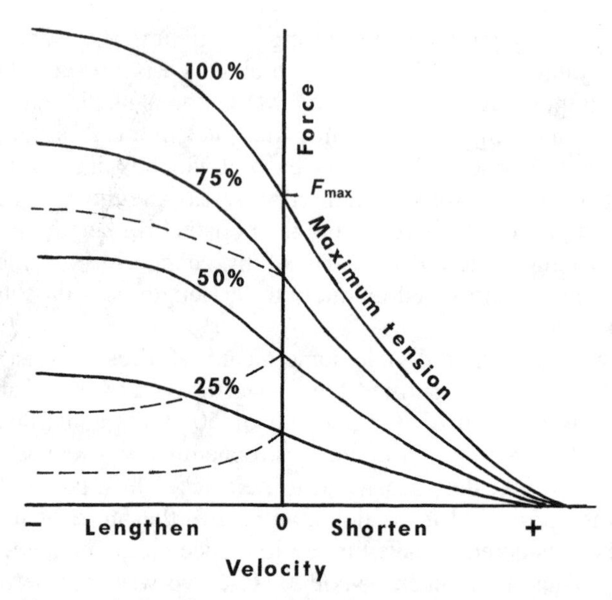

Figure 8.12 Force–velocity characteristics of skeletal muscle for different levels of muscle activation; shown are 25, 50, 75, and 100% levels of activation. All such measures must be taken as the muscle shortens or lengthens at a given length, and the length must be reported. During shortening, the curves follow the hyperbolic Hill model, but during lengthening, the curves depend on the experimental protocol; the solid lines are for isotonic activity while the dashed lines are for isovelocity activity.

that this curve represents the characteristics at a certain muscle length. To incorporate length as a variable as well as velocity requires a three-dimensional plot, as is discussed in Section 8.2.3.

The decrease of tension as the shortening velocity increases has been attributed to two main causes. A major reason appears to be the loss in tension as the cross bridges in the contractile element break and then reform in a shortened condition. A second cause appears to be the fluid viscosity in both the contractile element and the connective tissue. Such viscosity results in friction that requires an internal force to overcome and therefore results in a reduced tendon force. Whichever the cause of loss of tension, it is clear that the total effect is similar to that of viscous friction in a mechanical system and can therefore be modeled as some form of fluid damper. The loss of tension related to a combination of the number of cross bridges breaking and reforming and passive viscosity makes the force–velocity curve somewhat complicated to describe. If all the viscosity were passive, the slope of the force–velocity curve would be independent of activation. Conversely, if all the viscosity were related to the number of active cross bridges, the slope would be proportional to the activation. Green (1969) has analyzed families of force–velocity curves to determine the relative contribution of each of these mechanisms.

A curve fit of the force–velocity curve was demonstrated by Fenn and Marsh (1935), who used the equation

$$V = V_0 e^{P/B} - KP \tag{8.2}$$

where V = shortening velocity at any force
V_0 = shortening velocity of unloaded muscle
P = force
B, K = constants

A few years later Hill (1938) proposed a different mathematical relationship that bore some meaning with regard to the internal thermodynamics. Hill's curve was in the form of a hyperbola,

$$(P + a)(V + b) = (P_0 + a)b \tag{8.3}$$

where P_0 = maximum isometric tension
a = coefficient of shortening heat
b = $a \cdot V_0/P_0$
V_0 = maximum velocity (when $P = 0$)

More recently this hyperbolic form has been found to fit the empirical constant only during isotonic contractions near resting length.

The maximum velocity of shortening is often expressed as a function of l_0, the resting fiber length of the contractile elements of the muscle. Animal experiments have shown the maximum shortening velocity to be about 6 l_0/s (Faulkner et al., 1980; Close, 1965). However, it appears that this must be low for some human activity. Woittiez (1984) has calculated the shortening velocity in soleus, for example, to be above 10 l_0/s (based on plantarflexor velocity in excess of 8 rad/s and a moment arm length of 5 cm).

8.2.2 Eccentric Contractions

The vast majority of research done on isolated muscle during in vivo experiments has involved concentric contractions. As a result, there is relatively limited knowledge about the details of the force–velocity curve as the muscle lengthens. The curve certainly does not follow the detailed mathematical relationships that have been developed for concentric contractions.

This lack of information about eccentric contractions is unfortunate because normal human movement usually involves as much eccentric as concentric activity. If we neglect air and ground friction, level walking involves equal amounts of positive and negative work, and in downhill gait, negative work dominates. Figure 8.12 shows the general shape of the force–velocity curve during eccentric contractions. It can be seen that this curve is an extension of the concentric curve. During isotonic eccentric action, the solid line curves apply (Winters, 1990), but during isovelocity eccentric activity, the relationship follows the dotted lines shown in Figure 8.12 (Zahalak, 1990; Sutarno and McGill, 1995). If the maximum isometric force is F_{max}, the plateau reached in the eccentric phase varies from 1.1 F_{max} to 1.8 F_{max} (Winters, 1990). The reason given for the force increasing as the lengthening velocity increases was that within the cross bridges the force required to break the links is greater than that required to hold it at its isometric length. The plateau is reached at higher velocities when the cross-bridge links simply "give way" to produce no further increase in force. Also, there is viscous friction in the fluid surrounding the muscle fibers, and this friction force must be overcome as well. However, the DM (distributed moment) state variable model that approximates the Huxley-type cross-bridge theories (Zahalak and Ma, 1990) predicted the drop in tension reported in the eccentric region for constant velocity stretch.

Experimentally it is somewhat more difficult to conduct experiments involving eccentric work because an external device must be available to do the work on the human muscle. Such a requirement means that a motor is needed to provide an external force that will always exceed that of the muscle. Experiments on isolated muscle are safe to conduct, but in vivo experiments on humans are difficult because such a machine could cause lengthening even past the safe range of movement of a joint. The excessive force could tear the limb apart at the joint. Foolproof safety mechanisms have to be installed to prevent such an occurrence.

8.2.3 Combination of Length and Velocity Versus Force

In Section 8.1 and in this section, it is evident that the tendon force is a function of both length and velocity. Therefore, a proper representation of both these effects requires a three-dimensional plot like that shown in Figure 8.13. The resultant curve is actually a surface, which here is represented only for the maximum contraction condition. The more normal contractions are at a fraction of this maximum, so that surface plots would be required for each level of muscle activation, say at 75%, 50%, and 25%.

8.2.4 Combining Muscle Characteristics with Load Characteristics: Equilibrium

Muscles are the only motors in the human system, and when they are active, they must be in equilibrium with their load. The load can be static, such as holding a weight against gravity, or applying a static force against a fixed object (e.g., a wall, the floor), or in an isometric cocontraction where one muscle provides the load of the other. Or the load may be dynamic and the muscles are accelerating or decelerating an inertial load or overcoming the friction of a viscous load. In the vast majority of voluntary movements, there is a mixture of static and dynamic load. The one condition that is satisfied during the entire movement is that of equilibrium: at any given point in time

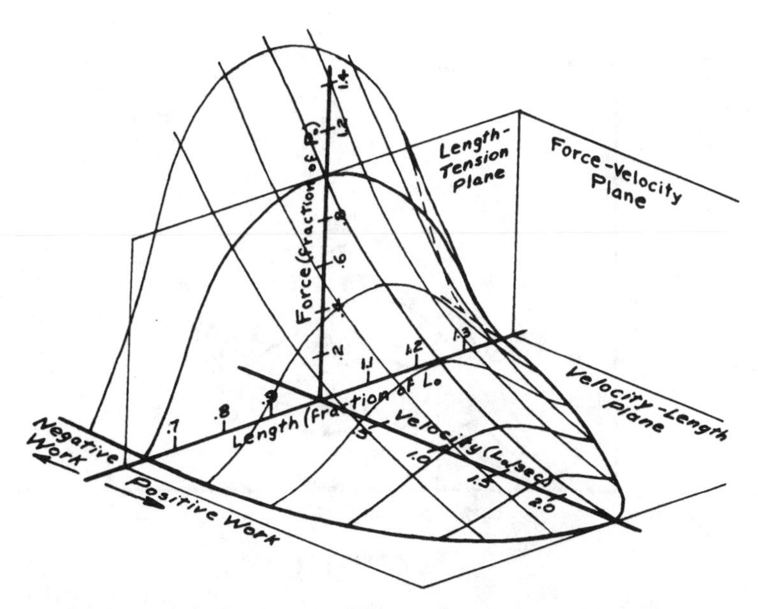

Figure 8.13 Three-dimensional plot showing the change in contractile element tension as a function of both velocity and length. Surface shown is for maximum muscle activation; a new "surface" will be needed to describe each level of activation. Influence of parallel elastic element is not shown.

the operating point will be the intersection of the muscle and load characteristics. In a dynamic movement, this operating or equilibrium point will be continuously changing. In Chapter 4, an equilibrium was recognized every time the equations of motion were written: Equation (4.3) and subsequent Examples 4.1, 4.2, 4.3, and 4.4. However, such equations recognized the net effect of all muscles acting at each joint and represented them as a torque motor without any concern about the internal characteristics of the muscles themselves. These are now examined.

Consider a muscle that is contracting against a springlike load that can have a linear or a nonlinear force–length characteristic. Figure 8.14*a* plots the force–length characteristics of a muscle at four different levels of activation, along with the linear and nonlinear characteristics of the spring loads. Assume that the muscle is at a length l_1 longer than the resting length, l_0 when the spring is at rest. The linear spring is compressed with a 50% muscle contraction, the equilibrium point a is reached, and the muscle will shorten from l_1 to l_2. When the nonlinear spring is compressed with a 50% contraction, the muscle shortens from l_1 to l_3 and the equilibrium point b is reached. If a gravitational load is lifted by the elbow flexors, the flexor muscles have the characteristics shown in Figure 8.14*b*. Starting with the forearm lowered to the side so that it is vertical, the initial equilibrium point is a. Then, as the flexors contract and shorten, equilibrium points b through e are reached. Fi-

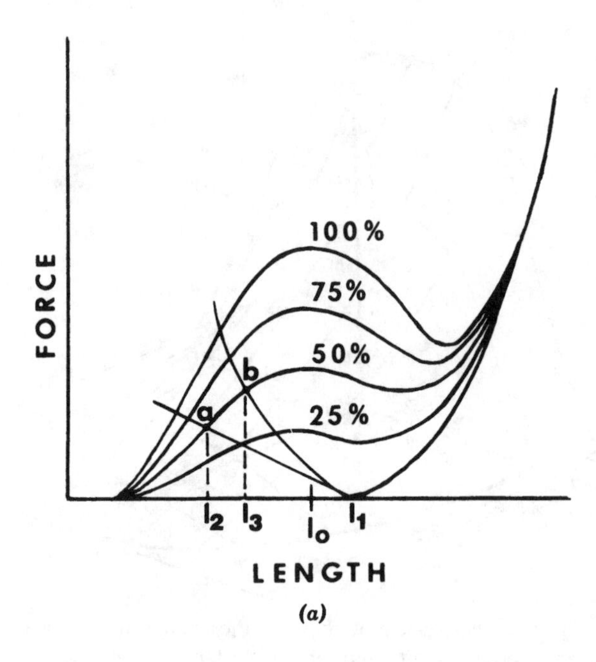

(a)

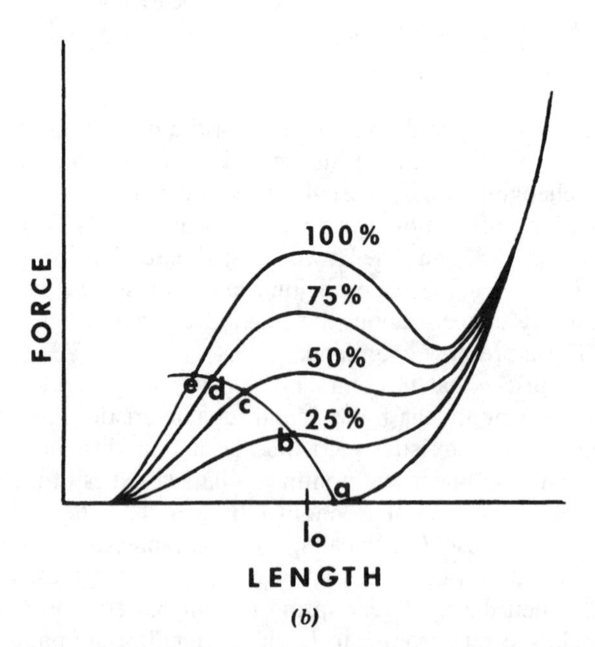

(b)

Figure 8.14 (*a*) Force–length characteristics of a muscle at four different levels of activation along with the characteristics of a linear and nonlinear spring. Intersection of spring (load) and muscle characteristics is the equilibrium point. (*b*) Equilibrium points of a load and elbow flexor muscle as the elbow is flexed against a gravitational load.

nally, as the muscles are activated 100%, the equilibrium point *e* is reached. During the transient conditions en route from *a* to *e*, the equilibrium point will also move about on the force–velocity characteristics, tracing a three-dimensional locus. The idea of dynamic equilibrium is now addressed.

We use the data from a typical walking stride to plot the time course of the contractions of the muscles about the ankle. Because the angular changes of the ankle are relatively small, we can consider that the muscle lengths are proportional to the ankle angle, and these length changes are also small. Thus, we can plot the event on a two-dimensional force–velocity curve. The lengthening or shortening velocity can be considered proportional to the angular velocity and the tendon forces are considered proportional to the muscle moment. Thus, using in vivo data, the traditional force–velocity curve becomes a moment–angular velocity curve. Figure 8.15 is the resultant plot for the period of time from heel contact to toe-off.

It can be seen from this time course of moment and angular velocity that this common movement is, in fact, quite complex. Contrary to what might be implied by force–velocity curves, a muscle does not operate along any simple curve, but actually goes through a complex combination of force and velocity

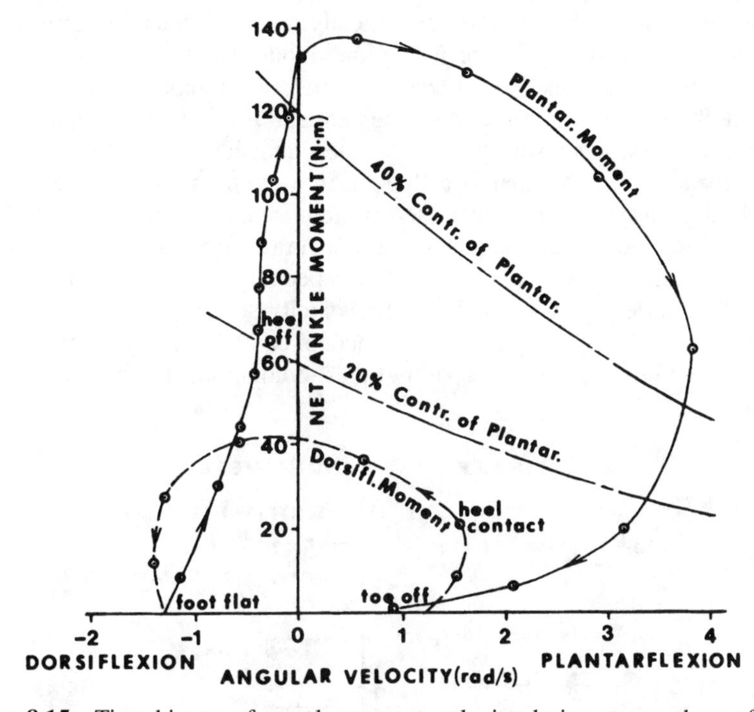

Figure 8.15 Time history of muscle moment–velocity during stance phase of walking. Dorsiflexor moment is shown as a dashed line; plantarflexor moment is a solid line. Stance begins with negative work, then alternating positive and negative, and finally a major positive work burst later in push-off, ending at toe-off.

changes. Initially the dorsiflexor muscles are on as the foot plantarflexes between the time of heel contact and flat foot (when the ground reaction force lies behind the ankle joint). Negative work is being done by the dorsiflexors as they lower the foot to the ground. After flat foot, the plantarflexors dominate and create a moment that tends to slow down the leg as it rotates over the foot, which is now fixed on the floor. Again, this is negative work. A few frames after, heel-off positive work begins, as indicated by the simultaneous plantarflexor muscle moment and plantarflexor velocity. This period is the active push-off phase, when most of the "new" energy is put back into the body.

8.3 MUSCLE MODELING

A variety of mechanical models of muscle have evolved to describe and predict tension, based on some input stimulation. Crowe (1970) and Gottlieb and Agarwal (1971) proposed a contractile component in conjunction with a linear series and a parallel elastic component plus a linear viscous damper. Glantz (1974) proposed nonlinear elastic components plus a linear viscous component. Winter (1976) has used a mass and a linear spring and damper system to simulate the second-order critically damped twitch. The purpose of this section is not to criticize or justify one model versus another, but rather to go over the principles behind the modeling and the components.

Figure 8.16 shows the force–displacement and force–velocity relationships of linear and nonlinear springs and dampers that have been proposed. The symbol for a viscous damper is a "piston in a cylinder," which can be considered full of a fluid of suitable viscosity represented by the constant K. The more common nonlinear models are exponential in form or to a power a, which is usually greater than 1. This is especially true of viscous friction, which often varies approximately as the velocity squared.

The total model of the passive components can take many configurations, as shown in Figure 8.17a. The parallel elastic component can be considered

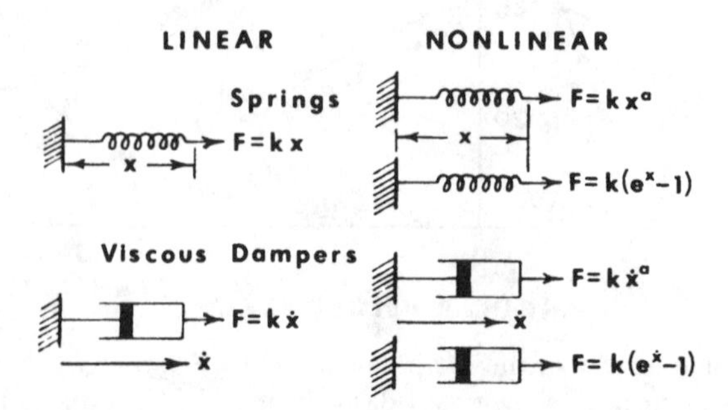

Figure 8.16 Schematic diagram of linear and nonlinear spring and viscous damper elements used to represent passive viscoelastic characteristics of muscle.

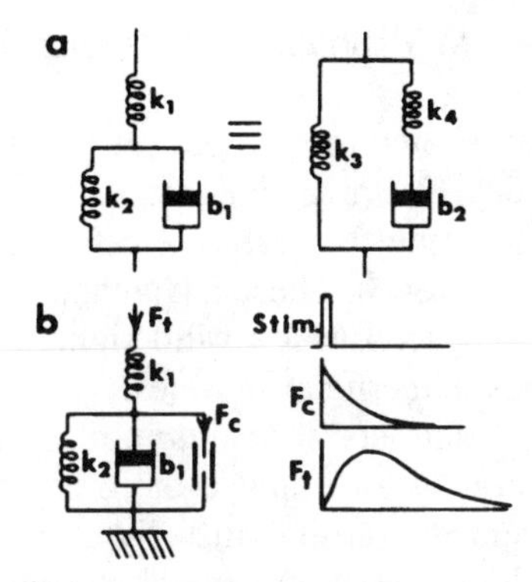

Figure 8.17 (*a*) Two equivalent series/parallel arrangements of linear elements of a muscle model. (*b*) Model showing contractile element acting on viscous elastic elements. Twitch tension of the tendon F_t results only if we assume an exponential activation tension from the contractile element F_c.

to be in parallel with the damper or a series of combinations of the damper and the series elastic component. For linear components, it does not make the slightest difference which configuration is used, because they can be made equivalent. Fung (1971) has shown that

$$k_1 = k_3 + k_4, \qquad \frac{k_1 k_2}{k_1 + k_2} = k_3, \qquad \frac{b_1}{k_1 + k_2} = \frac{b_2}{k_4} \tag{8.4}$$

This means that if either model is known, the other equivalent model can replace it, and it will have the same dynamic characteristics.

The total model requires the active contractile component to be represented by some form of force generator. The time course of the tension from the contractile component is sometimes referred to as its *active state,* and this is quite often assumed to be an exponential response to a stimulus. Figure 8.17*b* shows the contractile component combined with the passive components along with the time course of the active state F_c and the resultant tendon tension F_t. More complex models have been developed, one of which is presented in the next section.

8.3.1 Example of a Model—EMG Driven

Any realistic model of muscle must have a valid input to represent the motoneuron drive. It would be desirable to record the output of the motoneuron pool (summation of all recruited motor units) to any given muscle. Unfortunately this is impossible to do experimentally, so the best compromise is to record the EMG from the muscle. Students not familiar with the recording and interpretation of the EMG should read Chapter 9 before reviewing this example of muscle modeling.

Surface EMG has been shown to be more reliable than the EMG recorded from indwelling electrodes (Komi and Buskirk, 1970), and the number of motor units in the pick-up zone for surface electrodes is considerably larger than that for indwelling electrodes. These advantages, combined with the ease of application, make the surface EMG a valid signal to represent the average motor unit activity of most superficial muscles.

When each motor unit contracts, it results in a characteristic action potential and also the generation of an impulse of force by the activated muscle fibers. There is controversy regarding the shape of the tension waveform generated at the cross bridges versus the twitch waveform seen at the tendon. Some researchers assume the internal force waveform (sometimes called *active state*) to be a first-order exponential (Hatze, 1978; Gottlieb and Agarwal, 1971). However, research on muscles (Hill, 1953) showed that the time course of the biochemical thermal events was impulsive in nature. Using frog muscle at 0°C, the rate of heat production reaches a maximum 30 ms after stimulation. If we recall that reaction rates increase 2–3 times every 10°C, we could predict an increase of 15–40 in the reaction rates at muscle temperatures of 37°C. Therefore, the rate of heat production (indicating the release of chemical energy at the cross bridges) would be over in less than 2 ms. Thus, the associated mechanical energy release must also be assumed to be a short-duration impulse. Fortunately larger motor units also generate larger action potentials. Thus, the absolute value of the EMG is a series of short-duration impulses that can be considered to represent the contractile element's impulsive force. Thus, a full-wave rectified EMG followed by a nonlinearity (to model any nonlinearity between the EMG and muscle force) is used to represent the relationship between the EMG and the contractile element force.

Figure 8.18 shows the complete model, which incorporates the contractile element, a linear damper B to model any viscous or velocity-dependent elements (in the contractile element and connective tissue), a linear spring K to represent the series elastic elements (in the tendon, fascia, and cross bridges), and a mass element M to represent the effective mass of muscle tissue that must be accelerated as the force impulse wave travels from the motor end plate region toward both tendons. For an isometric contraction, the length L_0 of the muscle remains constant, but x_2 shortens, causing a lengthening of x_1 in the muscle's series elastic element. The impulsive contractile element force F_{ce} acts on the mass to accelerate it, but to do so the mass acts on the damper and spring elements. Thus, the equations of motion are

$$x_1 + x_2 = L_0 = \text{constant}$$

$$F_{ce} = M\ddot{x}_1 + B\dot{x}_1 + Kx_1 \tag{8.5a}$$

The tendon force F_t is seen only if the series elastic spring increases its length beyond resting length, or

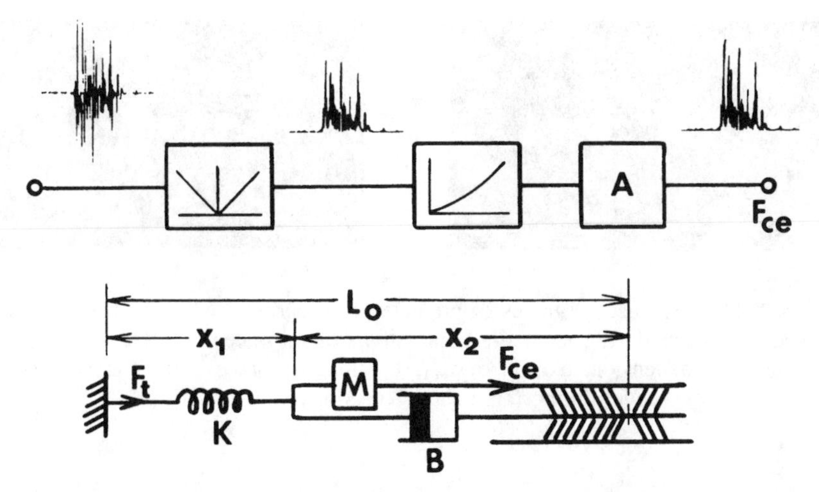

Figure 8.18 Biomechanical model of a muscle during an isometric contraction of varying tension. F_{ce} is modeled as an impulsive contractile element force acting on an equivalent mass M, a linear damping element B, and a series elastic element K. F_{ce} as input to the model is assumed to be the same shape as the full-wave rectified EMG with an empirically curve-fitted nonlinearity. See text for full details and justification for the assumptions.

$$F_t = Kx_1 \tag{8.5b}$$

In analog or digital form, this second-order differential equation can be solved for x_1 (Figure 8.19) to predict the tendon force for any given EMG. The final part of the model is to determine K, B, and M for the muscle. Fortunately we do not have to measure them separately. Rather, we can make use of the fact that the twitch waveform is quite close to that of a critically damped second-order system (Milner-Brown et al., 1973*a*). The duration of the twitches is sufficiently long, such that, in comparison, the motor unit action potential can be considered as an impulse. For a mass–spring–damper system that is critically damped, the twitch time T (see Section 8.0.5) allows us to calculate the correct ratios, $T = 2M/B$ or $B/M = 2/T$. Also, $B =$

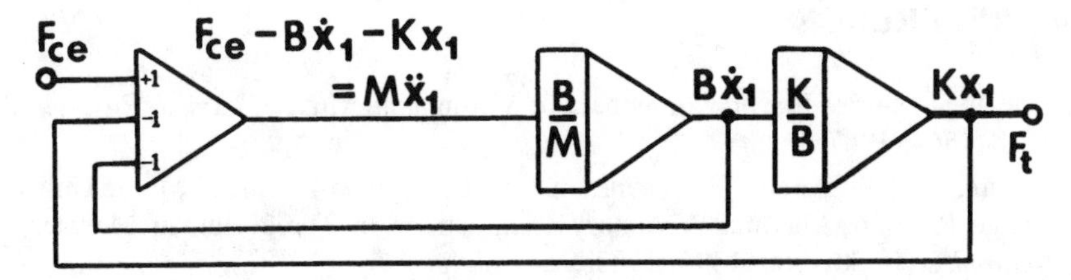

Figure 8.19 Analog or digital solution to solve for tendon force F_t for any given F_{ce} and for any muscle. B/M and K/B ratios are known if the twitch time T for that muscle is known. See text for complete details.

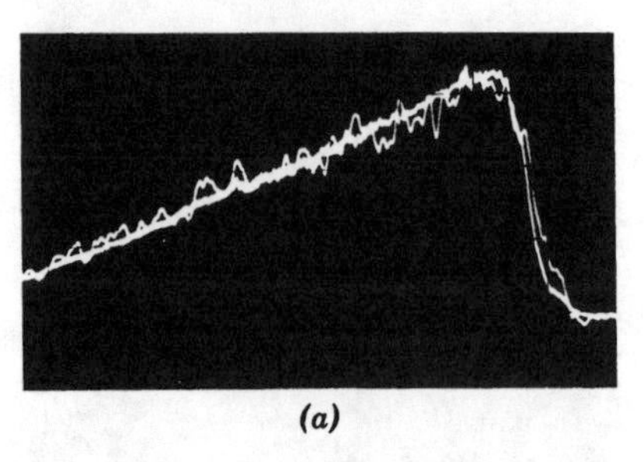

(a)

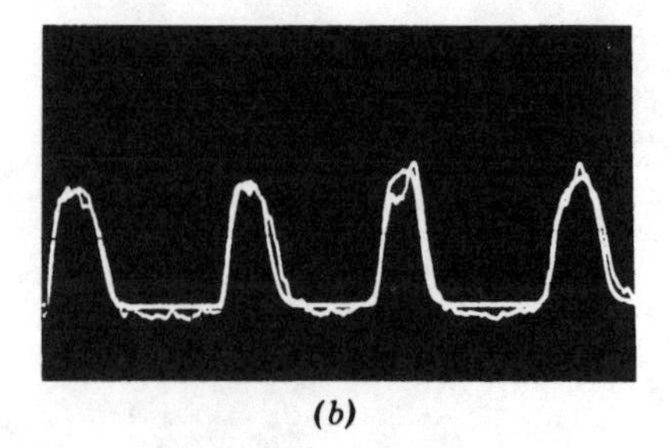

(b)

Figure 8.20 Oscilloscope traces of predicted F_t superimposed on that measured by a force transducer during isometric contractions of the biceps muscle. (*a*) During a ramp increase of tension. Noisier trace is F_t, smoother trace is the force transducer record. (*b*) Same model during rapid short bursts of force.

$2\sqrt{MK}$. Therefore, $K/B = B/4M = 1/2T$. Therefore, the differential equation as modeled in Figure 8.19 can be solved by knowing the values for B/M and K/B, which are automatically known if we know T for any given muscle. The only remaining parameter that is needed to curve-fit the model is a "gain" A to model the relationship between the EMG (in microvolts) and the contractile element tension F_{ce} (in newtons).

Figure 8.20 shows the results of comparisons between the predicted muscle tension F_t and that measured experimentally. The biceps muscle EMG was recorded along with elbow flexor tension from a force transducer attached firmly to the wrist via a cuff. The smoother curve is the transducer output, while the noisier signal is the predicted F_t. The difference between measured and predicted curves is minimal and can be partially attributed to several measurement and assumption errors. First, we are not measuring the tendon tension directly, but via a transducer attached to the wrist. The recorded tension will be further damped by the cuff/wrist tissue and in the synovial fluid and soft tissue in the elbow joint capsule. Second, certain unstated assumptions are inferred in this model. Other agonists (brachioradialis, brachialis) are assumed to have exactly the same activation profile as the biceps, and any antagonists (triceps) are assumed to be silent.

8.4 REFERENCES

Bahler, A. S. "Series Elastic Component of Mammalian Muscle," *Am. J. Physiol.* **213**: 1560–1564, 1967.

Bellemare, F., J. J. Woods, R. Johansson, and B. Bigland-Ritchie. "Motor Unit Discharge Rates in Maximal Voluntary Contractions in Three Human Muscles," *J. Neurophysiol.* **50**:1380–1392, 1983.

Buchthal, F. and H. Schmalbruch. "Contraction Times and Fibre Types in Intact Human Muscle," *Acta Physiol. Scand.* **79**:435–452, 1970.

Close, R. "The Relation Between Intrinsic Speed of Shortening and Duration of the Active State of Muscles," *J. Physiol.* **180**:542–559, 1965.

Crowe, A. "A Mechanical Model of Muscle and its Application to the Intrafusal Fibres of Mammalian Muscle Spindle," *J. Biomech.* **3**:583–592, 1970.

DeLuca, C. J., R. S. LeFever, M. P. McCue, and A. P. Xenakis. "Control Scheme Governing Concurrently Active Motor Units during Voluntary Contractions," *J. Physiol.* **329**:129–142, 1982.

Desmedt, J. E. and E. Godaux. "Ballistic Contractions in Fast and Slow Human Muscles: Discharge Patterns of Single Motor Units," *J. Physiol.* **285**:185–196, 1978.

Erim, Z., C. J. DeLuca, and K. Mineo. "Rank-Ordered Regulation of Motor Units," *Muscle & Nerve* **19**:563–573, 1996.

Faulkner, J. A., J. H. Niemeyer, L. C. Maxwell, and T. P. White. "Contractile Properties of Transplanted Extensor Digitorum Longus Muscle of Cats," *Am. J. Physiol.* **238**:120–126, 1980.

Feinstein, B., B. Lindegard, E. Nyman, and G. Wohlfart. "Morphological Studies of Motor Units in Normal Human Muscles," *Acta Anat.* **23**:127–142, 1955.

Fenn, W. O. and B. S. Marsh. "Muscular Force at Different Speeds of Shortening," *J. Physiol. London* **85**:277–297, 1935.

Fuglevand, A. J., D. A. Winter, and A. E. Patla. "Models of Recruitment and Rate Coding Organization in Motor Unit Pools," *J. Neurophysiol.* **70**:2470–2488, 1993.

Fung, Y. C. "Comparison of Different Models of the Heart Muscle," *J. Biomech.* **4**: 289–295, 1971.

Glantz, S. A. "A Constitutive Equation for the Passive Properties of Muscle," *J. Biomech.* **7**:137–145, 1974.

Gordon, A. M., A. F. Huxley, and F. J. Julian. "The Variation Is Isometric Tension with Sarcomere Length in Vertebrate Muscle Fibres," *J. Physiol.* **184**:170, 1966.

Gottlieb, G. L. and G. C. Agarwal. "Dynamic Relationship between Isometric Muscle Tension and the Electromyogram in Man," *J. Appl. Physiol.* **30**:345–351, 1971.

Green, D. G. "A Note on Modelling in Physiological Regulators," *Med. Biol. Eng.* **7**: 41–47, 1969.

Hatze, H. "A General Myocybernetic Control Model of Skeletal Muscle," *Biol. Cybern.* **28**:143–157, 1978.

Henneman, E. "Organization of the Spinal Cord," in *Medical Physiology,* vol. 1, 13th ed., V. B. Mountcastle, Ed. (C. V. Mosby, St. Louis, MO, 1974*a*).

Henneman, E. "Peripheral Mechanism Involved in the Control of Muscle," in *Medical Physiology,* vol. 1, 13th ed., V. B. Montcastle, Ed. (C. V. Mosby, St. Louis, MO, 1974*b*).

Hill, A. V. "The Heat of Shortening and Dynamic Constants of Muscle," *Proc. R. Soc. B.* **126**:136–195, 1938.

Hill, A. V. "Chemical Change and Mechanical Response in Stimulated Muscle," *Proc. R. Soc. B.* **141**:314–320, 1953.

Komi, P. V. and E. R. Buskirk. "Reproducibility of Electromyographic Measures with Inserted Wire Electrodes and Surface Electrodes," *Electromyography* **10**:357–367, 1970.

Milner-Brown, H. S. and R. B. Stein. "The Relation between the Surface Electromyogram and Muscular Force," *J. Physiol.* **246**:549–569, 1975.

Milner-Brown, H. S., R. B. Stein, and R. Yemm. "The Contractile Properties of Human Motor Units during Voluntary Isometric Contractions," *J. Physiol.* **228**:285–306, 1973*a*.

Milner-Brown, H. S., R. B. Stein, and R. Yemm. "The Orderly Recruitment of Human Motor Units during Voluntary Isometric Contractions," *J. Physiol.* **230**:359–370, 1973*b*.

Sale, D., J. Quinlan, E. Marsh, A. J. McComas, and A.Y. Belanger. "Influence of Joint Position on Ankle Plantarflexion in Humans," *J. Appl. Physiol.* **52**:1636–1642, 1982.

Sutarno, C. G. and S. M. McGill. "Isovelocity Lengthening Behaviour of Erector Spinae Muscles," *Europ. J. Appl. Physiol.* **70**:146–153, 1995.

van Ingen Schenau, G. J. "An Alternate View of the Concept of Utilization of Elastic Energy in Human Movement," *Human Movement Sci.* **3**:301–336, 1984.

Wani, A. M. and S. K. Guha. "A Model for Gradation of Tension Recruitment and Rate Coding," *Med. Biol. Eng.* **13**:870–875, 1975.

Winter, D. A. "Biomechanical Model Related EMG to Changing Isometric Tension," in *Dig. 11th Int. Conf Med. Biol. Eng.,* pp. 362–363, 1976.

Winters, J. M. "Hill-Based Muscle Models: A System Engineering Perspective," in *Multiple Muscle Systems: Biomechanics and Movement Organization,* J. M. Winters and S. L. J. Woo, Eds. (Springer, New York, 1990), pp. 69–93.

Woittiez, R. O. "A Quantitative Study of Muscle Architecture and Muscle Function," Ph.D. dissertation, Free University (Amsterdam), The Netherlands, 1984.

Wormolts, J. R. and W. K. Engel. "Correlation of Motor Unit Behaviour with Histochemical-Myofiber Type in Humans by Open-Biopsy Electromyography," in *New Developments in Electromyography and Clinical Neurophysiology,* vol. 1, J. E. Desmedt, Ed. (Karger, Basel, Switzerland, 1973).

Zahalak, G. I. "Modelling Muscle Mechanics (and Energetics)," in *Multiple Muscle Systems: Biomechanics and Movement Organization,* J. M. Winters and S. L. J. Woo, Eds. (Springer, New York, 1990), pp. 1–23.

Zahalak, G. I. and S.-P. Ma. "Muscle Activation and Contraction: Constitutive Relations Based Directly on Cross-Bridge Kinetics," *J. Biomech. Eng.* **112**:52–62, 1990.

9

KINESIOLOGICAL ELECTROMYOGRAPHY

9.0 INTRODUCTION

The electrical signal associated with the contraction of a muscle is called an *electromyogram* or, by its shorthand name, EMG. The study of EMGs, called *electromyography,* has revealed some basic information; however, much remains to be learned. Voluntary muscular activity results in an EMG that increases in magnitude with the tension. However, there are many variables that can influence the signal at any given time: velocity of shortening or lengthening of the muscle, rate of tension buildup, fatigue, and reflex activity. An understanding of the electrophysiology and the technology of recording is essential to the appreciation of the biomechanical relationships that follow.

9.1 ELECTROPHYSIOLOGY OF MUSCLE CONTRACTION

It is important to realize that muscle tissue conducts electrical potentials somewhat similarly to the way axons transmit action potentials. The name given to this special electrical signal generated in the muscle fibers as a result of the recruitment of a motor unit is a *motor unit action potential* (m.u.a.p.). Electrodes placed on the surface of a muscle or inside the muscle tissue (indwelling electrodes) will record the algebraic sum of all m.u.a.p.'s being transmitted along the muscle fibers at that point in time. Those motor units far away from the electrode site will result in a smaller m.u.a.p. than those of similar size near the electrode.

9.1.1 Motor End Plate

For a given muscle, there can be a variable number of motor units, each controlled by a motor neuron through special synaptic junctions called *motor end plates.* An action potential transmitted down the motoneuron (sometimes called the *final common pathway*) arrives at the motor end plate and triggers a sequence of electrochemical events. A quantum of ACh is released; it crosses the synaptic gap (200–500 Å wide) and causes a depolarization of the postsynaptic membrane. Such a depolarization can be recorded by a suitable microelectrode and is called an *end plate potential* (EPP). In normal circumstances, the EPP is large enough to reach a threshold level, and an action potential is initiated in the adjacent muscle fiber membrane. In disorders of neuromuscular transmission (e.g., depletion of ACh), there may not be a one-to-one relationship between each motor nerve action potential and a m.u.a.p. The end plate block may be complete, or may occur only at high stimulation rates, or intermittently.

9.1.2 Sequence of Chemical Events Leading to a Twitch

The beginning of the m.u.a.p. starts at the *Z* line of the contractile element (see Section 8.1) by means of an inward spread of the stimulus along the transverse tubular system. This results in a release of Ca^{2+} in the sarcoplasmic reticulum. Ca^{2+} rapidly diffuses to the contractile filaments of actin and myosin, where ATP is hydrolyzed to produce ADP plus heat plus mechanical energy (tension). The mechanical energy manifests itself as an impulsive force at the cross bridges of the contractile element. The time course of the contractile element's force has been the subject of considerable speculation and has been modeled mathematically, as described in Section 8.05.

9.1.3 Generation of a Muscle Action Potential

The depolarization of the transverse tubular system and the sarcoplasmic reticulum results in a depolarization "wave" along the direction of the muscle fibers. It is this depolarization wave front and the subsequent repolarization wave that are "seen" by the recording electrodes.

Many types of EMG electrodes have developed over the years, but generally they can be divided into two groups: surface and indwelling (intramuscular). Basmajian (1973) gives a detailed review of the use of different types along with their connectors. Surface electrodes consist of disks of metal, usually silver/silver chloride, of about 1 cm in diameter. These electrodes detect the average activity of superficial muscles and give more reproducible results than do indwelling types (Komi and Buskirk, 1970; Kadaba et al., 1985). Smaller disks can be used for smaller muscles. Indwelling electrodes are required, however, for the assessment of fine movements or to record from deep muscles. A needle electrode is nothing more than a fine hypodermic needle with an insulated conductor located inside and bared to the muscle tissue at the open end of the needle; the needle itself forms the other conductor. For research purposes, multielectrode types have been developed to investigate the "territory" of a motor unit (Buchthal et al., 1959), which has been found to vary from 2 to 15 mm in diameter. Fine-wire electrodes, having about the diameter of human hairs, are now widely used. They require a hypodermic needle to insert. After removal of the needle, the fine wires with their uninsulated tips remain inside in contact with the muscle tissue. A comparison of this experimental investigation of motor unit territory was seen to agree with theoretical predictions (Boyd et al., 1978).

Indwelling electrodes are influenced not only by waves that actually pass by their conducting surfaces, but also by waves that pass within a few millimeters of the bare conductor. The same is true for surface electrodes. The field equations that describe the electrode potential were originally derived by Lorente de No (1947) and were extended in rigorous formulations by Plonsey (1964, 1974) and Rosenfalck (1969). These equations were complicated by the formulation of current density functions to describe the temporal and spatial depolarization and repolarization of the muscle membrane. Thus, simplification of the current density to a dipole or tripole (Rosenfalck, 1969) yielded a reasonable approximation when the active fiber was more than 1 mm from the electrode surface (Andreassen and Rosenfalck, 1981).

In the dipole model (Figure 9.1), the current is assumed to be concentrated at two points along the fiber: a source of current I representing the depolar-

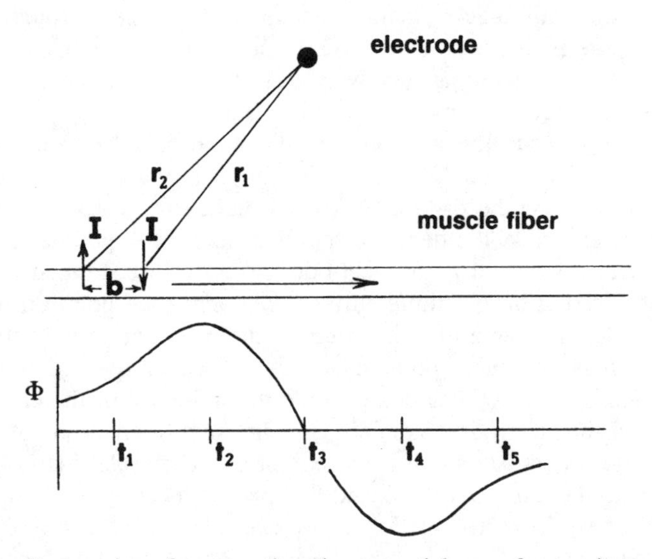

Figure 9.1 Propagation of motor unit action potential wave front as it passes beneath a recording electrode on the skin surface. The electrode voltage is a function of the magnitude of the dipole and the distances r_1 and r_2 from the electrode to the depolarizing and repolarizing currents.

ization and a sink of current $-I$ representing the repolarization, both separated by a distance b. The potential Φ, recorded at a point electrode at a distance r from the current source, is given by

$$\Phi = \frac{I}{4\Pi\sigma} \cdot \frac{1}{r} \tag{9.1}$$

where σ is the conductivity of the medium, which is assumed to be isotropic (uniform in all spatial directions).

The net potential recorded at the point electrode from both source and sink currents is

$$\begin{aligned} \Phi &= \frac{I}{4\Pi\sigma} \cdot \frac{1}{r_1} - \frac{I}{4\Pi\sigma} \cdot \frac{1}{r_2} \\ &= \frac{I}{4\Pi\sigma}\left(\frac{1}{r_1} - \frac{1}{r_2}\right) \end{aligned} \tag{9.2}$$

where r_1 and r_2 are the distances to the source and sink currents.

The time history of the action potential then depends on r_1 and r_2, which vary with time as the wave propagates along the muscle fiber. At t_1, as the wave approaches the electrode ($r_1 < r_2$), the potential will thus be positive and will increase. It will reach a maximum at t_2; then as r_1 becomes nearly equal to r_2, the amplitude suddenly decreases and passes through zero at t_3 when the dipole is directly beneath the electrode ($r_1 = r_2$). Then it becomes negative as the dipole propagates away from the electrode ($r_1 > r_2$). Thus, a biphasic wave is recorded by a single electrode.

A number of recording and biological factors affect the magnitude and shape of the biphasic signal. The duration of each phase is a function of the propagation velocity, the distance b (which varies from 0.5 to 2.0 mm) between source and sink, the depth of the fiber, and the electrode surface area. Equation (9.2) assumes a point electrode. Typical surface electrodes are not point electrodes but have a finite surface area, and each point on the surface can be considered an area of point sources; the potential on the surface is the average of all point source potentials. Figure 9.2 is presented from a study by Fuglevand et al. (1992) to demonstrate the influence of the electrode size and shape on the electrode potential. Two different orientations of strip electrodes are shown: A—along an axis parallel to the muscle fiber—and D—at right angles to the muscle fiber. B and E show the location of a strip of point-source electrodes to represent these two strip electrodes, each 10 mm long with 10 point sources; C and F show the action potentials at each point source. For the strip electrode in A, the distance from the points on the electrode to the fiber, r_f, is constant; thus, each action potential has the same amplitude and shape. However, there is a phase difference between each of the point-

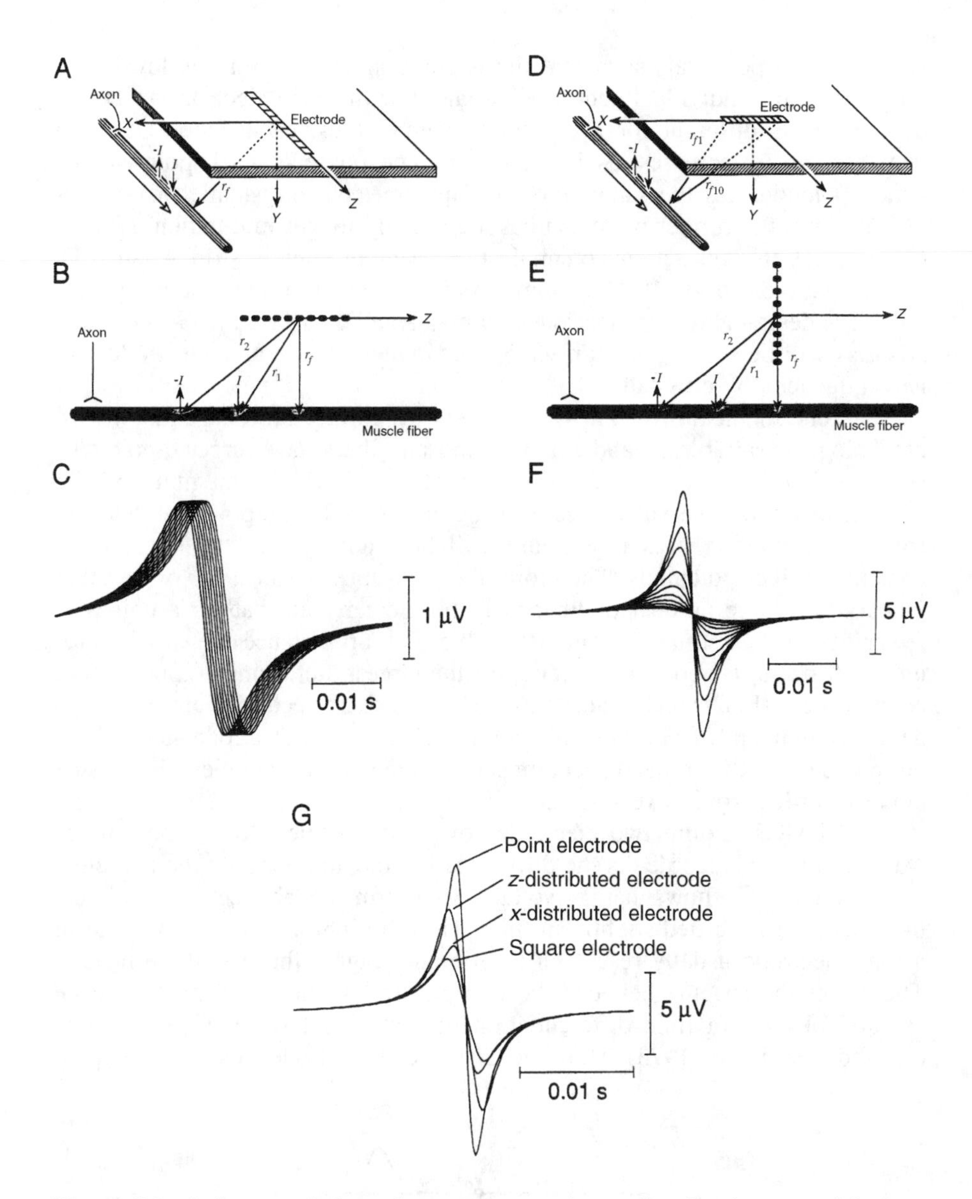

Figure 9.2 Influence of electrode geometry on the predicted action potentials from a single electrode with a variety of shapes and sizes. The potential was computed as the algebraic mean of the potentials detected by an array of point-source electrodes. See text for detailed discussion. (Reproduced by permission from *Biological Cybernetics*, "Detection of Motor Unit Action Potentials with Surface Electrodes: Influence of Electrode Size and Shaping," A. F. Fuglevand, D. A. Winter, A. E. Patla, and D. Stashuk, **67**:143–153, 1992; Fig. 3A–G.)

source action potentials such that the average action potential is lower and longer than the individual action potentials. For the strip electrode in D, the distance r_f is different for every point source; thus, each action potential decreases as r_f increases, as is seen in F. The phase of each point-source action potential has the same zero-crossing time, but the smaller action potentials from the further point sources are slightly longer in duration. Finally, G compares the net action potentials from a point source, strip A, strip D, and a square electrode 10×10 mm^2. As is evident from these action potentials, it is desirable to use electrodes with as small a surface area as possible. Also, as will be seen in Section 9.2.5, the smaller surface area electrodes are less influenced by crosstalk.

The zone of pickup for individual fibers by an electrode depends on two physiological variables, I and r. Larger muscle fibers have larger dipole currents, which increase with the diameter of the fiber. For an m.u.a.p., the number of fibers innervated must also be considered. The potential detected from the activation of a motor unit will be equivalent to the sum of the constituent fiber potentials. Therefore, the pick-up zone for a motor unit that innervates a large number of fibers will be greater than that for a unit with fewer fibers. An estimate of the detectable pick-up distance for small motor units (50 fibers) is about 0.5 cm and, for the largest motor units (2500 fibers), about 1.5 cm (Fuglevand et al., 1992). Thus, surface electrodes are limited to detecting m.u.a.p.'s from those fibers quite close to the electrode site and are not prone to pickup from adjacent muscles (called *crosstalk*) unless the muscle being recorded from is very small.

Most EMGs require two electrodes over the muscle site, so the voltage waveform that is recorded is the difference in potential between the two electrodes. Figure 9.3 shows that the voltage waveform at each electrode is almost the same, but is shifted slightly in time. Thus, a biphasic potential waveform at one electrode usually results in a difference signal that has three phases. The closer the spacing between the two electrodes, the more the difference signal looks like a time differentiation of that signal recorded at a single electrode (Kadefors, 1973). Thus, closely spaced electrodes result in the spec-

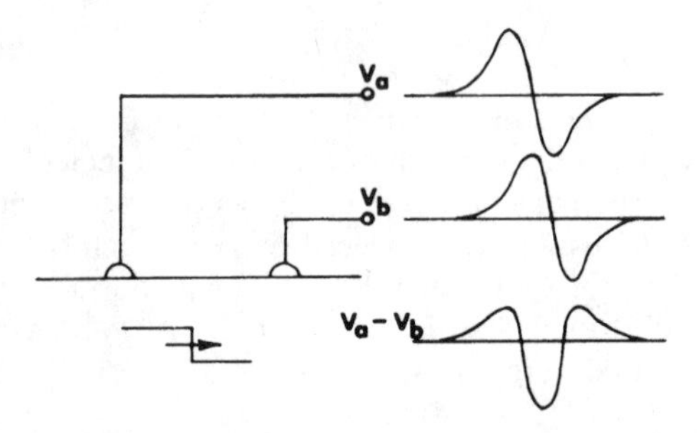

Figure 9.3 Voltage waveform present at two electrodes due to a single propagating wave. Voltage recorded is the difference voltage $V_a - V_b$, which is triphasic in comparison with the biphasic waveform seen at a single electrode.

trum of the EMG having higher frequency components than those recorded from widely spaced electrodes.

9.1.4 Duration of the Motor Unit Action Potential

As indicated in the previous section, the larger the surface area, the longer the duration of the m.u.a.p. Thus, surface electrodes automatically record longer duration m.u.a.p.'s than indwelling electrodes (Kadefors, 1973; Basmajian, 1973). Needle electrodes record durations of 3–20 ms, while surface electrodes record durations about twice that long. However, for a given set of electrodes the duration of the m.u.a.p.'s is a function of the velocity of the propagating wave front and the depth of the motor unit below the electrode surface. The velocity of propagation of the m.u.a.p. in normals has been found to be about 4 m/s (Buchthal et al., 1955). The faster the velocity, the shorter the duration of the m.u.a.p. Such a relationship has been used to a limited extent in the detection of velocity changes. In fatigue and in certain myopathies (muscle pathologies), the average velocity of the m.u.a.p.'s that are recruited is reduced; thus, the duration of the m.u.a.p.'s increases (Johansson et al., 1970; Gersten et al., 1965; Kadefors, 1973). Because the peak amplitude of each phase of the m.u.a.p. remains the same, the area under each phase will increase. Thus, when we measure the average amplitude of the EMG (from the full-wave rectified waveform), it will appear to increase (Fuglevand, et al., 1992). During voluntary contractions, it has been possible, under special laboratory conditions (Milner-Brown and Stein, 1975), to detect the duration and amplitude of muscle action potentials directly from the EMG. However, during unconstrained movements, a computer analysis of the total EMG would be necessary to detect a shift in the frequency spectrum (Kwatny et al., 1970), or an autocorrelation analysis can yield the average m.u.a.p. duration (Person and Mishin, 1964).

The distance between the motor unit and the surface of the electrode severely influences the amplitude of the action potential, as is predicted in Equation (9.2), and also influences the duration of the action potential. Figure 9.4, taken from Fuglevand et al. (1992), predicts the duration and amplitude of the action potential from a 50-fiber motor unit with electrode-unit distances out to 20 mm. It is evident that the duration of the action potential increases as the amplitude decreases with distance. Thus, the frequency content of the m.u.a.p.'s from the more distant sites decreases. Fuglevand et al. (1992) showed that the mean power frequency of the m.u.a.p. would decrease from 160 Hz at an electrode-motor unit distance of 1 mm to 25 Hz at an electrode-motor unit distance of 20 mm.

9.1.5 Detection of Motor Unit Action Potentials from Electromyogram During Graded Contractions

Using recordings from multiple indwelling electrodes, DeLuca et al. (1982) pioneered techniques to identify individual m.u.a.p.'s during a low-level

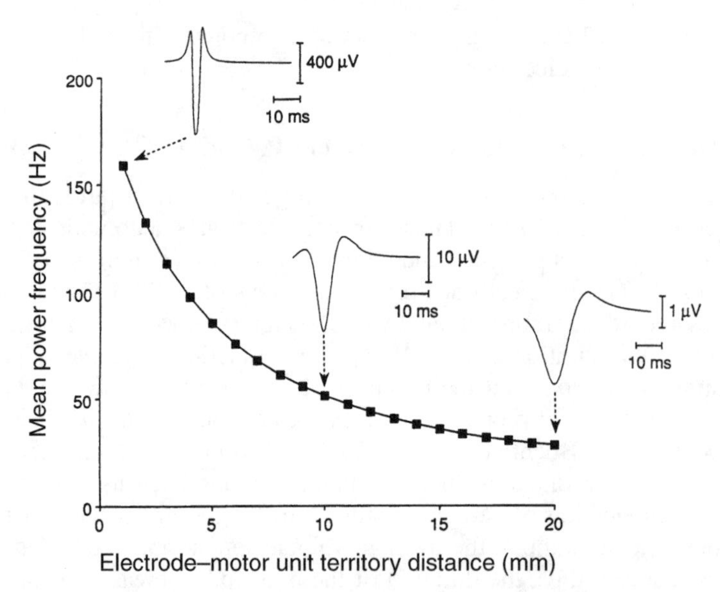

Figure 9.4 Variation in the amplitude and frequency content of the motor unit action potential with increased electrode–motor unit distance. Predictions were for a small motor unit (50 fibers) with 4-mm^2 bipolar electrodes with an interelectrode spacing of 11 mm. Shown are the m.u.a.p.'s at 1-, 10-, and 20-mm distances. (Reproduced by permission from *Biological Cybernetics,* "Detection of Motor Unit Action Potentials with Surface Electrodes: Influence of Electrode Size and Shaping," A. F. Fuglevand, D. A. Winter, A. E. Patla, and D. Stashuk, **67**:143–153, 1992; Fig. 8.)

graded contraction. More recent techniques use a quadrifilar needle electrode recording of three channels followed by decomposition algorithms involving template matching, template updating, and motor unit firing statistics (De Luca, 1993). As many as four motor units were detected and their firing rates tracked from 0 to 100% of maximum voluntary contraction (MVC) of the tibialis anterior muscle (Erim et al., 1996). It should be noted that only a fraction of all active motor units can be tracked—only those within the pick-up region of the electrode array. EMG recordings from surface electrodes have been decomposed in order to determine the firing profile of those motor units detected (McGill et al., 1987).

9.2 RECORDING OF THE ELECTROMYOGRAM

A biological amplifier of certain specifications is required for the recording of the EMG, whether from surface or from indwelling electrodes. It is valuable to discuss the reasons behind these specifications with respect to considerable problems in getting a "clean" EMG signal. Such a signal is the summation of m.u.a.ps and should be undistorted and free of noise or artifacts. Undistorted means that the signal has been amplified linearly over the range of the amplifier and recording system. The larger signals (up to 5 mV) have been amplified as much as the smaller signals (100 μV and below). The most common distortion is overdriving of the amplifier system such that the larger signals appear to be clipped off. Every amplifier has a dynamic range and should be such that the largest EMG signal expected will not exceed that range. Noise can be introduced from sources other than the muscle, and can be biological in origin or man-made. An electrocardiogram (ECG) signal picked up by EMG electrodes on the thoracic muscles can be considered unwanted biological noise. Man-made noise usually comes from power lines (hum), from machinery, or is generated within the components of the amplifier. Artifacts generally refer to false signals generated by the electrodes themselves or the cabling system. Anyone familiar with EMG recording will recall the lower frequency baseline jumps, called *movement artifacts,* which result from touching of the electrodes and movement of the cables.

The major considerations to be made when specifying the EMG amplifier are:

1. Gain and dynamic range
2. Input impedance
3. Frequency response
4. Common-mode rejection

9.2.1 Amplifier Gain

Surface EMGs have a maximum amplitude of 5 mV peak to peak, as recorded during a MVC. Indwelling electrodes can have a larger amplitude, up to 10 mV. A single m.u.a.p. has an amplitude of about 100 μV. The noise level of the amplifier is the amplitude of the higher frequency random signal seen when the electrodes are shorted together, and should not exceed 50 μV, preferably 20 μV. The gain of the amplifier is defined as the ratio of the output voltage to the input voltage. For a 2-mV input and a gain of 1000, the output will be 2 V. The exact gain chosen for any given situation will depend on what is to be done with the output signal. The EMG can be recorded on a pen recorder or magnetic tape, viewed on an oscilloscope, or fed straight into a computer. In each case, the amplified EMG should not exceed the range of input signals expected by this recording equipment. Fortunately most of these recording systems have internal amplifiers that can be adjusted to accommodate a wide range of input signals. In general, a good bioamplifier should have a range of gains selectable from 100 to 10,000. Independent of the amplifier gain, the amplitude of the signal should be reported as it appears at the electrodes, in millivolts.

9.2.2 Input Impedance

The input impedance (resistance) of a biological amplifier must be sufficiently high so as not to attenuate the EMG signal as it is connected to the input terminals of the amplifier. Consider the amplifier represented in Figure 9.5*a*. The active input terminals are 1 and 2, with a common terminal *c*. The need for a three-input terminal amplifier (differential amplifier) is explained in Section 9.2.4.

Each electrode–skin interface has a finite impedance that depends on many factors: thickness of the skin layer, cleaning of the skin prior to the attachment of the electrodes, area of the electrode surface, and temperature of the electrode paste (it warms up from room temperature after attachment). Indwelling electrodes have a higher impedance because of the small surface area of bare wire that is in contact with the muscle tissue.

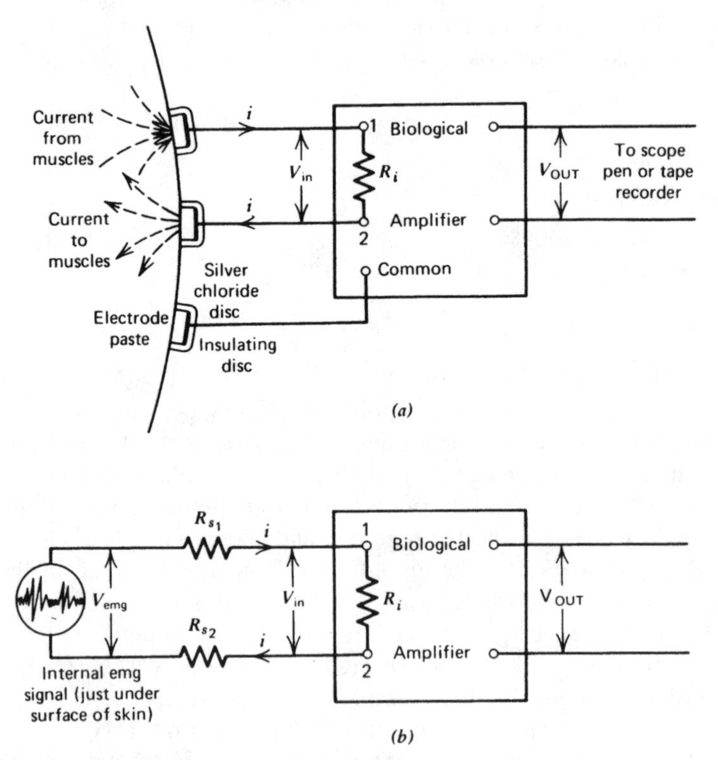

Figure 9.5 Biological amplifier for recording electrode potentials. (*a*) Current resulting from muscle action potentials flows across skin–electrode interface to develop a voltage V_{in} at the input terminals of the amplifier. A third, common electrode is normally required because the amplifier is a differential amplifier. (*b*) Equivalent circuit showing electrodes replaced by series resistors R_{s1} and R_{s2}. V_{in} will be nearly equal to V_{EMG} if $R_i \gg R_s$.

In Figure 9.5*b*, the electrode–skin interface has been replaced with an equivalent resistance. This is a simplification of the actual situation. A correct representation is a more complex impedance to include the capacitance effect between the electrode and the skin. As soon as the amplifier is connected to the electrodes, the minute EMG signal will cause current to flow through the electrode resistances R_{s1} and R_{s2} to the input impedance of the amplifier R_i. The current flow through the electrode resistances will cause a voltage drop so that the voltage at the input terminals V_{in} will be less than the desired signal V_{EMG}. For example, if $R_{s1} = R_{s2} = 10{,}000\ \Omega$ and $R_i = 80{,}000\ \Omega$, a 2-mV EMG signal will be reduced to 1.6 mV at V_{in}. A voltage loss of 0.2 mV occurs across each of the electrodes. If R_{s1} and R_{s2} were decreased by better skin preparation to 1000 Ω, and R_i were increased to 1 MΩ, the 2-mV EMG signal would be reduced only slightly, to 1.998 mV. Thus, it is desirable to have input impedances of 1 MΩ or higher, and to prepare the skin to reduce the impedance to 1000 Ω or less. For indwelling electrodes, the electrode impedance can be as high as 50,000 Ω, so an amplifier with at least 5 MΩ input impedance should be used.

9.2.3 Frequency Response

The frequency bandwidth of an EMG amplifier should be such as to amplify, without attenuation, all frequencies present in the EMG. The bandwidth of any amplifier, as shown in Figure 9.6, is the difference between the upper cutoff frequency f_2 and the lower cutoff frequency f_1. The gain of the amplifier at these cutoff frequencies is 0.707 of the gain in the midfrequency region. If we express the midfrequency gain as 100%, the gain at the cutoff frequencies has dropped to 70.7%, or the power has dropped to $(0.707)^2$ =

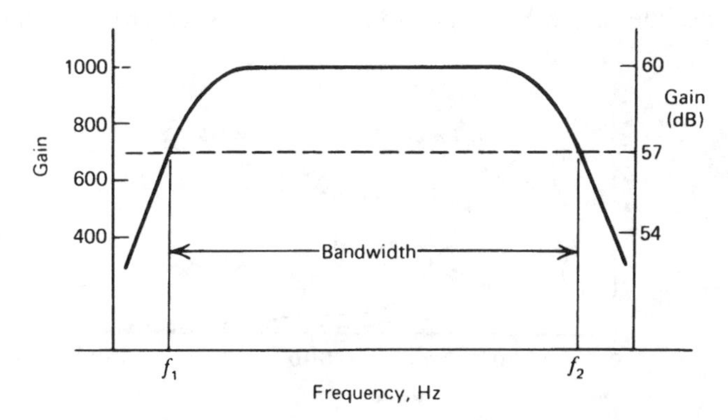

Figure 9.6 Frequency response of the biological amplifier showing a gain of 1000 (60 dB), and lower and upper cutoff frequencies f_1 and f_2.

0.5. These are also referred to as the half-power points. Often amplifier gain is expressed in logarithmic form and expressed in decibels,

$$\text{gain (dB)} = 20 \log_{10} (\text{linear gain}) \tag{9.3}$$

If the linear gain were 1000, the gain in decibels would be 60, and the gain at the cutoff frequency would be 57 dB (3 dB less than that at midfrequency).

In a high-fidelity amplifier used for music reproduction, f_1 and f_2 are designed to accommodate the range of human hearing, from 50 to 20,000 Hz. All frequencies present in the music will be amplified equally, producing a true undistorted sound at the loudspeakers. Similarly the EMG should have all its frequencies amplified equally. The spectrum of the EMG has been widely reported in the literature, with a range from 5 Hz at the low end to 2000 Hz at the upper end. For surface electrodes, the m.u.a.p.'s are longer in duration and thus have negligible power beyond 1000 Hz. A recommended range for surface EMG is 10–1000 Hz, and 20–2000 Hz for indwelling electrodes. If computer pattern recognition of individual m.u.a.p.'s is being done, the upper cutoff frequencies should be increased to 5 and 10 kHz, respectively. Figure 9.7 shows a typical EMG spectrum, and it can be seen that most of the signal is concentrated in the band between 20 and 200 Hz, with a lesser component extending up to 1000 Hz.

The spectra of other physiological and noise signals must also be considered. ECG signals contain power out to 100 Hz, so it may not be possible to eliminate such interference, especially when monitoring muscle activity around the thoracic region. The major interference comes from power line hum: in North America it is 60 Hz, in Europe 50 Hz. Unfortunately hum lies right in the middle of the EMG spectrum, so nothing can be done to filter it out. Movement artifacts, fortunately, lie in the 0–10 Hz range and normally should not cause problems. Unfortunately some of the lower quality cabling systems can generate large low-frequency artifacts that can seriously interfere with the baseline of the EMG recording. Usually such artifacts can be elim-

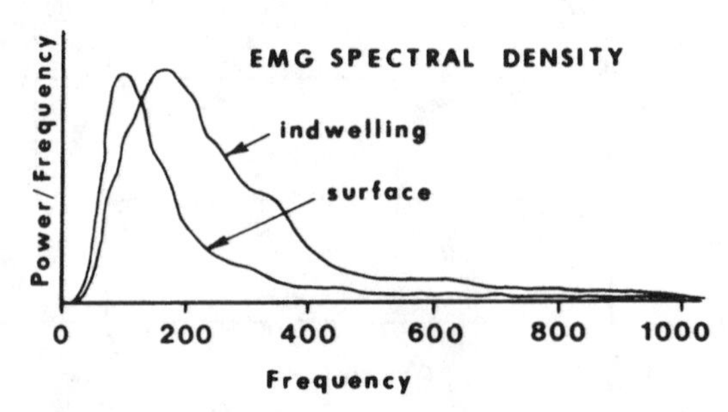

Figure 9.7 Frequency spectrum of EMG as recorded via surface and indwelling electrodes. Higher frequency content of indwelling electrodes is due to closer spacing between electrodes and their closer proximity to the active muscle fibers.

inated by good low-frequency filtering, by setting f_1 to about 20 Hz. If this fails, the only solution is to replace the cabling or go to the expense of using microamplifiers right at the skin surface.

It is valuable to see the EMG signal as it is filtered using a wide range of bandwidths. Figure 9.8 shows the results of such filtering and the obvious distortion of the signal when f_1 and f_2 are not set properly.

9.2.4 Common-Mode Rejection

The human body is a good conductor, and therefore will act as an antenna to pick up any electromagnetic radiation that is present. The most common radiation comes from domestic power: power cords, fluorescent lighting, and electric machinery. The resulting interference may be so large as to prevent recording of the EMG. If we were to use an amplifier with a single-ended input, we would see the magnitude of this interference. Figure 9.9 depicts hum interference on the active electrode. It appears as a sinusoidal signal, and if the muscle is contracting, its EMG is added. However, hum could be 100 mV, and would drown out even the largest EMG signal.

If we replace the single-ended amplifier with a differential amplifier (Figure 9.10), we can possibly eliminate most of the hum. Such an amplifier takes the difference between the signals on the active terminals. As can be seen, this hum interference appears as an equal amplitude on both active terminals. Because the body acts as an antenna, all locations pick up the same hum signal. Because this unwanted signal is common to both active terminals, it is called a *common-mode signal.* At terminal 1, the net signal is $V_{\text{hum}} + \text{emg}_1$; at terminal 2 it is $V_{\text{hum}} + \text{emg}_2$. The amplifier has a gain of A. Therefore, the ideal output signal is

$$\begin{aligned} e_o &= A(e_1 - e_2) \\ &= A(V_{\text{hum}} + \text{emg}_1 - V_{\text{hum}} - \text{emg}_2) \\ &= A(\text{emg}_1 - \text{emg}_2) \end{aligned} \tag{9.4}$$

The output e_o is an amplified version of the difference between the EMGs on electrodes 1 and 2. No matter how much hum is present at the individual electrodes, it has been removed by a perfect subtraction within the differential amplifier. Unfortunately a perfect subtraction never occurs, and the measure of how successfully this has been done is given by the common-mode rejection ratio (CMRR). If CMRR is 1000:1, then all but 1/1000 of the hum will be rejected. Thus, the hum at the output is given by

$$V_o(\text{hum}) = \frac{A \times V_{\text{hum}}}{\text{CMRR}} \tag{9.5}$$

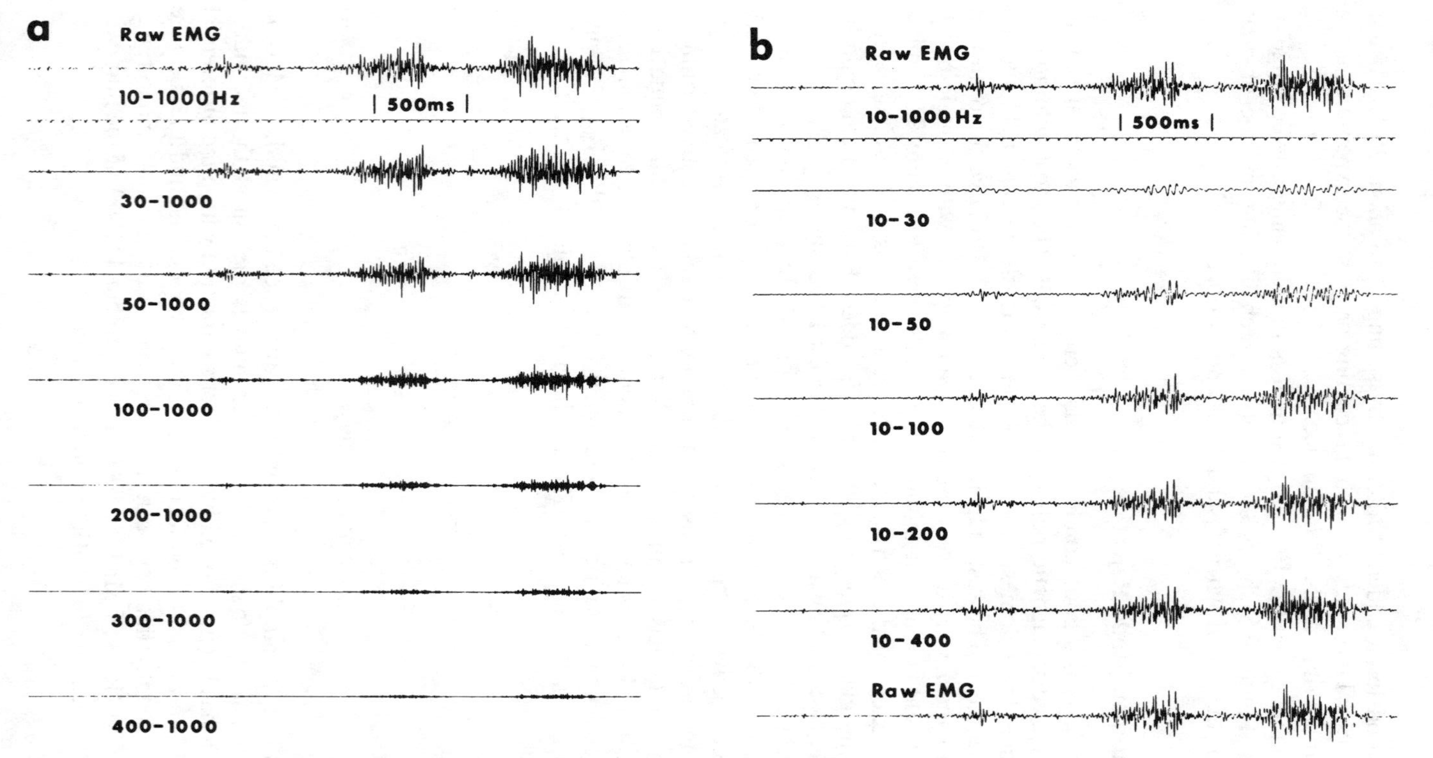

Figure 9.8 Surface EMG filtered with varying cutoff frequencies. (*a*) With lower cutoff frequency varied from 30 to 400 Hz, showing the effect of rejecting lower frequencies. (*b*) With upper cutoff frequency varied from 30 to 400 Hz, showing the loss of higher frequencies.

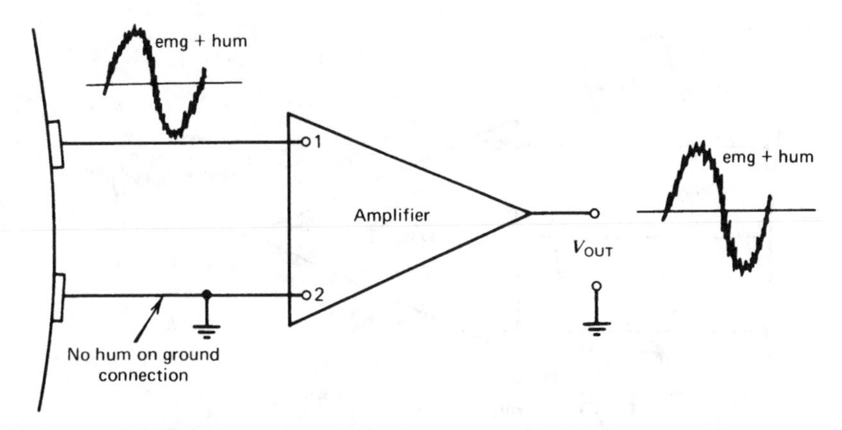

Figure 9.9 Single-ended amplifier showing lack of rejection of hum present on non-grounded active terminal.

Example 9.1. Consider the EMG on the skin to be 2 mV in the presence of hum of 500 mV. CMRR is 10,000:1 and the gain is 2000. Calculate the output EMG and hum.

$$e_o = A(\text{emg}_1 - \text{emg}_2)$$

$$= 2000 \times 2 \text{ MV} = 4 \text{ V}$$

$$\text{output hum} = 2000 \times 500 \text{ mV} \div 10{,}000 = 100 \text{ mV}$$

Hum is always present to some extent unless the EMG is being recorded with battery-powered equipment not in the presence of domestic power. Its magnitude can be seen in the baseline when no EMG is present. Figure 9.11 shows two EMG records, the first with hum quite evident, the latter with negligible hum

CMRR is often expressed as a logarithmic ratio rather than a linear ratio. The units of this ratio are decibels,

$$\text{CMRR (dB)} = 20 \log_{10} \text{CMRR (linear)} \tag{9.6}$$

If CMRR = 10,000:1, then CMRR (dB) = 20 $\log_{10}$ 10,000 = 80 dB. In good-quality biological amplifiers, CMRR should be 80 dB or higher.

9.2.5 Crosstalk in Surface Electromyograms

The detectable pick-up distance for small motor units was estimated to be about 0.5 cm and about 1.5 cm for the largest units (Fuglevand et al., 1992). In Figure 9.12, for example, we see a number of surface electrodes over

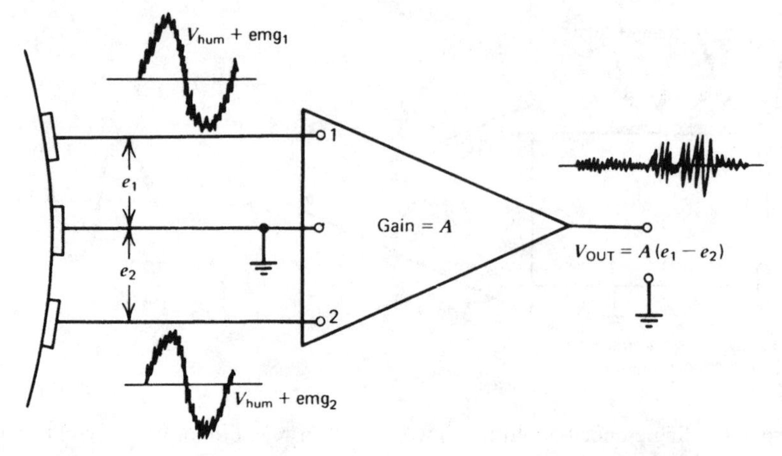

Figure 9.10 Biological amplifier showing how the differential amplifier rejects the common (hum) signal by subtracting the hum signal that is present at an equal amplitude at each active terminal. Different EMG signals are present at each electrode; thus, the subtraction does not result in a cancellation.

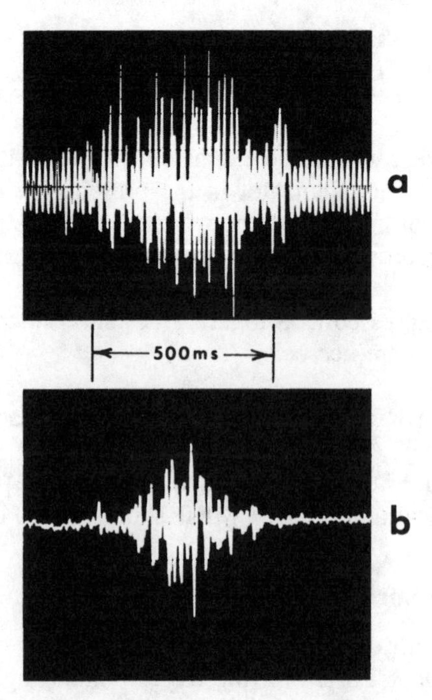

Figure 9.11 Storage oscilloscope recordings of an EMG signal. (*a*) With hum present. (*b*) Without hum.

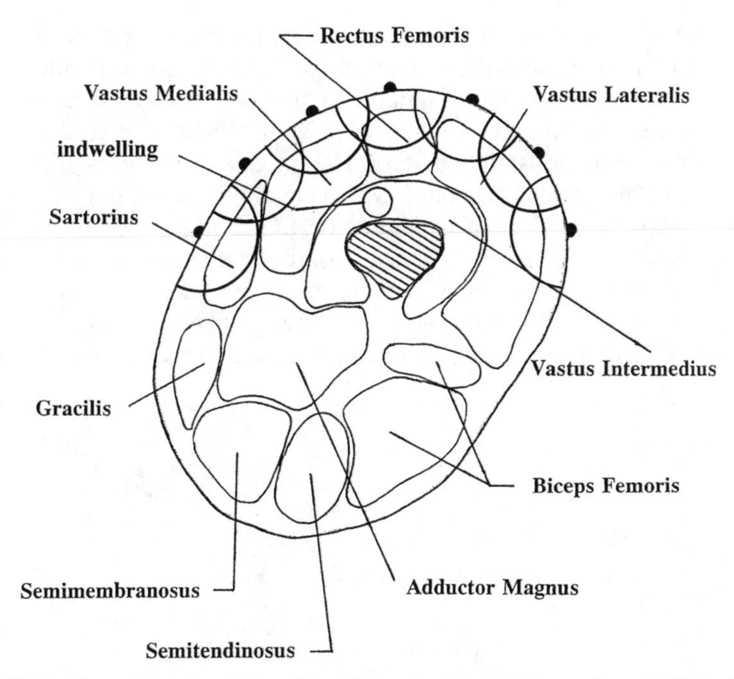

Figure 9.12 Location of seven surface electrodes placed across the midthigh showing the region of pick-up for each site relative to the underlying muscles. The zone of pick-up for each is shown as an arc beneath electrode; there is an overlapping zone between adjacent electrodes that will result in some crosstalk. Deeper muscles, such as the vastus intermedius and adductor magnus, require indwelling electrodes whose zone of pick-up is small because of the small surface area and close spacing of the electrode tips.

several thigh muscles in the midthigh region. The range of pick-up is shown as an arc beneath each electrode. Those m.u.a.p.'s whose fibers are close to each electrode are not subject to crosstalk; however, there is an overlapping zone of pick-up where both electrodes detect m.u.a.p.'s from the same active motor units. This common pick-up is called crosstalk.

Experienced researchers who are knowledgeable about their surface anatomy chose sites that minimize crosstalk. However, because of close proximity of the desired muscles, it may be necessary to check that crosstalk is minimal. The most common way to test for crosstalk is to conduct a manual resistance test prior to any given experiment. Consider two adjacent muscles A and B that, during the course of a given movement, have common periods of activity. The purpose of the manual resistance tests is to get a contraction on muscle A without any activity on B and vice versa. Consider electrodes placed on the leg on three muscles that are fairly close together: tibialis anterior (TA), peroneus longus (PER), and lateral gastrocnemius (LG). Functionally the TA is a dorsiflexor and invertor, PER is an evertor and plantarflexor, and LG is a plantarflexor. The results of a functional test for crosstalk from the PER to the TA and LG is presented in Figure 9.13. Because it is difficult for many subjects to voluntarily create a unique contraction (such as eversion for the PER), it is very helpful for the subject to view the EMG recording and use visual feedback to assist in the test. Here we see two very distinct PER contractions. The first was eversion with a small amount of dorsiflexion: note the minor activity on TA and negligible on LG. The second contraction was eversion with a minor amount of plantarflexion: note the minor activity on the LG and none on TA. It is not critical to be successful on both tests simultaneously, or even on the very first attempt. The student is referred to Winter et al. (1994) for more details and examples of manual resistance tests.

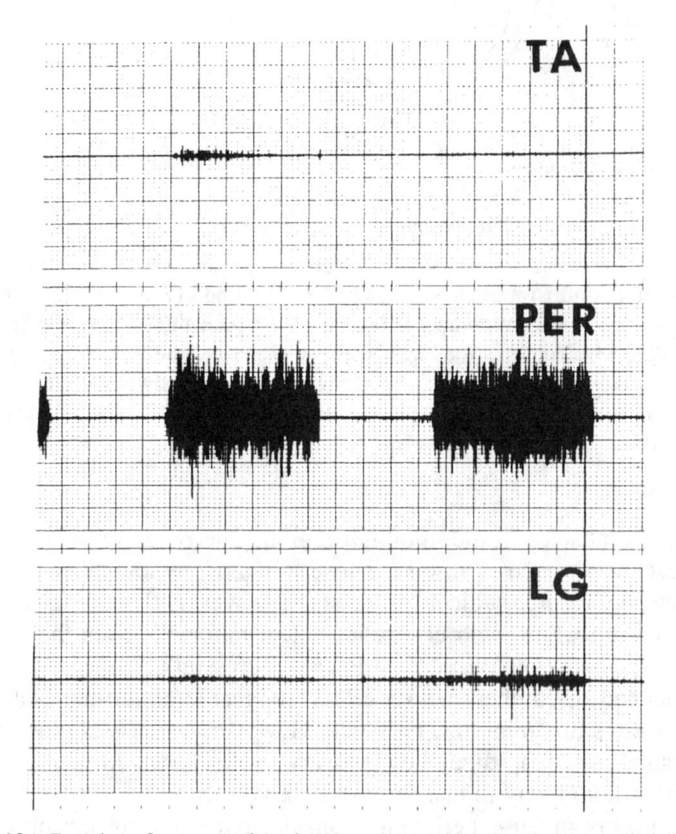

Figure 9.13 Results of a manual resistance test to ensure no crosstalk between the peroneii (PER) muscle and the tibialis anterior (TA) and lateral gastrocnemius (LG) muscles. The first trial showed negligible crosstalk between PER and TA; the second contraction showed negligible crosstalk between PER and LG. See text for further details. (Reprinted from the *Journal of Electromyography and Kinesiology*, Vol. 4, Winter, D. A., A. J. Fuglevand, and S. Archer. "Crosstalk in Surface Electromyography: Theoretical and Practical Estimates," pp. 15–26, 1994, with permission of Elsevier.)

However, there are many situations when a manual resistance test cannot be done and there is a good chance that adjacent channels may have some common EMG signal. The question is how much do they have in common? This question was initially addressed by Winter et al. (1994) using the well-recognized signal-processing technique called cross-correlation. The adjacent channels, $x(t)$ and $y(t)$, are isometrically contracted for about 10 s at a moderate contraction level. The general equation for cross-correlation, $R_{xy}(\tau)$, between signals $x(t)$ and $y(t)$ is

$$R_{xy}(\tau) = \frac{\frac{1}{T}\int_0^T x(t) \cdot y(t - \tau)\, dt}{\sqrt{R_{xx}(0) \cdot R_{yy}(0)}} \tag{9.7}$$

where $x(t)$ and $y(t)$, respectively, are at a phase shift of 0. $R_{xx}(0)$ and $R_{yy}(0)$ are the mean squares of $x(t)$ and $y(t)$, respectively, over the time T. τ is normally varied ± 30 ms, and a typical cross-correlation for two electrode pairs located 2 cm apart is presented in Figure 9.14. Note that the peak correlation is 0.6 at zero phase-shift. Because the electrode pairs were placed parallel with the muscle fibers, each pair was the same distance from the motor point; thus, any common m.u.a.p.'s will be virtually in phase. With the peak $R_{xy} = 0.6$, the net correlation, $R^2_{xy} = 0.36$, telling us the two EMG had

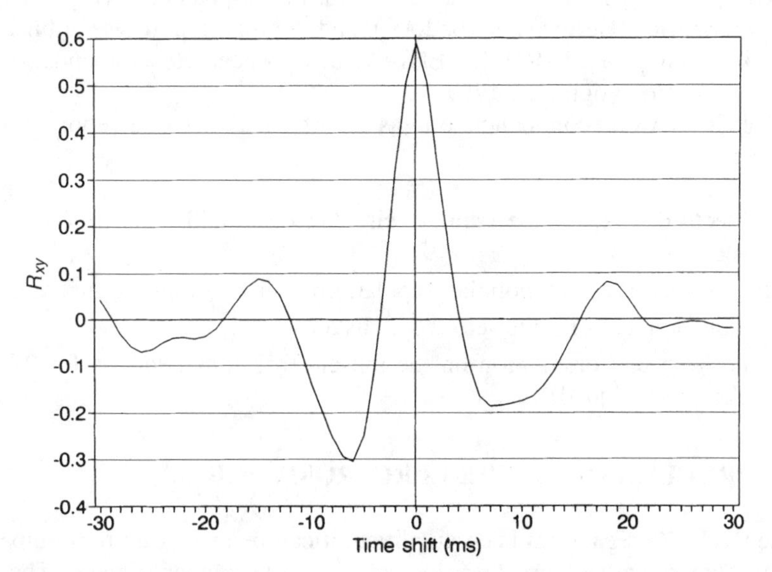

Figure 9.14 Results of a cross-correlation, R_{xy}, between two adjacent electrodes in Figure 9.12. Here $R_{xy} = 0.6$, indicating a common pick-up of $R^2_{xy} = 0.36$ or 36% crosstalk. (Reprinted from the *Journal of Electromyography and Kinesiology*, Vol. 4, Winter, D. A., A. J. Fuglevand, and S. Archer. "Crosstalk in Surface Electromyography: Theoretical and Practical Estimates," pp. 15–26, 1994, with permission of Elsevier.)

36% of their signals in common. R_{xy} drops off sharply as the distance between electrode pairs is increased (Winter et al., 1994). At 2.5 cm, $R_{xy} = 0.48$ (23% crosstalk); at 5.0 cm, $R_{xy} = 0.24$ (6% crosstalk); and at 7.5 cm, $R_{xy} = 0.14$ (2% crosstalk). Reducing the size of the electrodes reduces the pick-up zone of the electrode, and thus the crosstalk. The motor units at a greater distance from the electrodes contribute most to the crosstalk, and the m.u.a.p.'s from those units are longer duration (lower frequency). Thus, by processing the EMG through a differentiator, the higher frequency (closer) m.u.a.p.'s are emphasized, while the lower frequency (further) m.u.a.p.'s are attenuated.

9.2.6 Recommendations for Surface Electromyogram Reporting and Electrode Placement Procedures

Over the past 30 years, the availability of EMG recording equipment and the number of laboratories using such equipment has exploded dramatically. Most labs have developed their own protocols regarding the details needed for reporting results and for selection of the electrode sites over the muscles. The first attempt at developing recommended standards was done by the International Society of Electrophysiological Kinesiology, which resulted in an Ad Hoc Committee Report (Winter et al., 1980). More recently, a considerably more in-depth report was published with the support of the European Union and was called SENIAM (Surface EMG for a Non-Invasive Assessment of Muscles). During 1996–1999, the report was developed and reviewed by over 100 laboratories (Hermens et al., 2000), and the final report was published as a booklet and as a CD-ROM: SENIAM 8: European Recommendations for Surface Electromyography, 1999.

The SENIAM recommendations as to information to be reported are as follows:

1. Electrode size, shape (square, circular, etc.), and material (Ag, AgCl, Ag/AgCl, etc.)
2. Electrode type (monopolar, bipolar, one- or two-dimensional array)
3. Skin preparation, inter-electrode distance
4. Position and orientation on the muscle (recommendations for 27 muscles are included)

9.3 PROCESSING OF THE ELECTROMYOGRAM

Once the EMG signal has been amplified, it can be processed for comparison or correlation with other physiological or biomechanical signals. The need for changing the EMG into another processed form is caused by the fact that the raw EMG may not be suitable for recording or correlation. For example, because of the higher frequencies present in the EMG, it is impossible to record it directly on a pen recorder. The frequency response of most recorders

(0–60 Hz) means that most of the higher frequency components of the EMG are not seen.

The more common types of on-line processing are:

1. Half- or full-wave rectification (the latter also called *absolute value*)
2. Linear envelope detector (half- or full-wave rectifier followed by a low-pass filter)
3. Integration of the full-wave rectified signal over the entire period of muscle contraction
4. Integration of the full-wave rectified signal for a fixed time, reset to zero, then integration cycle repeated
5. Integration of the full-wave rectified signal to a preset level, reset to zero, then integration repeated

These various processing methods are shown schematically in Figure 9.15, together with a sample record of a typical EMG processed all five ways. This is now discussed in detail.

9.3.1 Full-Wave Rectification

The full-wave rectifier generates the absolute value of the EMG, usually with a positive polarity. The original raw EMG has a mean value of zero because it is recorded with a biological amplifier with a low-frequency cutoff around 10 Hz. However, the full-wave rectified signal does not cross through zero and therefore has an average or bias level that fluctuates with the strength of the muscle contraction. The quantitative use of the full-wave rectified signal by itself is somewhat limited; it serves as an input to the other processing schemes. The main application of the full-wave rectified signal is in semi-quantitative assessments of the phasic activity of various muscle groups. A visual examination of the amplitude changes of the full-wave rectified signal gives a good indication of the changing contraction level of the muscle. The proper unit for the amplitude of the rectified signal is the millivolt, the same as for the original EMG.

9.3.2 Linear Envelope

If we filter the full-wave rectified signal with a low-pass filter, we have what is called the *linear envelope.* It can be described best as a moving average because it follows the trend of the EMG and quite closely resembles the shape of the tension curve. It is reported in millivolts. There is considerable confusion concerning the proper name for this signal. Many researchers call it an *integrated EMG* (IEMG). Such a term is quite wrong because it can be confused with the mathematical term *integrated,* which is a different form of processing.

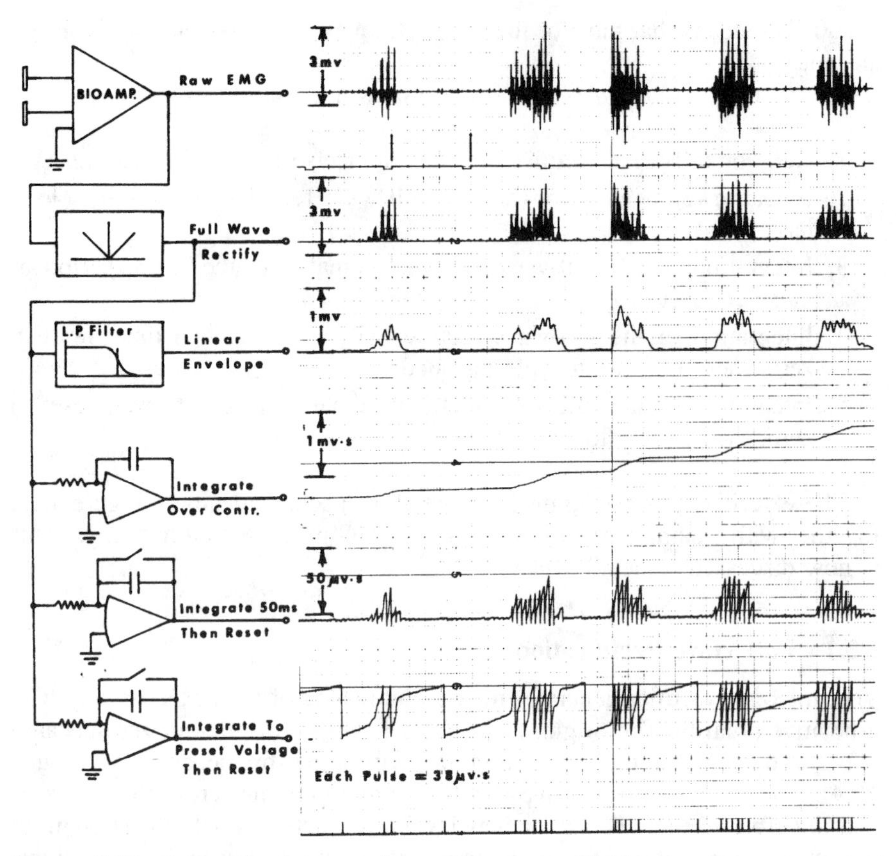

Figure 9.15 Schematic diagram of several common EMG processing systems and the results of simultaneous processing of EMG through these systems. See text for details.

There is a need to process the signal to provide the assessor with a pattern that can be justified on some biophysical basis. Some researchers have full-wave rectified the raw EMG and low-pass filtered at a high frequency but with no physiological basis (Forssberg, 1985; Murray et al., 1985). If it is desired that the linear envelope bear some relationship with the muscle force or the joint moment of force, the processing should model the biomechanics of muscle tension generation. The basic unit of muscle tension is the muscle twitch, and the summation of muscle twitches as a result of recruitment is matched by a superposition of m.u.a.p.'s. There is an inherent delay between the m.u.a.p. and the resultant twitch waveform. If we consider the full-wave rectified signal to be an impulse and the twitch to be the response, we can define the transfer function of the desired processing. The duration of the full-wave rectified m.u.a.p. is about 10 ms, while the twitch waveform peaks at 50–110 ms and lasts up to 300 ms, so this impulse/response relationship is close. Twitch waveforms have been analyzed and have been found to be a second-order system, critically damped or slightly overdamped (Crosby, 1978; Milner-Brown et al., 1973). The cut-off frequency of these second-order responses ranged from 2.3 Hz to 7.8 Hz. Figure 9.16 is presented to show the linear envelope processing of the EMG to model the relationship between the m.u.a.p. and the twitch. A critically damped second-order low-pass filter has a cut-off frequency f_c, which is related to the twitch time T as follows:

$$f_c = 1/2\pi T \tag{9.8}$$

Thus, the table in Figure 9.16 shows the relationship between f_c and T for the range of twitch times reported in the literature. The soleus muscle with a twitch time $\approx$ 106 ms would require a filter with f_c = 1.5 Hz. The reader is referred back to Section 8.0.5 for information on the twitch shape and times.

Good correlations have been reported between the muscle force and linear envelope waveforms during isometric anisotonic contractions (Calvert and Chapman, 1977; Crosby, 1978). Readers are also referred back to Section 8.3.1 and Figure 8.20, which modeled the muscle contraction as a mass-spring-damper system that was critically damped. The critically damped low-pass filters described above have exactly the same response as this previously

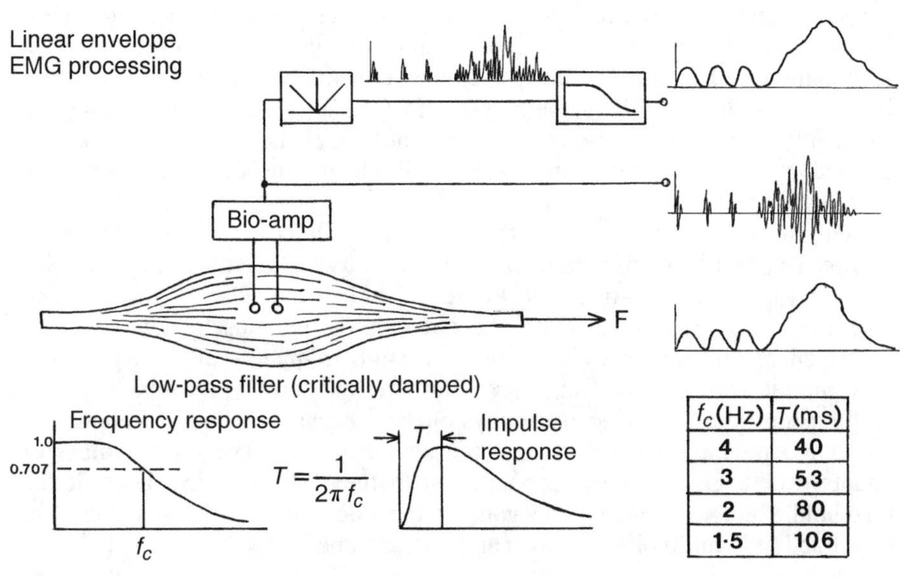

f_c(Hz)	T(ms)
4	40
3	53
2	80
1·5	106

Figure 9.16 Linear envelope processing of the EMG with a critically damped low-pass filter to match the impulse response of the muscle being recorded. The full-wave rectified EMG acts as a series of impulses that, when filtered, mimic the twitch response of the muscle and in a graded contraction mimic the superposition of muscle twitches.

described mechanical system. Thus, the matching of the muscle force waveform with that from the model is exactly the same as would be achieved with a linear envelope detector with the same cut-off frequency as used in the mechanical model.

9.3.3 True Mathematical Integrators

There are several different forms of mathematical integrators as shown in Figure 9.15. The purpose of an integrator is to measure the "area under the curve." Thus, the integration of the full-wave rectified signal will always increase as long as any EMG activity is present. The simplest form starts its integration at some preset time and continues during the time of the muscle activity. At the desired time, which could be a single contraction or a series of contractions, the integrated value can be recorded. The unit of a properly integrated signal is millivolt-seconds (mV · s). The only true way to find the average EMG during a given contraction is to divide the integrated value by the time of the contraction; this will yield a value in millivolts.

A second form of integrator involves a resetting of the integrated signal to zero at regular intervals of time (40–200 ms). Such a scheme yields a series of peaks that represent the trend of the EMG amplitude with time. Each peak represents the average EMG over the previous time interval, and the series of peaks is a "moving" average. Each peak has units of millivolt-seconds so that the sum of all the peaks during a given contraction yields the total integrated signal, as described in the previous paragraph. There is a close similarity between this reset, integrated curve and the linear envelope. Both follow the trend of muscle activity. If the reset time is too high, it will not be able to follow rapid fluctuations of EMG activity, and if it is reset too frequently, noise will be present in the trendline. If the integrated peaks are divided by the integration time, the amplitude of the signal can again be reported in millivolts.

A third common form of integration uses a voltage level reset. The integration begins before the contraction. If the muscle activity is high, the integrator will rapidly charge up to the reset level; if the activity is low, it will take longer to reach rest. Thus, the strength of the muscle contraction is measured by the frequency of the resets. High frequency of reset pulses is high muscle activity; low frequency is low muscle activity. Intuitively such a relationship is attractive to neurophysiologists because it has a similarity to the action potential rate present in the neural system. The total number of counts over a given time is proportional to the EMG activity. Thus, if the threshold voltage level and the gain of the integrater are known, the total EMG activity (millivolt-seconds) can be determined.

9.4 RELATIONSHIP BETWEEN ELECTROMYOGRAM AND BIOMECHANICAL VARIABLES

The major reason for processing the basic EMG is to derive a relationship between it and some measure of muscle function. A question that has been posed for years is: "How valuable is the EMG in predicting muscle tension?" Such a relationship is very attractive because it would give an inexpensive and noninvasive way of monitoring muscle tension. Also, there may be information in the EMG concerning muscle metabolism, power, fatigue state, or contractile elements recruited.

9.4.1 Electromyogram Versus Isometric Tension

Bouisset (1973) has presented an excellent review of the state of knowledge regarding the EMG and muscle tension in normal isometric contractions. The EMG processed through a linear envelope detector has been widely used to compare the EMG–tension relationship, especially if the tension is changing with time. If constant tension experiments are done, it is sufficient to calculate the average of the full-wave rectified signal, which is the same as that derived from a long time-constant linear envelope circuit. Both linear and nonlinear relationships between EMG amplitude and tension have been discovered. Typical of the work reporting linear relationships is an early study by Lippold (1952) on the calf muscles of humans. Zuniga and Simons (1969) and Vredenbregt and Rau (1973), on the other hand, found quite nonlinear relationships between tension and EMG in the elbow flexors over a wide range of joint angles. Both these studies were in effect static calibrations of the muscle under certain length conditions; a reproduction of their results is presented in Figure 9.17.

Another way of representing the level of EMG activity is to count the action potentials over a given period of time. Close et al. (1960) showed a

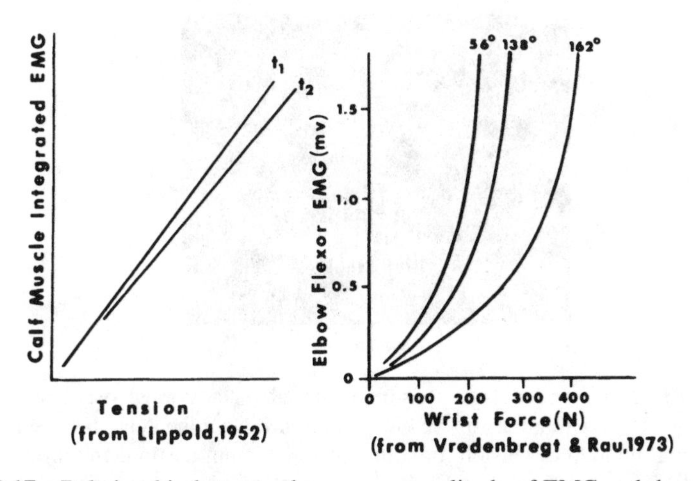

Figure 9.17 Relationship between the average amplitude of EMG and the tension in the muscle in isometric contraction. Linear relationships have been shown by some researchers while others report the EMG amplitude to increase more rapidly than the tension.

linear relationship between the count rate and the integrated EMG, so it was not surprising that the count rate increased with muscle tension in an almost linear fashion.

The relationship between force and linear envelope EMG also holds during dynamic changes of tension. Inman et al. (1952) first demonstrated this by a series of force transducer signals that were closely matched by the envelope EMG from the muscle generating the isometric force. Gottlieb and Agarwal (1971) mathematically modeled this relationship with a second-order low-pass system (e.g., a low-pass filter). Under dynamic contraction conditions, the tension is seen to lag behind the EMG signal, as shown in Figure 9.18. The delay is due to the fact that the twitch corresponding to each m.u.a.p. reaches its peak 40–100 ms afterward. Thus, as each motor unit is recruited, the resulting summation of twitch forces will also have a similar delay behind the EMG.

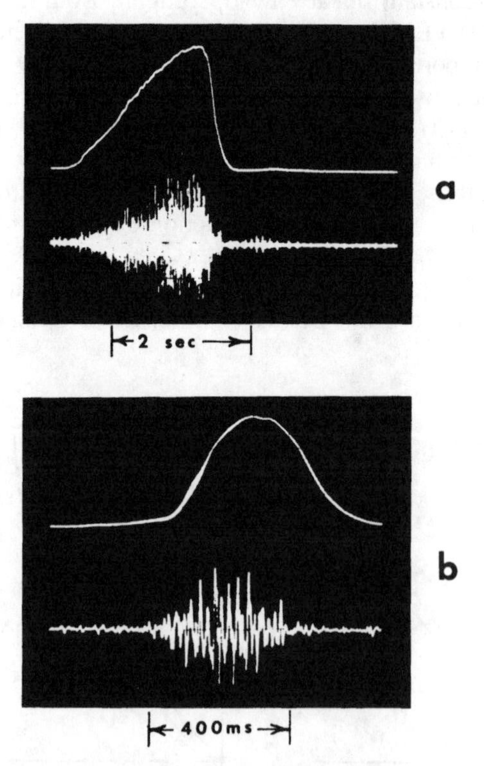

Figure 9.18 EMG and muscle tension recorded on a storage oscilloscope during varying isometric contractions of the biceps muscles. Note the delay between the EMG and the initial buildup of tension, time to reach maximum tension, and drop of tension after the EMG has ceased. (*a*) During a gradual buildup and rapid relaxation. (*b*) During a short 400-ms contraction.

In spite of the reasonably reproducible relationships, the question still arises as to how valid these relationships are for dynamic conditions when many muscles act across the same joint:

1. How does the relationship change with length? Does the length change merely alter the mechanical advantage of the muscle, or does the changing overlap of the muscle fibers affect the EMG itself (Vredenbregt and Rau, 1973)?
2. How do other agonist muscles share the load at that joint, especially if some of the muscles have more than one function (Vredengregt and Rau, 1973)?
3. In many movements, there is antagonist activity. How much does this alter the force being predicted by creating an extra unknown force?

With the present state of knowledge, it appears that a suitably calibrated linear envelope EMG can be used as a coarse predictor of muscle tension for muscles whose length is not changing rapidly.

9.4.2 Electromyogram During Muscle Shortening and Lengthening

In order for a muscle to do positive or negative work, it must also undergo length changes while it is creating tension. Thus, it is important to see how well the EMG can predict tension under these more realistic conditions. A major study has been reported by Komi (1973). In it the subject did both positive and negative work on an isokinetic muscle-testing machine. The subject was asked to generate maximum tension while the muscle lengthened or shortened at controlled velocities. The basic finding was that the EMG amplitude remained fairly constant in spite of decreased tension during shortening and increased tension during eccentric contractions (see Section 8.2.1). Such results support the theory that the EMG amplitude indicates the state of activation of the contractile element, which is quite different from the tension recorded at the tendon. Also, these results combined with later results (Komi et al., 1987) indicate that the EMG amplitude associated with negative work is considerably less than that associated with the same amount of positive work. Thus, if the EMG amplitude is a relative measure of muscle metabolism, such a finding supports the experiments that found negative work to have somewhat less metabolic cost than positive work.

9.4.3 Electromyogram Changes During Fatigue

Muscle fatigue occurs when the muscle tissue cannot supply the metabolism at the contractile element, because of either ischemia (insufficient oxygen) or local depletion of any of the metabolic substrates. Mechanically, fatigue manifests itself by decreased tension, assuming that the muscle activation remains constant, as indicated by a constant surface EMG or stimulation rate. Conversely, the maintenance of a constant tension after onset of fatigue requires increased motor unit recruitment of new motor units to compensate for a decreased firing rate of the already recruited units (Vredenbregt and Rau, 1973). Such findings also indicate that all or some of the motor units are decreasing their peak twitch tensions but are also increasing their contraction times. The net result of these changes is a decrease in tension.

Fatigue not only reduces the muscle force but also may alter the shape of the motor action potentials. It is not possible to see the changes of shape of the individual m.u.a.p.'s in a heavy voluntary contraction. However, an autocorrelation shows an increase in the average duration of the recruited m.u.a.p. (see Section 9.1.4). Also, the EMG spectrum shifts to reflect these duration changes; Kadefors et al. (1973) found that the higher frequency components decreased. The net result is a decrease in the EMG frequency spectrum, which has been attributed to the following:

1. Lower conduction velocity of the action potentials along the muscle fibers below the nonfatigued velocity of 4.5 m/s (Mortimer et al., 1970; Krogh-Lund and Jørgensen, 1991).
2. Some of the larger and faster motor units with shorter duration m.u.a.p.'s dropping out.
3. A tendency for the motor units to fire synchronously, which increases the amplitude of the EMG. Normally each motor unit fires independently of others in the same muscle so that the EMG can be considered to be the summation of a number of randomly timed m.u.a.p.'s. However, during fatigue, a tremor (8–10 Hz) is evident in the EMG and force records. These fluctuations are neurological in origin and are caused by motor units firing in synchronized bursts.

The major EMG measures of fatigue are an increase in the amplitude of the EMG and a decrease in its frequency spectrum. One of those measures is median power frequency, f_m, which is the frequency of the power spectral density function below which half the power lies and above which the other half of the power lies:

$$\int_0^{f_m} X^2(f)\,df = \int_{f_m}^{\infty} X^2(f)\,df = \frac{1}{2}\int_0^{\infty} X^2(f)\,df \tag{9.9}$$

where $X(f)$ is the amplitude of the harmonic at frequency f and $X^2(f)$ is the power at frequency f.

A representative paper that shows the majority of these measures (Krogh-Lund and Jørgensen, 1991) reports a 10% decrease in conduction velocity, a 45% decrease in the median power frequency, and a 250% increase in the root mean square (rms) amplitude. The median power frequency has also been shown to be highly correlated with the amplitude of the low-frequency bandwidth (15–45 Hz) and the ratio of the high-frequency bandwidth (>95 Hz) to the low-frequency bandwidth (Allison and Fujiwara, 2002). An alternate and very common statistical measure (Öberg et al., 1994) is mean power frequency (MPF):

$$\text{MPF} = \frac{\int_0^F f \cdot X^2(f)\, df}{\int_0^F X^2(f)\, df} \text{ Hz} \tag{9.10}$$

where F is the maximum frequency analyzed.

There is virtually no difference in these two frequency measures and their correlations with independent measures of fatigue (Kerr and Callaghan, 1999).

9.5 REFERENCES

Allison, G. T. and T. Fujiwara. "The Relationship Between EMG Median Frequency and Low Frequency Band Amplitude Changes at Different Levels of Muscle Capacity," *Clin. Biomech.* **17**:464–469, 2002.

Andreassen, S. and A. Rosenfalck. "Relationship of Intracellular and Extracellular Action Potentials of Skeletal Muscle Fibers," *CRC Crit. Rev. Bioeng.* **6**:267–306.

Basmajian, J. V. "Electrodes and Electrode Connectors," in *New Developments in Electromyography and Clinical Neurophysiology,* Vol. 1, J. E. Desmedt, Ed. (Karger, Basel, Switzerland, 1973), pp. 502–510.

Bouisset, S. "EMG and Muscle Force in Normal Motor Activities," in *New Developments in Electromyography and Clinical Neurophysiology,* Vol. 1, J. E. Desmedt, Ed. (Karger, Basel, Switzerland, 1973), pp. 547–583.

Boyd, D. C., P. D. Lawrence, and P. J. A. Bratty. "On Modelling the Single Motor Unit Action Potential," *IEEE Trans. Biomed. Eng.* **BME-25**:236–242, 1978.

Buchthal, F., C. Guld, and P. Rosenfalck. "Propagation Velocity in Electrically Activated Fibers in Man," *Acta Physiol. Scand.* **34**:75–89, 1955.

Buchthal, F., F. Erminio, and P. Rosenfalck. "Motor Unit Territory in Different Human Muscles," *Acta Physiol. Scand.* **45**:72–87, 1959.

Calvert, T. W. and A. B. Chapman. "Relationship Between Surface EMG and Force Transients in Muscle: Simulation and Experimental Results," *Proc. IEEE* **65**:682–689, 1977.

Close, J. R., E. D. Nickel, and A. B. Todd. "Motor Unit Action Potential Counts," *J. Bone Jt. Surg.* **42-A**:1207–1222, 1960.

Crosby, P. A. "Use of Surface Electromyography as a Measure of Dynamic Force in Human Limb Muscles," *Med. Biol. Eng. Comput.* **16**:519–524, 1978.

DeLuca, C. J. "Precision Decomposition of EMG Signals," *Methods Clin. Neurophysiol.* **4**:1–28, 1993.

DeLuca, C. J., R. S. LeFever, M. P. McCue, and A. P. Xenakis. "Control Scheme Governing Concurrently Active Motor Units during Voluntary Contractions," *J. Physiol.* **329**:129–142, 1982.

Erim, Z., C. J. DeLuca, K. Mineo, and T. Aoki. "Rank-Ordered Regulation of Motor Units," *Muscle & Nerve* **19**:563–573, 1996.

Forssberg, H. "Ontogeny of Human Locomotor Control," *Exp. Brain Res.* **57**:480–493, 1985.

Fuglevand, A. J., D. A. Winter, A. E. Patla, and D. Stashuk. "Detection of Motor Unit Action Potentials with Surface Electrodes: Influence of Electrode Size and Spacing," *Biol. Cybernetics* **67**:143–153, 1992.

Gersten, J. W., F. S. Cenkovich, and G. D. Jones. "Harmonic Analysis of Normal and Abnormal Electromyograms," *Am. J. Phys. Med.* **44**:235–240, 1965.

Gottlieb, G. L., and G. C. Agarwal. "Dynamic Relationship between Isometric Muscle Tension and the Electromyogram in Man," *J. Appl. Physiol.* **30**:345–351, 1971.

Hermens, H. J., B. Freriks, C. Disselhorst-Klug, and G. Rau. "Development of Recommendations for SEMG Sensors and Sensor Placement Procedures," *J. Electromyogr. Kinesiol.* **10**:361–374, 2000.

Inman, V. T., H. J. Ralston, J. B. Saunders, B. Feinstein, and E. W. Wright. "Relation of Human Electromyogram to Muscular Tension," *Electroencephalogr. Clin. Neurophysiol.* **4**:187–194, 1952.

Johansson, S., L. E. Larsson, and R. Ortengren. "An Automated Method for the Frequency Analysis of Myoelectric Signals Evaluated by an Investigation of the Spectral Changes Following Strong Sustained Contractions," *Med. Biol. Eng.* **8**:257–264, 1970.

Kadaba, M. P., M. E. Wootten, J. Gainey, and G. V. B. Cochran. "Repeatability of Phasic Muscle Activity: Performance of Surface and Intramuscular Wire Electrodes in Gait Analysis," *J. Orthop. Res.* **3**:350–359, 1985.

Kadefors, R. "Myo-electric Signal Processing as an Estimation Problem," in *New Developments in Electromyography and Clinical Neurophysiology,* Vol. 1, J. E. Desmedt, Ed. (Karger, Basel, Switzerland, 1973), pp. 519–532.

Kadefors, R., I. Petersen, and H. Broman. "Spectral Analysis of Events in the Electromyogram," in *New Developments in Electromyography and Clinical Neurophysiology,* Vol. 1, J. E. Desmedt, Ed. (Karger, Basel, Switzerland, 1973), pp. 628–637.

Kerr, D. and J. P. Callaghan. "Establishing a Relationship Between Spectral Indicators of Fatigue and Ratings of Perceived Discomfort," *Proc. 31st Conf. Human Factors Assoc. Canada:* 301–307, 1999.

Komi, P. V. "Relationship Between Muscle Tension, EMG, and Velocity of Contraction under Concentric and Eccentric Work," in *New Developments in Electromyography and Clinical Neurophysiology,* Vol. 1, J. E. Desmedt, Ed. (Karger, Basel, Switzerland, 1973), pp. 596–606.

Komi, P. V. and E. R. Buskirk. "Reproducibility of Electromyographic Measures with Inserted Wire Electrodes and Surface Electrodes," *Electromyography* **10**:357–367, 1970.

Komi, P. V., M. Kaneko, and O. Aura. "EMG Activity of the Leg Extensor Muscles with Special Reference to Mechanical Efficiency in Concentric and Eccentric Exercise," *Int. J. Sports Med.* **8**:22–29, Suppl., 1987.

Krogh-Lund, C. and K. Jørgensen. "Changes in Conduction Velocity, Median Frequency, and Root-Mean-Square Amplitude of the Electromyogram During 25% Maximal Voluntary Contraction of the Triceps Brachii Muscle, to Limit of Endurance," *Eur. J. Physiol.* **63**:60–69, 1991.

Kwatny, E., D. H. Thomas, and H. G. Kwatny. 'An Application of Signal Processing Techniques to the Study of Myoelectric Signals," *IEEE Trans. Biomed. Eng.* **BME-17**:303–312, 1970.

Lippold, O. C. J. "The Relationship Between Integrated Action Potentials in a Human Muscle and its Isometric Tension," *J. Physiol.* **177**:492–499, 1952.

Lorente de No, R. "A Study of Nerve Physiology: Analysis of the Distribution of Action Currents of Nerve in Volume Conductors," *Studies from Rockefeller Inst. Med. Res.* **132**:384–477, 1947.

McGill, K. C., L. J. Dorfman, J. E. Howard, and E. V. Valaines. "Decomposition Analysis of the Surface Electromyogram," in *Proc. 9th Ann. Conf. IEEE Eng. Med. Biol. Soc.,* pp. 2001–2003, 1987.

Milner-Brown, H. S. and R. B. Stein. "The Relation between Surface Electromyogram and Muscular Force," *J. Physiol.* **246**:549–569, 1975.

Milner-Brown, H. S., R. B. Stein, and R. Yemm. "Contractile Properties of Human Motor Units During Voluntary Isometric Contractions," *J. Physiol.* **228**:285–306, 1973.

Mortimer, J. T., R. Magnusson, and I. Petersen. "Conduction Velocity in Ischemic Muscle: Effect on EMG Frequency Spectrum," *Am. J. Physiol.* **219**:1324–1329, 1970.

Murray, M. P., G. B. Spurr, S. B. Sepic, G. M. Gardner, and L. A. Mollinger. "Treadmill vs. Floor Walking: Kinematics, Electromyogram and Heart Rate," *J. Appl. Physiol.* **59**:87–91, 1985.

Öberg, T., L. Sandsjö, and R. Kadefors. "EMG Mean Power Frequency: Obtaining a Reference Value," *Clin. Biomech.* **9**:253–257, 1994.

Person, R. S. and L. N. Mishin. "Auto and Crosscorrelation Analysis of the Electrical Activity of Muscles," *Med. Biol. Eng.* **2**:155–159, 1964.

Plonsey, R. "Volume Conductor Fields of Action Currents," *Biophys. J.* **4**:317–328, 1964.

Plonsey, R. "The Active Fiber in a Volume Conductor," *IEEE Trans. Biomed. Eng.* **BME-21**:371–381, 1974.

Rosenfalck, P. "Intra and Extracellular Potential Fields of Active Nerve and Muscle Fibres," *Acta Physiol. Scand.* **321**:1–165, Suppl., 1969.

Vredenbregt, J. and G. Rau. "Surface Electromyography in Relation to Force, Muscle Length and Endurance," in *New Developments in Electromyography and Clinical Neurophysiology,* Vol. 1, J. E. Desmedt, Ed. (Karger, Basel, Switzerland, 1973), pp. 607–622.

Winter, D. A., A. J. Fuglevand, and A. J. Archer. "Crosstalk in Surface Electromyography: Theoretical and Practical Estimates," *J. Electromyogr. Kinesiol.* **4**:15–26, 1994.

Winter, D. A., G. Rau, R. Kadefors, R. Broman, and C. DeLuca. "Units, Terms and Standards in Reporting EMG research," *Report of an AD Hoc Committee of the Internat. Soc. Electrophysiol. Kinesiol.,* Aug. 1980.

Zuniga, E. N. and D. G. Simons. "Non-linear Relationship between Averaged Electromyogram Potential and Muscle Tension in Normal Subjects," *Arch. Phys. Med.* **50**:613–620, 1969.

APPENDIX A

KINEMATIC, KINETIC, AND ENERGY DATA

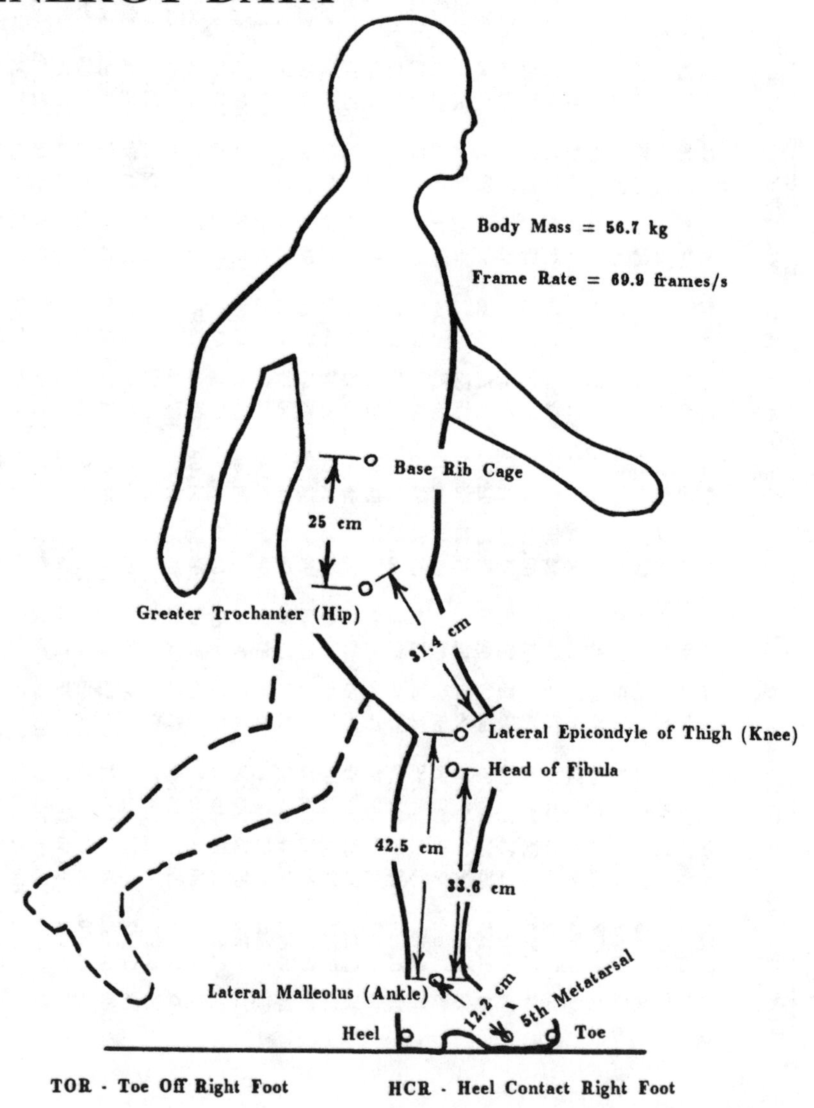

Figure A.1 Walking trial—marker locations and mass and frame rate information.

TABLE A.1 Raw Coordinate Data (cm)

FRAME	TIME	BASE RIB CAGE		RIGHT HIP		RIGHT KNEE		RIGHT FIBULA		RIGHT ANKLE		RIGHT HEEL		RIGHT METAT.		RIGHT TOE	
	S	X	Y	X	Y	X	Y	X	Y	X	Y	X	Y	X	Y	X	Y
1	0.000	46.98	104.41	44.94	78.58	41.00	47.40	35.91	40.53	9.31	21.44	2.95	24.24	7.53	9.35	11.73	3.63
2	0.014	49.22	104.79	47.31	78.58	45.02	46.89	40.06	40.02	12.70	22.46	7.23	26.02	10.54	10.63	13.72	4.77
3	0.029	51.10	105.17	49.57	78.71	48.68	47.27	44.23	40.15	16.49	23.73	10.64	27.30	14.20	12.03	17.63	6.43
4	0.043	53.13	105.30	51.74	79.21	52.50	47.53	48.43	40.15	20.81	24.37	14.71	27.55	18.78	12.53	22.47	6.94
5	0.057	54.86	105.43	53.33	79.09	56.13	47.91	52.70	40.78	24.96	24.24	18.72	27.42	23.17	12.66	27.12	7.32
6	0.072	56.81	106.06	55.41	79.98	59.87	48.67	56.56	41.29	29.33	24.62	23.09	27.17	28.31	12.41	32.51	6.81
7	0.086	58.25	106.32	56.73	80.49	63.34	49.44	60.29	41.80	33.57	23.73	26.95	26.02	33.44	12.03	38.02	6.81
8	0.100	60.03	106.95	58.89	81.00	66.90	50.84	64.36	42.69	38.78	23.22	31.27	25.01	39.55	11.26	44.38	6.30
9	0.114	61.56	107.08	60.79	81.12	69.96	51.09	67.79	43.20	43.11	22.21	35.73	22.84	45.01	10.24	50.36	6.04
10	0.129	63.54	107.46	62.78	82.01	73.22	51.73	71.31	43.84	48.15	21.06	40.64	20.93	51.46	8.97	57.05	5.28
11	0.143	65.20	107.85	64.69	82.40	76.27	53.00	74.74	44.86	53.23	20.04	45.73	19.02	57.56	8.21	63.54	5.15
12	0.157	66.92	107.85	66.67	82.78	79.01	53.51	77.86	45.11	58.27	18.52	51.27	16.86	63.99	7.44	70.23	5.03
13	0.172	68.77	107.59	68.65	82.65	81.62	54.15	80.73	45.49	63.68	16.86	56.56	14.44	70.30	6.30	76.79	4.52
14	0.186	70.37	107.59	70.88	83.16	83.99	54.53	83.48	45.87	69.22	15.59	62.23	12.66	76.73	5.79	83.48	4.77
15	0.200	72.43	107.59	73.19	82.90	86.56	54.78	86.30	46.00	74.72	14.44	68.23	11.01	83.63	5.66	90.12	5.54
16	0.215	74.00	107.59	74.89	83.03	88.51	54.91	88.89	46.13	80.36	13.43	74.13	9.10	89.53	5.28	96.14	5.54
17	0.229	75.68	107.59	76.95	82.90	90.57	54.91	90.95	45.87	85.86	12.66	80.01	7.57	95.53	5.79	102.40	6.55
18	0.243	77.67	107.08	79.19	82.65	93.06	54.65	93.32	46.00	91.54	12.03	86.32	6.55	101.72	5.79	108.84	7.70
19	0.257	79.55	106.95	81.20	82.40	94.56	54.02	95.45	45.62	96.85	11.64	92.02	5.41	107.79	6.17	114.16	8.97
20	0.272	81.47	107.21	83.12	82.14	96.10	54.02	96.99	45.24	101.45	11.77	97.50	5.15	113.28	7.70	119.39	10.88
21	0.286	83.53	106.45	85.69	81.50	98.16	53.25	99.18	44.73	106.56	11.77	103.13	4.77	118.78	8.21	124.38	12.15
22	0.300	85.86	105.68	87.77	80.49	99.73	51.86	101.13	43.84	110.68	11.52	108.39	4.14	123.15	9.35	128.24	13.30
23	0.315	87.74	105.81	89.91	80.61	101.36	51.73	103.14	43.71	114.85	12.15	112.43	4.90	126.94	10.88	131.77	15.33
24	0.329	90.34	105.17	92.25	80.49	103.57	51.09	105.36	43.58	118.33	12.15	116.30	5.28	130.55	11.39	135.39	16.23
25	0.343	92.25	104.79	94.41	80.10	105.23	50.96	107.26	43.46	120.75	12.28	118.84	4.65	133.22	11.64	137.80	16.73
26	0.357	94.36	104.16	97.03	79.72	107.59	50.84	109.50	42.69	122.99	12.03	121.34	4.14	135.21	11.39	139.79	16.23
27	0.372	96.57	103.77	99.37	79.98	109.93	50.58	111.58	42.95	124.31	11.90	122.14	3.75	136.27	10.37	140.85	15.21
28	0.386	98.73	103.52	101.53	79.34	112.34	50.46	113.87	41.80	124.94	10.63	122.65	3.25	137.16	8.97	141.99	13.68
29	0.400	101.40	103.26	104.45	79.60	114.89	50.33	116.28	41.80	125.83	9.61	123.03	3.50	137.66	7.95	142.88	11.90
30	0.415	103.60	103.65	106.78	79.60	116.96	50.46	118.11	41.80	127.02	9.74	123.20	4.14	138.34	6.30	144.45	10.24
31	0.429	105.94	103.65	108.86	79.60	118.53	50.33	119.81	41.55	127.82	9.48	123.12	4.01	138.77	5.41	145.13	8.59
32	0.443	108.22	103.52	111.28	80.10	120.44	50.58	121.71	42.18	128.33	9.61	123.24	4.14	139.27	4.65	145.38	6.81
33	0.458	110.75	104.03	113.80	80.49	122.84	50.96	124.24	42.44	129.07	9.35	123.60	4.14	139.51	4.14	146.38	6.30

34	0.472	112.91	104.03	115.58	81.38	125.25	51.86	126.14	42.57	129.45	9.61	123.98	4.39	139.63	3.37	146.25	5.41
35	0.486	115.20	104.28	117.36	81.89	127.54	51.86	128.05	42.44	129.83	9.74	123.98	4.26	140.01	3.63	147.14	5.41
36	0.500	116.74	104.54	118.78	81.63	128.32	51.86	128.83	42.69	129.46	9.10	123.48	4.14	139.39	3.75	146.52	4.77
37	0.515	119.05	105.43	120.96	81.63	129.87	51.86	130.13	42.95	129.74	9.35	123.89	4.65	139.80	3.63	146.54	4.90
38	0.529	121.31	105.30	123.10	82.27	130.86	51.60	131.49	42.18	130.10	9.23	124.24	4.52	140.02	3.75	147.27	4.52
39	0.543	123.28	106.19	124.93	83.03	132.18	52.36	132.31	42.82	130.53	9.61	124.55	4.90	140.45	3.75	147.33	5.28
40	0.558	125.03	106.19	126.94	82.78	133.04	51.35	132.79	42.44	130.37	9.35	124.64	4.26	140.55	3.50	147.04	4.77
41	0.572	127.21	106.83	128.61	83.03	133.58	51.73	132.81	42.44	130.27	9.61	124.67	4.77	139.94	3.63	147.19	4.52
42	0.586	128.71	106.95	130.24	83.03	134.31	51.35	133.17	43.20	129.98	8.97	123.75	4.77	139.91	3.37	146.91	4.77
43	0.601	130.90	107.34	132.04	82.90	134.72	50.96	133.70	42.57	130.39	9.61	124.54	4.14	140.19	3.50	147.31	4.52
44	0.615	132.52	107.97	133.03	83.41	135.70	52.62	134.56	43.20	130.23	9.61	124.12	4.52	139.90	4.01	146.78	4.26
45	0.629	134.48	107.85	134.86	83.54	136.26	51.86	134.73	42.95	129.89	9.48	123.91	4.77	139.82	3.63	146.95	4.26
46	0.643	136.33	108.10	136.20	83.80	137.09	51.86	135.18	43.07	130.22	10.24	124.37	4.90	140.02	3.37	147.02	4.77
47	0.658	138.33	108.23	138.08	82.90	137.95	51.47	135.66	42.69	130.06	9.99	124.08	4.77	140.11	3.75	146.99	4.26
48	0.672	140.14	108.48	139.38	82.90	138.74	51.60	136.19	42.69	130.47	10.12	124.36	5.03	140.01	3.25	147.01	4.01
49	0.686	142.20	107.97	140.67	82.78	139.15	51.09	136.73	42.57	130.49	9.99	124.64	5.28	140.16	3.12	147.04	4.52
50	0.701	144.05	107.46	142.02	82.65	140.11	51.09	137.05	42.57	130.56	9.99	124.20	4.90	139.60	3.63	146.85	4.01
51	0.715	146.01	107.08	143.85	82.01	140.92	50.58	137.87	42.06	130.49	10.24	124.76	4.77	140.03	3.50	147.28	3.75
52	0.729	148.58	107.46	146.04	82.40	142.22	50.84	138.66	42.44	131.27	10.24	125.17	5.28	140.18	3.63	147.18	4.14
53	0.744	149.92	106.57	147.51	81.63	143.05	50.58	139.23	41.93	130.96	10.12	125.24	5.03	140.25	3.37	147.25	4.01
54	0.758	152.34	106.32	148.90	81.50	143.69	50.71	140.12	42.57	131.47	10.63	125.49	5.66	140.50	3.37	147.50	3.88
55	0.772	153.94	105.81	150.50	81.12	144.27	50.46	140.45	42.95	131.03	10.63	124.80	6.17	140.07	2.99	146.94	3.75
56	0.786	156.15	105.30	152.46	81.00	145.59	50.71	141.52	42.31	131.34	10.63	125.61	5.92	140.50	3.37	147.24	3.88
57	0.801	158.65	104.92	154.57	81.00	147.07	50.58	142.61	42.18	131.80	10.75	125.94	7.06	140.70	3.50	147.45	3.75
58	0.815	160.50	104.66	156.30	80.61	148.41	50.71	143.83	43.20	132.12	12.15	125.89	7.70	140.65	4.01	147.39	4.14
59	0.829	162.61	104.41	157.90	80.10	149.75	50.33	144.66	42.31	131.94	12.03	125.96	7.83	140.34	3.75	146.95	3.75
60	0.844	164.58	104.03	159.88	80.23	151.22	50.33	146.26	42.82	132.26	12.66	126.03	8.46	140.41	3.75	147.02	3.88
61	0.858	166.66	103.77	162.08	80.23	153.30	50.33	147.96	42.69	132.69	12.92	126.58	9.61	140.20	4.26	146.69	4.26
62	0.872	168.90	104.03	164.19	79.60	155.41	50.07	150.06	42.57	133.39	13.93	126.90	10.50	140.01	3.88	146.76	4.01
63	0.887	171.00	103.90	166.42	79.85	157.01	50.07	152.30	42.69	133.97	14.83	127.48	12.28	140.21	3.88	146.83	3.75
64	0.901	173.53	103.14	169.07	78.96	160.29	49.06	154.95	42.06	135.48	14.95	128.73	13.04	140.95	4.14	147.06	3.50
65	0.915	175.81	103.52	171.74	79.47	163.22	49.06	157.62	42.31	136.75	16.48	129.62	14.83	141.08	4.65	147.18	3.75
66	0.929	179.23	103.14	175.16	78.83	166.89	48.55	161.03	41.04	139.14	17.37	132.15	16.10	141.94	5.28	148.31	2.99
67	0.944	181.49	103.14	177.80	78.58	170.04	48.16	164.69	41.42	140.90	18.64	134.02	18.52	142.42	5.79	148.40	2.86
68	0.958	183.65	103.77	180.60	79.09	174.11	48.04	168.25	41.55	143.69	19.79	136.95	21.57	143.69	7.57	149.04	3.63
69	0.972	184.86	103.39	182.44	77.94	176.97	47.02	171.50	40.15	145.54	20.42	139.43	23.35	144.14	7.95	148.59	3.50

TABLE A.1 *(Continued)*

FRAME	TIME	BASE RIB CAGE		RIGHT HIP		RIGHT KNEE		RIGHT FIBULA		RIGHT ANKLE		RIGHT HEEL		RIGHT METAT.		RIGHT TOE	
	S	X	Y	X	Y	X	Y	X	Y	X	Y	X	Y	X	Y	X	Y
70	0.987	187.71	104.28	185.67	78.83	181.73	47.53	176.00	40.15	149.53	21.82	143.81	25.13	146.86	9.74	150.17	4.26
71	1.001	188.82	104.16	187.29	78.20	184.36	46.89	179.40	39.51	152.68	22.21	146.95	25.90	149.11	10.75	151.91	4.14
72	1.015	191.13	104.92	189.48	78.83	188.33	47.27	184.00	39.89	156.39	23.73	150.54	27.17	153.34	12.03	156.52	5.66
73	1.030	192.42	105.55	190.89	79.09	191.78	47.66	187.71	40.53	159.84	24.24	153.73	27.68	157.17	12.28	160.73	5.92
74	1.044	194.40	106.19	193.00	79.60	195.29	48.16	191.47	41.04	163.99	24.37	157.50	27.93	161.82	12.66	165.38	6.43
75	1.058	196.18	105.94	195.16	79.72	199.10	48.93	195.79	41.42	168.44	24.50	161.95	27.04	167.29	12.15	170.98	6.68
76	1.072	197.69	106.06	196.42	80.10	202.66	49.31	199.47	41.55	172.50	23.61	165.88	25.77	171.99	10.88	176.57	6.17
77	1.087	199.57	106.70	198.68	81.00	206.19	50.33	203.39	42.44	177.30	22.84	170.56	25.01	177.81	10.12	182.65	6.17
78	1.101	201.17	107.21	200.53	81.63	209.57	51.22	206.90	43.20	182.08	21.95	175.08	23.10	183.74	9.61	189.21	4.90
79	1.115	202.99	107.59	202.48	82.14	212.66	52.11	210.63	43.84	187.09	21.06	179.83	21.57	189.76	8.97	195.61	4.77
80	1.130	204.67	107.59	204.80	82.65	215.74	52.62	213.83	44.47	192.20	19.66	184.95	19.28	196.02	7.19	201.75	4.39
81	1.144	206.27	107.46	206.78	82.52	218.49	53.13	216.96	44.47	197.36	17.88	190.11	16.73	202.45	6.81	208.94	3.75
82	1.158	208.30	107.72	209.07	83.03	221.28	53.89	219.88	45.24	202.96	16.86	195.58	14.57	209.19	5.41	215.94	4.26
83	1.173	210.18	107.72	211.07	83.03	223.80	54.02	222.91	45.24	208.15	15.59	201.27	12.53	215.14	4.90	222.65	3.50
84	1.187	211.95	107.46	212.97	82.78	226.20	54.40	225.70	45.49	213.86	13.68	207.37	10.12	221.88	4.39	229.13	3.88
85	1.201	213.83	107.21	215.10	82.90	228.59	54.65	228.21	45.49	219.43	12.79	213.45	8.59	228.72	3.75	235.59	4.90
86	1.215	215.63	107.08	217.03	82.65	230.52	54.65	230.52	45.11	224.92	11.39	219.32	6.68	234.84	4.26	241.72	5.28
87	1.230	217.58	106.70	219.11	82.65	232.85	54.65	232.60	45.24	230.94	11.39	225.85	5.66	241.38	4.52	247.74	6.30
88	1.244	219.36	106.45	220.89	82.27	234.63	53.64	234.76	44.47	236.28	11.01	232.09	4.52	247.86	4.52	254.10	7.32
89	1.258	221.06	105.94	222.71	81.63	236.07	53.13	236.84	44.35	241.80	10.37	237.35	4.01	253.00	6.05	259.36	9.48
90	1.273	223.53	105.55	225.18	81.12	238.29	52.75	239.18	43.46	247.07	10.63	243.89	3.63	259.41	7.06	265.01	11.26
91	1.287	225.31	104.92	227.60	81.00	239.81	51.73	241.47	43.33	252.03	11.01	249.23	3.63	264.24	8.84	269.33	12.92
92	1.301	227.33	104.66	229.75	80.10	241.33	50.84	243.36	42.82	255.83	11.01	253.54	3.63	268.31	9.86	273.40	14.32
93	1.316	229.28	104.41	231.57	79.98	243.03	50.46	245.32	42.82	259.31	11.26	257.53	4.14	271.78	10.75	276.49	15.84
94	1.330	231.49	104.03	234.16	79.60	245.23	50.33	247.40	42.82	262.03	12.15	260.50	4.01	274.75	11.26	278.95	16.48
95	1.344	233.80	103.52	236.47	79.21	247.67	49.82	249.57	42.31	263.95	10.88	262.55	3.63	276.42	10.88	281.39	15.72
96	1.358	235.62	103.01	238.80	78.83	250.00	49.82	251.91	41.68	265.27	10.24	263.49	3.12	277.74	9.99	282.70	14.70
97	1.373	238.04	103.01	241.60	79.21	252.80	50.33	254.32	42.06	266.41	9.86	264.12	2.99	278.63	8.59	283.72	13.43
98	1.387	240.34	103.14	244.29	79.21	255.23	50.20	256.50	41.55	267.06	9.23	264.14	3.37	279.15	7.44	284.50	11.39
99	1.401	242.88	103.01	246.83	79.09	257.39	50.07	258.92	41.42	268.72	9.48	264.52	3.63	280.42	6.05	286.15	9.10
100	1.416	245.09	102.63	249.16	79.34	259.60	50.07	260.74	41.68	269.78	9.23	264.69	3.63	281.10	4.90	287.08	7.95
101	1.430	247.53	103.26	251.60	79.72	261.90	50.96	263.30	42.57	270.18	9.48	265.21	4.26	281.25	4.39	287.86	6.68
102	1.444	249.91	102.88	253.73	80.36	264.42	50.96	265.95	42.44	270.78	9.10	265.18	3.63	281.47	3.50	288.09	5.54
103	1.459	251.81	103.39	255.63	81.00	266.70	51.60	267.59	42.82	271.15	9.23	265.30	4.39	281.84	3.63	288.20	5.15
104	1.473	253.96	104.16	257.65	81.38	268.47	51.47	269.23	42.69	271.40	9.23	265.16	3.88	281.96	3.25	288.45	5.03
105	1.487	256.07	104.28	259.51	81.25	269.56	51.35	270.19	42.31	271.34	8.59	265.36	3.50	281.52	2.74	288.14	4.39
106	1.501	257.61	105.05	260.92	81.63	270.34	51.60	270.85	42.44	270.85	8.84	264.99	3.88	281.15	2.35	287.77	4.52

TABLE A.2(*a*) Filtered Marker Kinematics—Rib Cage and Greater Trochanter (Hip)

TABLE A.2(*a*) Filtered Marker Kinematics—Rib Cage and Greater Trochanter (Hip)

			BASE RIB CAGE						RIGHT HIP					
	FRAME	TIME S	X M	VX M/S	AX M/S/S	Y M	VY M/S	AY M/S/S	X M	VX M/S	AX M/S/S	Y M	VY M/S	AY M/S/S
TOR	1	0.000	0.4695	1.43	0.3	1.0435	0.22	1.0	0.4474	1.64	-2.2	0.7870	0.01	3.3
	2	0.014	0.4900	1.41	-1.6	1.0467	0.23	0.5	0.4706	1.59	-4.6	0.7875	0.07	4.3
	3	0.029	0.5100	1.38	-2.9	1.0501	0.24	0.4	0.4928	1.51	-5.6	0.7889	0.14	4.8
	4	0.043	0.5294	1.33	-3.4	1.0535	0.24	0.6	0.5138	1.43	-5.3	0.7914	0.21	4.7
	5	0.057	0.5481	1.28	-3.3	1.0570	0.25	0.6	0.5336	1.36	-3.9	0.7948	0.27	3.9
	6	0.072	0.5661	1.24	-2.5	1.0607	0.26	0.2	0.5526	1.31	-2.0	0.7991	0.32	2.6
	7	0.086	0.5836	1.21	-1.5	1.0644	0.26	-0.7	0.5712	1.30	-0.2	0.8039	0.34	1.2
	8	0.100	0.6007	1.20	-0.5	1.0680	0.24	-1.8	0.5898	1.31	1.0	0.8089	0.35	-0.2
	9	0.114	0.6178	1.20	0.2	1.0713	0.21	-2.8	0.6086	1.33	1.7	0.8139	0.34	-1.5
	10	0.129	0.6349	1.20	0.5	1.0739	0.16	-3.5	0.6278	1.36	1.9	0.8185	0.31	-2.8
	11	0.143	0.6521	1.21	0.5	1.0759	0.11	-3.7	0.6474	1.38	1.9	0.8226	0.26	-3.8
	12	0.157	0.6695	1.22	0.5	1.0770	0.05	-3.4	0.6673	1.41	1.6	0.8259	0.20	-4.4
	13	0.172	0.6869	1.22	0.5	1.0774	0.01	-2.7	0.6876	1.43	1.1	0.8283	0.13	-4.6
	14	0.186	0.7044	1.23	0.6	1.0773	-0.02	-2.1	0.7082	1.44	0.5	0.8298	0.07	-4.7
	15	0.200	0.7221	1.24	1.0	1.0767	-0.05	-1.9	0.7288	1.44	0.2	0.8302	0.00	-4.7
	16	0.215	0.7400	1.26	1.6	1.0758	-0.08	-2.1	0.7495	1.45	0.3	0.8298	-0.07	-4.6
	17	0.229	0.7581	1.29	2.2	1.0745	-0.11	-2.3	0.7702	1.45	0.7	0.8283	-0.13	-4.4
	18	0.243	0.7768	1.32	2.8	1.0727	-0.14	-2.5	0.7910	1.47	1.1	0.8260	-0.19	-4.0
	19	0.257	0.7960	1.37	3.0	1.0704	-0.18	-2.8	0.8121	1.48	1.4	0.8228	-0.25	-3.1
	20	0.272	0.8158	1.41	2.9	1.0675	-0.22	-2.8	0.8335	1.51	1.7	0.8189	-0.28	-1.7
	21	0.286	0.8363	1.45	2.5	1.0640	-0.26	-2.4	0.8552	1.53	1.8	0.8147	-0.30	0.1
	22	0.300	0.8573	1.48	2.0	1.0600	-0.29	-1.6	0.8773	1.56	1.9	0.8105	-0.28	1.6
	23	0.315	0.8787	1.51	1.4	1.0557	-0.31	-0.7	0.8997	1.59	2.1	0.8067	-0.25	2.3
	24	0.329	0.9003	1.52	1.1	1.0512	-0.31	0.3	0.9226	1.62	2.2	0.8033	-0.21	2.6
	25	0.343	0.9222	1.54	1.3	1.0468	-0.30	1.7	0.9460	1.65	2.1	0.8005	-0.18	2.8
	26	0.357	0.9443	1.56	1.6	1.0427	-0.26	3.2	0.9698	1.68	1.6	0.7983	-0.13	3.1
	27	0.372	0.9668	1.58	1.8	1.0392	-0.21	4.4	0.9940	1.69	0.8	0.7967	-0.09	3.7
HCR	28	0.386	0.9896	1.61	1.5	1.0367	-0.14	5.0	1.0183	1.70	-0.4	0.7959	-0.03	4.4
	29	0.400	1.0128	1.63	0.9	1.0353	-0.06	4.9	1.0425	1.68	-1.7	0.7959	0.04	4.9
	30	0.415	1.0362	1.64	0.0	1.0349	0.00	4.3	1.0664	1.65	-3.0	0.7970	0.11	5.0
	31	0.429	1.0596	1.63	-0.9	1.0353	0.06	3.7	1.0897	1.60	-4.0	0.7991	0.18	4.5
	32	0.443	1.0828	1.61	-1.8	1.0366	0.11	3.2	1.1121	1.53	-4.7	0.8023	0.24	3.1

33	0.458	1.1056	1.58	-2.5	1.0384	0.15	2.9	1.1336	1.46	-4.7	0.8060	0.27	1.1
34	0.472	1.1279	1.54	-2.8	1.0408	0.19	2.7	1.1540	1.40	-3.8	0.8100	0.27	-0.6
35	0.486	1.1496	1.50	-2.6	1.0438	0.23	2.3	1.1736	1.36	-2.3	0.8139	0.25	-1.5
36	0.500	1.1707	1.46	-2.2	1.0473	0.26	1.8	1.1928	1.34	-1.0	0.8173	0.23	-1.4
37	0.515	1.1915	1.43	-2.0	1.0512	0.28	1.0	1.2118	1.33	-0.8	0.8205	0.21	-1.3
38	0.529	1.2118	1.41	-2.1	1.0552	0.29	0.3	1.2307	1.31	-1.4	0.8234	0.19	-1.6
39	0.543	1.2317	1.38	-2.1	1.0594	0.29	-0.2	1.2494	1.29	-2.4	0.8260	0.17	-2.0
40	0.558	1.2511	1.35	-1.8	1.0634	0.28	-0.7	1.2675	1.25	-3.2	0.8282	0.14	-2.0
41	0.572	1.2702	1.32	-1.3	1.0673	0.27	-1.1	1.2850	1.20	-3.6	0.8300	0.11	-1.7
42	0.586	1.2890	1.31	-0.7	1.0710	0.25	-1.6	1.3018	1.14	-3.3	0.8314	0.09	-1.6
43	0.601	1.3076	1.30	-0.2	1.0744	0.22	-2.4	1.3178	1.10	-2.6	0.8325	0.07	-1.9
44	0.615	1.3263	1.30	0.3	1.0773	0.18	-3.2	1.3332	1.07	-1.7	0.8333	0.03	-2.7
45	0.629	1.3450	1.31	0.7	1.0796	0.13	-4.0	1.3484	1.05	-0.9	0.8335	-0.01	-3.3
46	0.643	1.3638	1.33	1.0	1.0810	0.07	-4.6	1.3633	1.04	-0.3	0.8330	-0.06	-3.3
47	0.658	1.3829	1.34	1.1	1.0815	0.00	-5.0	1.3782	1.04	0.3	0.8318	-0.11	-2.6
48	0.672	1.4021	1.36	1.1	1.0809	-0.08	-4.8	1.3931	1.05	1.1	0.8300	-0.14	-1.8
49	0.686	1.4217	1.37	1.1	1.0793	-0.14	-4.0	1.4083	1.07	1.9	0.8279	-0.16	-1.2
50	0.701	1.4414	1.39	0.9	1.0769	-0.19	-3.0	1.4239	1.11	2.2	0.8255	-0.17	-0.8
51	0.715	1.4614	1.40	0.6	1.0739	-0.23	-2.3	1.4400	1.14	1.8	0.8230	-0.18	-0.5
52	0.729	1.4814	1.40	0.4	1.0704	-0.26	-1.9	1.4564	1.16	1.2	0.8204	-0.18	-0.2
53	0.744	1.5015	1.41	0.5	1.0666	-0.28	-1.3	1.4731	1.17	1.1	0.8178	-0.18	0.2
54	0.758	1.5217	1.42	0.8	1.0624	-0.29	-0.5	1.4900	1.19	1.4	0.8151	-0.18	0.5
55	0.772	1.5421	1.43	0.9	1.0582	-0.29	0.5	1.5071	1.21	1.8	0.8127	-0.17	0.6
56	0.786	1.5626	1.44	0.7	1.0541	-0.28	1.3	1.5247	1.24	1.9	0.8103	-0.16	0.5
57	0.801	1.5833	1.45	0.5	1.0503	-0.25	1.9	1.5426	1.27	2.3	0.8081	-0.15	0.4
58	0.815	1.6041	1.46	0.7	1.0468	-0.22	2.1	1.5610	1.31	3.1	0.8059	-0.15	0.4
59	0.829	1.6250	1.47	1.5	1.0439	-0.19	2.1	1.5800	1.36	4.5	0.8038	-0.14	0.3
60	0.844	1.6461	1.50	2.6	1.0413	-0.16	1.9	1.5999	1.44	6.0	0.8018	-0.14	0.0
61	0.858	1.6678	1.54	3.6	1.0392	-0.14	1.5	1.6210	1.53	7.2	0.7998	-0.14	-0.3
62	0.872	1.6903	1.60	4.0	1.0373	-0.12	1.2	1.6436	1.64	7.7	0.7977	-0.15	-0.3
63	0.887	1.7136	1.66	3.6	1.0357	-0.10	1.5	1.6679	1.75	7.3	0.7955	-0.15	-0.1
64	0.901	1.7377	1.70	1.8	1.0343	-0.08	2.3	1.6937	1.85	5.5	0.7933	-0.15	0.2
65	0.915	1.7623	1.71	-1.0	1.0334	-0.04	3.2	1.7208	1.91	2.5	0.7911	-0.15	0.7
66	0.929	1.7866	1.67	-4.0	1.0332	0.01	4.0	1.7483	1.92	-1.1	0.7891	-0.13	1.3
67	0.944	1.8101	1.60	-6.0	1.0338	0.07	4.3	1.7756	1.88	-4.2	0.7873	-0.11	2.0
68	0.958	1.8323	1.50	-6.4	1.0353	0.14	4.2	1.8020	1.80	-6.4	0.7859	-0.08	2.9
69	0.972	1.8531	1.41	-5.8	1.0377	0.19	3.8	1.8270	1.69	-7.5	0.7851	-0.03	3.8

TABLE A.2(*a*) (*Continued*)

	FRAME	TIME S	BASE RIB CAGE X M	VX M/S	AX M/S/S	Y M	VY M/S	AY M/S/S	RIGHT HIP X M	VX M/S	AX M/S/S	Y M	VY M/S	AY M/S/S
TOR	70	0.987	1.8727	1.34	-4.7	1.0409	0.24	3.0	1.8505	1.58	-7.7	0.7851	0.03	4.5
	71	1.001	1.8914	1.28	-3.5	1.0447	0.28	1.8	1.8723	1.48	-6.8	0.7861	0.10	4.8
	72	1.015	1.9093	1.24	-2.4	1.0489	0.30	0.4	1.8927	1.39	-5.0	0.7880	0.17	4.6
	73	1.030	1.9268	1.21	-1.4	1.0532	0.29	-0.8	1.9120	1.33	-3.0	0.7909	0.23	4.1
	74	1.044	1.9440	1.20	-0.7	1.0572	0.27	-1.3	1.9308	1.30	-1.1	0.7947	0.29	3.5
	75	1.058	1.9610	1.19	-0.3	1.0610	0.26	-1.1	1.9493	1.30	0.5	0.7991	0.33	2.8
	76	1.072	1.9780	1.19	0.1	1.0645	0.24	-1.1	1.9680	1.32	1.7	0.8042	0.37	1.5
	77	1.087	1.9951	1.19	0.4	1.0679	0.22	-1.6	1.9870	1.35	2.4	0.8096	0.38	-0.4
	78	1.101	2.0122	1.20	0.7	1.0710	0.19	-2.6	2.0066	1.39	2.5	0.8149	0.36	-2.3
	79	1.115	2.0294	1.21	1.0	1.0735	0.15	-3.4	2.0267	1.42	2.0	0.8198	0.31	-3.9
	80	1.130	2.0469	1.23	1.3	1.0753	0.10	-3.7	2.0473	1.45	1.1	0.8238	0.24	-4.8
	81	1.144	2.0647	1.25	1.3	1.0763	0.04	-3.8	2.0681	1.45	0.1	0.8268	0.17	-5.0
	82	1.158	2.0827	1.27	1.0	1.0765	-0.01	-3.7	2.0889	1.45	-0.8	0.8287	0.10	-4.9
	83	1.173	2.1009	1.28	0.7	1.0760	-0.06	-3.6	2.1095	1.43	-1.1	0.8297	0.03	-4.7
	84	1.187	2.1193	1.29	0.5	1.0748	-0.11	-3.4	2.1298	1.42	-1.1	0.8296	-0.03	-4.4
	85	1.201	2.1378	1.30	0.6	1.0728	-0.16	-3.0	2.1500	1.40	-0.7	0.8287	-0.09	-4.2
	86	1.215	2.1563	1.30	0.8	1.0702	-0.20	-2.7	2.1699	1.40	0.1	0.8270	-0.15	-4.1
	87	1.230	2.1751	1.32	1.1	1.0671	-0.24	-2.3	2.1899	1.40	1.0	0.8244	-0.21	-3.8
	88	1.244	2.1940	1.34	1.5	1.0635	-0.27	-1.8	2.2100	1.43	2.0	0.8210	-0.26	-3.1
	89	1.258	2.2133	1.36	1.6	1.0595	-0.29	-1.1	2.2306	1.46	2.6	0.8169	-0.30	-2.0
	90	1.273	2.2329	1.38	1.6	1.0552	-0.30	-0.3	2.2518	1.50	2.6	0.8124	-0.32	-0.7
	91	1.287	2.2529	1.41	1.7	1.0510	-0.30	0.4	2.2735	1.54	2.6	0.8078	-0.32	0.6
	92	1.301	2.2731	1.43	1.9	1.0468	-0.29	0.8	2.2957	1.57	2.8	0.8033	-0.30	1.8
	93	1.316	2.2938	1.46	2.2	1.0428	-0.27	1.3	2.3185	1.62	3.3	0.7992	-0.27	3.0
	94	1.330	2.3149	1.49	2.4	1.0390	-0.25	1.9	2.3419	1.67	3.5	0.7957	-0.22	4.0
	95	1.344	2.3365	1.53	2.6	1.0356	-0.22	2.8	2.3661	1.72	3.1	0.7930	-0.15	4.8
	96	1.358	2.3587	1.57	2.6	1.0328	-0.17	3.5	2.3910	1.75	2.0	0.7913	-0.08	5.3
HCR	97	1.373	2.3813	1.60	2.2	1.0307	-0.12	3.7	2.4163	1.77	0.2	0.7908	0.00	5.3
	98	1.387	2.4045	1.63	1.3	1.0294	-0.06	3.9	2.4417	1.76	-1.7	0.7913	0.07	5.2
	99	1.401	2.4279	1.64	0.1	1.0289	-0.01	4.1	2.4667	1.72	-3.5	0.7929	0.15	4.8
	100	1.416	2.4514	1.63	-1.1	1.0292	0.05	4.5	2.4909	1.66	-4.7	0.7955	0.21	4.0
	101	1.430	2.4746	1.61	-2.3	1.0304	0.12	4.7	2.5142	1.59	-5.2	0.7989	0.26	2.4
	102	1.444	2.4974	1.57	-3.2	1.0327	0.19	4.4	2.5364	1.51	-5.3	0.8029	0.28	0.6
	103	1.459	2.5194	1.52	-4.0	1.0358	0.25	3.5	2.5575	1.44	-5.2	0.8069	0.28	-1.0
	104	1.473	2.5407	1.45	-4.9	1.0398	0.29	2.1	2.5775	1.36	-5.2	0.8108	0.25	-1.8
	105	1.487	2.5610	1.37	-6.4	1.0441	0.31	0.4	2.5965	1.29	-6.0	0.8142	0.22	-2.0
	106	1.501	2.5800	1.27	-8.6	1.0485	0.30	-1.1	2.6143	1.19	-7.9	0.8171	0.19	-2.1

TABLE A.2(*b*) Filtered Marker Kinematics—Femoral Lateral Epicondyle (Knee) and Head of Fibula

TABLE A.2(*b*) Filtered Marker Kinematics—Femoral Lateral Epicondyle (Knee) and Head of Fibula

	FRAME	TIME	RIGHT KNEE X	RIGHT KNEE VX	RIGHT KNEE AX	RIGHT KNEE Y	RIGHT KNEE VY	RIGHT KNEE AY	RIGHT FIBULA X	RIGHT FIBULA VX	RIGHT FIBULA AX	RIGHT FIBULA Y	RIGHT FIBULA VY	RIGHT FIBULA AY
		S	M	M/S	M/S/S	M	M/S	M/S/S	M	M/S	M/S/S	M	M/S	M/S/S
TOR	1	0.000	0.4075	2.66	6.0	0.4754	-0.13	5.9	0.3570	2.84	8.4	0.4052	-0.16	5.1
	2	0.014	0.4461	2.71	1.6	0.4741	-0.03	8.3	0.3985	2.93	3.7	0.4036	-0.07	7.1
	3	0.029	0.4850	2.71	-1.5	0.4746	0.10	9.5	0.4407	2.95	0.0	0.4034	0.05	8.2
	4	0.043	0.5235	2.67	-3.5	0.4771	0.24	9.4	0.4829	2.93	-2.8	0.4049	0.17	8.2
	5	0.057	0.5613	2.61	-4.8	0.4815	0.37	8.2	0.5244	2.87	-4.8	0.4082	0.28	7.1
	6	0.072	0.5980	2.53	-5.8	0.4877	0.48	6.1	0.5649	2.79	-5.9	0.4130	0.37	5.4
	7	0.086	0.6336	2.44	-6.5	0.4952	0.55	3.6	0.6041	2.70	-6.7	0.4189	0.44	3.4
	8	0.100	0.6678	2.34	-7.0	0.5034	0.58	1.2	0.6420	2.60	-7.2	0.4255	0.47	1.4
	9	0.114	0.7006	2.24	-7.4	0.5118	0.58	-0.7	0.6784	2.49	-7.7	0.4324	0.48	-0.7
	10	0.129	0.7319	2.13	-7.8	0.5200	0.56	-2.4	0.7133	2.38	-8.1	0.4391	0.45	-2.5
	11	0.143	0.7615	2.02	-8.0	0.5278	0.51	-4.1	0.7464	2.26	-8.3	0.4453	0.40	-4.1
	12	0.157	0.7896	1.90	-7.9	0.5347	0.44	-5.7	0.7779	2.14	-8.4	0.4506	0.34	-5.1
	13	0.172	0.8159	1.79	-7.6	0.5404	0.35	-7.0	0.8076	2.02	-8.3	0.4549	0.26	-5.7
	14	0.186	0.8407	1.68	-7.2	0.5447	0.24	-7.8	0.8357	1.90	-8.1	0.4580	0.17	-6.0
	15	0.200	0.8641	1.58	-6.7	0.5474	0.13	-8.3	0.8620	1.79	-7.9	0.4598	0.08	-6.1
	16	0.215	0.8860	1.49	-6.2	0.5483	0.01	-8.3	0.8868	1.68	-7.4	0.4604	0.00	-5.9
	17	0.229	0.9067	1.41	-5.6	0.5476	-0.11	-8.0	0.9100	1.57	-6.5	0.4597	-0.09	-5.6
	18	0.243	0.9263	1.33	-4.7	0.5451	-0.22	-7.1	0.9318	1.49	-5.1	0.4579	-0.16	-5.0
	19	0.257	0.9448	1.27	-3.2	0.5412	-0.31	-5.6	0.9526	1.43	-3.3	0.4551	-0.23	-3.9
	20	0.272	0.9627	1.24	-1.2	0.5361	-0.38	-3.6	0.9727	1.40	-1.4	0.4514	-0.27	-2.4
	21	0.286	0.9803	1.24	0.9	0.5303	-0.42	-0.9	0.9925	1.39	0.2	0.4472	-0.30	-0.6
	22	0.300	0.9981	1.27	2.8	0.5242	-0.41	1.8	1.0124	1.40	1.2	0.4430	-0.29	0.6
	23	0.315	1.0165	1.32	4.2	0.5186	-0.36	3.8	1.0326	1.42	1.7	0.4389	-0.28	1.1
	24	0.329	1.0358	1.39	5.0	0.5138	-0.30	4.7	1.0531	1.45	1.8	0.4350	-0.26	1.0
	25	0.343	1.0561	1.46	4.8	0.5100	-0.23	4.7	1.0741	1.48	1.6	0.4314	-0.25	1.0
	26	0.357	1.0776	1.52	3.7	0.5072	-0.17	4.4	1.0953	1.50	1.0	0.4279	-0.23	1.5
	27	0.372	1.0997	1.57	1.8	0.5053	-0.10	4.2	1.1169	1.51	0.2	0.4247	-0.21	2.6
HCR	28	0.386	1.1223	1.58	-0.2	0.5042	-0.05	4.1	1.1384	1.50	-0.8	0.4220	-0.16	4.0
	29	0.400	1.1448	1.56	-1.6	0.5039	0.01	4.2	1.1598	1.48	-1.5	0.4202	-0.09	4.9
	30	0.415	1.1669	1.53	-2.1	0.5046	0.07	4.1	1.1808	1.46	-1.9	0.4194	-0.02	4.8
	31	0.429	1.1885	1.50	-2.3	0.5061	0.13	3.5	1.2015	1.43	-2.6	0.4196	0.05	3.7
	32	0.443	1.2097	1.46	-3.2	0.5083	0.17	2.0	1.2216	1.38	-4.0	0.4207	0.09	1.9

33	0.458	1.2304	1.41	-5.3	0.5110	0.19	-0.1	1.2410	1.31	-6.2	0.4221	0.10	0.1
34	0.472	1.2500	1.31	-7.7	0.5138	0.17	-2.1	1.2592	1.21	-8.5	0.4235	0.09	-1.1
35	0.486	1.2679	1.19	-9.5	0.5160	0.13	-3.3	1.2755	1.07	-9.9	0.4247	0.07	-1.7
36	0.500	1.2839	1.04	-9.8	0.5175	0.08	-3.5	1.2898	0.92	-10.4	0.4255	0.04	-1.6
37	0.515	1.2977	0.90	-9.3	0.5182	0.03	-2.9	1.3019	0.77	-10.4	0.4260	0.02	-1.0
38	0.529	1.3097	0.78	-8.2	0.5183	-0.01	-2.0	1.3119	0.63	-9.8	0.4262	0.02	-0.1
39	0.543	1.3200	0.67	-7.0	0.5180	-0.03	-0.9	1.3198	0.49	-8.3	0.4265	0.02	0.6
40	0.558	1.3288	0.58	-5.4	0.5175	-0.03	0.3	1.3260	0.39	-6.0	0.4269	0.03	0.8
41	0.572	1.3365	0.51	-3.4	0.5171	-0.02	1.3	1.3309	0.32	-3.3	0.4274	0.04	0.4
42	0.586	1.3435	0.48	-1.5	0.5170	0.00	1.3	1.3352	0.29	-1.2	0.4281	0.05	-0.3
43	0.601	1.3502	0.47	-0.1	0.5172	0.02	0.2	1.3393	0.29	0.0	0.4287	0.04	-1.1
44	0.615	1.3570	0.48	0.7	0.5175	0.01	-1.4	1.3435	0.29	0.5	0.4291	0.01	-1.8
45	0.629	1.3639	0.49	1.0	0.5175	-0.02	-2.5	1.3477	0.30	0.8	0.4291	-0.02	-2.0
46	0.643	1.3710	0.51	1.0	0.5169	-0.06	-2.6	1.3521	0.32	1.1	0.4287	-0.04	-1.8
47	0.658	1.3784	0.52	1.0	0.5157	-0.10	-2.0	1.3567	0.33	1.4	0.4278	-0.07	-1.1
48	0.672	1.3859	0.54	1.2	0.5141	-0.12	-1.0	1.3617	0.35	1.6	0.4268	-0.08	-0.3
49	0.686	1.3937	0.55	1.4	0.5123	-0.13	0.1	1.3669	0.38	1.8	0.4256	-0.08	0.5
50	0.701	1.4018	0.58	1.4	0.5105	-0.12	1.1	1.3725	0.41	1.9	0.4246	-0.06	1.3
51	0.715	1.4102	0.59	1.1	0.5089	-0.10	1.7	1.3785	0.43	1.8	0.4238	-0.04	1.9
52	0.729	1.4188	0.61	1.1	0.5077	-0.07	1.8	1.3849	0.46	1.8	0.4235	-0.01	2.1
53	0.744	1.4275	0.62	1.8	0.5069	-0.04	1.6	1.3917	0.49	2.2	0.4236	0.02	1.6
54	0.758	1.4366	0.66	3.2	0.5065	-0.02	1.1	1.3988	0.52	3.0	0.4241	0.04	0.8
55	0.772	1.4463	0.72	4.7	0.5062	-0.01	0.5	1.4066	0.57	4.1	0.4247	0.04	0.2
56	0.786	1.4571	0.79	5.7	0.5061	-0.01	-0.3	1.4151	0.64	5.4	0.4254	0.04	-0.1
57	0.801	1.4690	0.88	6.5	0.5059	-0.02	-1.0	1.4248	0.73	6.8	0.4260	0.04	-0.4
58	0.815	1.4822	0.98	7.5	0.5055	-0.04	-1.5	1.4359	0.83	8.5	0.4265	0.03	-0.9
59	0.829	1.4969	1.09	8.9	0.5048	-0.06	-2.0	1.4486	0.97	10.4	0.4269	0.02	-1.5
60	0.844	1.5135	1.23	10.6	0.5037	-0.10	-2.5	1.4635	1.13	12.1	0.4270	-0.01	-2.1
61	0.858	1.5322	1.40	12.3	0.5021	-0.14	-3.1	1.4809	1.31	13.4	0.4267	-0.04	-2.7
62	0.872	1.5534	1.58	13.8	0.4999	-0.18	-3.2	1.5011	1.51	14.2	0.4257	-0.09	-3.2
63	0.887	1.5775	1.79	14.5	0.4969	-0.23	-2.8	1.5242	1.72	14.4	0.4242	-0.14	-3.3
64	0.901	1.6046	2.00	13.8	0.4933	-0.26	-1.9	1.5503	1.93	14.1	0.4219	-0.18	-2.9
65	0.915	1.6346	2.19	11.7	0.4894	-0.28	-0.8	1.5793	2.12	13.0	0.4189	-0.22	-2.2
66	0.929	1.6671	2.33	9.0	0.4853	-0.28	0.5	1.6110	2.30	11.4	0.4156	-0.24	-1.2
67	0.944	1.7014	2.44	6.1	0.4812	-0.27	2.0	1.6450	2.45	9.4	0.4119	-0.26	-0.1
68	0.958	1.7369	2.51	3.5	0.4776	-0.23	3.8	1.6810	2.57	7.6	0.4083	-0.25	1.8
69	0.972	1.7732	2.54	1.6	0.4747	-0.16	5.6	1.7185	2.66	5.9	0.4049	-0.20	4.3

TABLE A.2(*b*) (*Continued*)

	FRAME	TIME	RIGHT KNEE						RIGHT FIBULA					
			X	VX	AX	Y	VY	AY	X	VX	AX	Y	VY	AY
		S	M	M/S	M/S/S	M	M/S	M/S/S	M	M/S	M/S/S	M	M/S	M/S/S
TOR	70	0.987	1.8097	2.55	0.3	0.4730	-0.07	7.1	1.7572	2.74	4.1	0.4024	-0.12	6.8
	71	1.001	1.8462	2.55	-0.3	0.4728	0.04	8.1	1.7967	2.78	2.3	0.4014	-0.01	8.2
	72	1.015	1.8826	2.54	-0.5	0.4743	0.16	8.2	1.8368	2.80	0.6	0.4021	0.11	8.0
	73	1.030	1.9190	2.54	-0.8	0.4775	0.28	7.6	1.8769	2.80	-0.9	0.4045	0.22	6.7
	74	1.044	1.9552	2.52	-1.5	0.4823	0.38	6.4	1.9168	2.78	-2.2	0.4083	0.30	5.1
	75	1.058	1.9911	2.49	-2.8	0.4884	0.46	4.8	1.9563	2.74	-3.6	0.4132	0.36	3.7
	76	1.072	2.0265	2.44	-4.3	0.4955	0.52	3.0	1.9951	2.68	-4.9	0.4188	0.41	2.3
	77	1.087	2.0610	2.37	-5.7	0.5032	0.55	1.0	2.0329	2.60	-6.1	0.4249	0.43	0.7
	78	1.101	2.0943	2.28	-6.8	0.5112	0.55	-1.1	2.0694	2.50	-7.0	0.4311	0.43	-1.2
	79	1.115	2.1262	2.18	-7.4	0.5189	0.52	-2.8	2.1044	2.40	-7.7	0.4371	0.40	-2.9
	80	1.130	2.1565	2.07	-7.7	0.5260	0.47	-4.1	2.1379	2.28	-8.0	0.4424	0.34	-4.2
	81	1.144	2.1853	1.96	-7.7	0.5323	0.40	-5.0	2.1697	2.17	-8.0	0.4469	0.28	-5.0
	82	1.158	2.2124	1.85	-7.6	0.5375	0.32	-6.0	2.1999	2.05	-8.1	0.4504	0.20	-5.7
	83	1.173	2.2381	1.74	-7.4	0.5416	0.23	-6.9	2.2284	1.94	-8.0	0.4527	0.12	-6.1
	84	1.187	2.2622	1.63	-7.2	0.5442	0.13	-7.9	2.2553	1.82	-7.7	0.4537	0.03	-6.2
	85	1.201	2.2848	1.53	-6.9	0.5452	0.01	-8.7	2.2805	1.72	-6.9	0.4534	-0.06	-5.9
	86	1.215	2.3060	1.44	-6.3	0.5443	-0.12	-8.9	2.3043	1.63	-5.5	0.4519	-0.14	-5.1
	87	1.230	2.3259	1.35	-5.4	0.5416	-0.25	-8.1	2.3270	1.56	-3.9	0.4493	-0.21	-4.0
	88	1.244	2.3446	1.28	-4.0	0.5372	-0.35	-6.2	2.3489	1.51	-2.5	0.4459	-0.26	-2.4
	89	1.258	2.3625	1.24	-2.0	0.5315	-0.43	-3.6	2.3703	1.49	-1.5	0.4420	-0.28	-0.8
	90	1.273	2.3800	1.22	0.3	0.5250	-0.46	-0.7	2.3914	1.47	-0.9	0.4380	-0.28	0.7
	91	1.287	2.3975	1.25	2.9	0.5184	-0.45	2.2	2.4124	1.46	-0.3	0.4341	-0.26	1.6
	92	1.301	2.4156	1.31	5.4	0.5122	-0.39	4.7	2.4333	1.46	0.7	0.4305	-0.23	1.7
	93	1.316	2.4349	1.40	6.9	0.5071	-0.31	6.2	2.4542	1.48	1.7	0.4274	-0.21	1.6
	94	1.330	2.4556	1.51	7.1	0.5033	-0.22	6.7	2.4756	1.51	2.5	0.4246	-0.19	1.7
	95	1.344	2.4779	1.60	5.8	0.5009	-0.12	6.4	2.4975	1.55	2.6	0.4220	-0.16	2.3
	96	1.358	2.5014	1.67	3.7	0.4998	-0.03	5.4	2.5200	1.59	2.1	0.4200	-0.12	3.3
HCR	97	1.373	2.5257	1.71	1.3	0.4999	0.03	4.3	2.5429	1.61	1.2	0.4186	-0.07	4.1
	98	1.387	2.5502	1.71	-0.7	0.5008	0.09	3.4	2.5661	1.62	-0.1	0.4180	0.00	4.4
	99	1.401	2.5746	1.69	-2.4	0.5025	0.13	2.7	2.5893	1.61	-1.9	0.4185	0.06	3.9
	100	1.416	2.5984	1.64	-4.3	0.5047	0.17	1.6	2.6121	1.57	-4.4	0.4197	0.11	2.2
	101	1.430	2.6214	1.56	-6.8	0.5072	0.18	0.0	2.6341	1.48	-7.5	0.4215	0.12	0.0
	102	1.444	2.6431	1.45	-9.5	0.5098	0.17	-1.5	2.6545	1.35	-10.6	0.4232	0.11	-1.7
	103	1.459	2.6628	1.29	-11.7	0.5120	0.13	-2.5	2.6727	1.18	-12.7	0.4246	0.07	-2.5
	104	1.473	2.6800	1.11	-12.7	0.5136	0.10	-2.6	2.6883	0.99	-13.3	0.4253	0.04	-2.2
	105	1.487	2.6945	0.93	-12.3	0.5147	0.06	-2.2	2.7010	0.80	-12.6	0.4256	0.01	-1.6
	106	1.501	2.7065	0.76	-11.4	0.5153	0.03	-1.7	2.7111	0.63	-11.4	0.4256	-0.01	-1.0

TABLE A.2(*c*) Filtered Marker Kinematics—Lateral Malleolus (Ankle) and Heel

TABLE A.2(*c*) Filtered Marker Kinematics—Lateral Malleolus (Ankle) and Heel

			RIGHT ANKLE						RIGHT HEEL					
	FRAME	TIME S	X M	VX M/S	AX M/S/S	Y M	VY M/S	AY M/S/S	X M	VX M/S	AX M/S/S	Y M	VY M/S	AY M/S/S
TOR	1	0.000	0.0939	2.24	19.1	0.2143	0.80	-6.3	0.0300	2.39	17.2	0.2360	1.32	-15.9
	2	0.014	0.1279	2.50	16.4	0.2251	0.68	-10.0	0.0659	2.59	12.0	0.2531	1.03	-22.6
	3	0.029	0.1653	2.71	13.5	0.2337	0.51	-13.2	0.1042	2.73	8.0	0.2655	0.67	-26.3
	4	0.043	0.2054	2.88	10.8	0.2396	0.30	-15.2	0.1440	2.82	5.7	0.2723	0.28	-27.2
	5	0.057	0.2477	3.02	8.7	0.2423	0.07	-16.0	0.1849	2.89	4.8	0.2734	-0.11	-26.1
	6	0.072	0.2917	3.13	7.4	0.2417	-0.16	-15.5	0.2267	2.96	5.4	0.2692	-0.47	-23.7
	7	0.086	0.3372	3.23	6.7	0.2379	-0.37	-14.1	0.2696	3.05	7.0	0.2600	-0.79	-20.4
	8	0.100	0.3841	3.32	6.4	0.2311	-0.56	-12.1	0.3138	3.16	9.0	0.2466	-1.05	-16.4
	9	0.114	0.4322	3.41	6.5	0.2219	-0.72	-9.7	0.3599	3.30	10.5	0.2299	-1.26	-11.9
	10	0.129	0.4817	3.51	6.7	0.2106	-0.84	-7.1	0.4083	3.46	11.3	0.2107	-1.39	-7.3
	11	0.143	0.5326	3.60	6.8	0.1979	-0.92	-4.0	0.4590	3.63	11.2	0.1900	-1.47	-3.0
	12	0.157	0.5848	3.70	6.4	0.1844	-0.95	-0.5	0.5120	3.78	10.5	0.1687	-1.48	1.0
	13	0.172	0.6384	3.79	5.4	0.1707	-0.93	3.0	0.5671	3.93	9.3	0.1477	-1.44	4.5
	14	0.186	0.6931	3.85	3.7	0.1577	-0.86	6.2	0.6242	4.05	7.5	0.1276	-1.35	7.3
	15	0.200	0.7486	3.89	1.3	0.1460	-0.76	8.6	0.6829	4.14	5.2	0.1090	-1.23	9.7
	16	0.215	0.8045	3.89	-1.6	0.1360	-0.62	10.1	0.7427	4.20	2.1	0.0925	-1.07	11.6
	17	0.229	0.8600	3.85	-5.1	0.1282	-0.47	10.7	0.8029	4.20	-1.7	0.0783	-0.90	12.9
	18	0.243	0.9145	3.75	-8.8	0.1226	-0.31	10.3	0.8628	4.15	-6.2	0.0668	-0.70	13.3
	19	0.257	0.9672	3.60	-12.5	0.1192	-0.17	8.9	0.9215	4.02	-11.4	0.0582	-0.51	12.6
	20	0.272	1.0173	3.39	-16.2	0.1177	-0.06	6.7	0.9779	3.82	-17.1	0.0521	-0.34	10.8
	21	0.286	1.0641	3.13	-20.0	0.1175	0.02	3.9	1.0309	3.53	-23.2	0.0483	-0.21	8.1
	22	0.300	1.1068	2.82	-23.6	0.1182	0.05	0.7	1.0790	3.16	-28.9	0.0462	-0.11	4.6
	23	0.315	1.1447	2.46	-26.3	0.1190	0.04	-2.5	1.1212	2.71	-33.1	0.0451	-0.07	1.1
	24	0.329	1.1771	2.07	-27.4	0.1193	-0.02	-5.1	1.1565	2.21	-35.2	0.0441	-0.08	-1.1
	25	0.343	1.2038	1.67	-26.3	0.1185	-0.11	-6.5	1.1844	1.70	-34.7	0.0428	-0.11	-1.2
	26	0.357	1.2249	1.31	-23.1	0.1162	-0.21	-6.1	1.2051	1.22	-31.6	0.0411	-0.12	0.4
	27	0.372	1.2413	1.01	-18.3	0.1126	-0.28	-3.7	1.2192	0.80	-26.1	0.0395	-0.09	2.4
HCR	28	0.386	1.2539	0.79	-13.3	0.1081	-0.31	-0.2	1.2280	0.47	-19.3	0.0384	-0.05	3.5
	29	0.400	1.2639	0.63	-9.2	0.1037	-0.29	3.0	1.2327	0.25	-12.5	0.0381	0.01	3.3
	30	0.415	1.2720	0.53	-6.7	0.0999	-0.23	4.6	1.2350	0.11	-6.8	0.0385	0.05	2.1
	31	0.429	1.2789	0.44	-5.5	0.0972	-0.16	4.7	1.2360	0.05	-2.7	0.0394	0.07	0.9
	32	0.443	1.2846	0.37	-5.1	0.0955	-0.09	3.7	1.2365	0.04	-0.4	0.0404	0.07	0.1

33	0.458	1.2894	0.30	-4.8	0.0945	-0.05	2.3	1.2370	0.04	0.5	0.0414	0.07	-0.3
34	0.472	1.2931	0.23	-4.3	0.0940	-0.03	1.2	1.2377	0.05	0.8	0.0424	0.06	-0.4
35	0.486	1.2960	0.18	-3.3	0.0937	-0.02	0.6	1.2384	0.06	0.9	0.0432	0.06	-0.3
36	0.500	1.2982	0.14	-2.4	0.0935	-0.01	0.5	1.2394	0.08	0.9	0.0440	0.05	-0.5
37	0.515	1.2999	0.11	-1.9	0.0934	0.00	0.5	1.2406	0.09	0.2	0.0448	0.05	-0.8
38	0.529	1.3012	0.08	-2.1	0.0935	0.01	0.4	1.2419	0.08	-1.1	0.0453	0.03	-1.0
39	0.543	1.3022	0.05	-2.2	0.0936	0.01	0.3	1.2429	0.06	-2.2	0.0457	0.02	-0.9
40	0.558	1.3026	0.02	-2.0	0.0938	0.01	0.5	1.2435	0.02	-2.5	0.0458	0.00	-0.6
41	0.572	1.3026	-0.01	-1.3	0.0940	0.02	1.0	1.2435	-0.01	-1.8	0.0458	0.00	0.0
42	0.586	1.3023	-0.02	-0.5	0.0945	0.04	1.3	1.2431	-0.03	-0.7	0.0458	0.00	0.6
43	0.601	1.3020	-0.02	0.3	0.0952	0.06	1.2	1.2425	-0.04	0.3	0.0459	0.02	1.1
44	0.615	1.3017	-0.01	0.9	0.0962	0.08	0.6	1.2421	-0.02	1.1	0.0463	0.04	1.0
45	0.629	1.3016	0.00	1.5	0.0974	0.08	-0.2	1.2418	0.00	1.6	0.0469	0.05	0.5
46	0.643	1.3018	0.03	1.7	0.0985	0.07	-0.9	1.2420	0.02	1.8	0.0477	0.05	-0.1
47	0.658	1.3024	0.05	1.6	0.0994	0.05	-1.1	1.2425	0.05	1.7	0.0484	0.04	-0.5
48	0.672	1.3034	0.07	1.3	0.1000	0.04	-0.9	1.2433	0.07	1.6	0.0489	0.03	-0.5
49	0.686	1.3046	0.09	0.9	0.1005	0.03	-0.3	1.2445	0.09	1.3	0.0494	0.03	0.1
50	0.701	1.3059	0.10	0.5	0.1009	0.03	0.3	1.2460	0.11	0.9	0.0498	0.04	1.2
51	0.715	1.3074	0.11	0.1	0.1014	0.04	0.8	1.2477	0.12	0.2	0.0504	0.06	2.3
52	0.729	1.3090	0.10	-0.4	0.1020	0.05	1.4	1.2494	0.12	-0.6	0.0516	0.10	3.1
53	0.744	1.3104	0.10	-0.5	0.1029	0.08	1.9	1.2510	0.10	-0.9	0.0534	0.15	3.4
54	0.758	1.3117	0.09	-0.2	0.1042	0.11	2.6	1.2523	0.09	-0.6	0.0559	0.20	3.5
55	0.772	1.3129	0.09	0.3	0.1061	0.15	3.2	1.2536	0.08	-0.3	0.0591	0.25	3.5
56	0.786	1.3142	0.10	0.8	0.1086	0.20	3.6	1.2547	0.08	-0.1	0.0631	0.30	3.5
57	0.801	1.3157	0.11	1.4	0.1118	0.25	3.6	1.2559	0.08	0.3	0.0677	0.35	3.8
58	0.815	1.3174	0.14	2.7	0.1158	0.30	3.2	1.2571	0.09	1.5	0.0731	0.41	4.6
59	0.829	1.3196	0.19	4.7	0.1205	0.35	3.2	1.2585	0.12	3.5	0.0794	0.48	5.9
60	0.844	1.3227	0.27	7.2	0.1257	0.39	3.5	1.2606	0.19	6.1	0.0869	0.58	7.4
61	0.858	1.3274	0.39	9.8	0.1317	0.45	4.0	1.2640	0.30	9.1	0.0960	0.69	8.5
62	0.872	1.3340	0.55	12.2	0.1385	0.51	4.4	1.2692	0.45	12.3	0.1068	0.82	9.3
63	0.887	1.3432	0.74	14.0	0.1463	0.57	4.5	1.2769	0.65	15.4	0.1195	0.96	9.9
64	0.901	1.3552	0.95	15.0	0.1549	0.64	4.4	1.2878	0.89	17.8	0.1343	1.11	10.0
65	0.915	1.3704	1.17	15.3	0.1645	0.70	3.6	1.3024	1.16	19.1	0.1511	1.25	9.1
66	0.929	1.3887	1.39	15.2	0.1749	0.74	2.2	1.3210	1.44	19.2	0.1700	1.37	6.1
67	0.944	1.4101	1.61	15.1	0.1857	0.76	0.6	1.3436	1.71	18.0	0.1902	1.42	0.8
68	0.958	1.4346	1.82	14.9	0.1967	0.76	-1.1	1.3699	1.95	15.8	0.2106	1.39	-5.7
69	0.972	1.4622	2.03	14.4	0.2074	0.73	-3.0	1.3995	2.16	12.9	0.2299	1.26	-11.7

TABLE A.2(*c*) (*Continued*)

	FRAME	TIME S	RIGHT ANKLE X M	VX M/S	AX M/S/S	Y M	VY M/S	AY M/S/S	RIGHT HEEL X M	VX M/S	AX M/S/S	Y M	VY M/S	AY M/S/S
TOR	70	0.987	1.4928	2.23	13.4	0.2175	0.67	-5.3	1.4317	2.32	9.9	0.2467	1.05	-16.4
	71	1.001	1.5261	2.41	12.2	0.2266	0.58	-8.2	1.4659	2.44	7.8	0.2600	0.79	-19.6
	72	1.015	1.5619	2.58	11.3	0.2340	0.44	-11.3	1.5016	2.55	7.3	0.2693	0.49	-21.8
	73	1.030	1.5999	2.74	10.7	0.2391	0.25	-13.9	1.5387	2.65	7.8	0.2741	0.17	-22.9
	74	1.044	1.6401	2.89	10.2	0.2413	0.04	-15.2	1.5774	2.77	8.5	0.2741	-0.16	-22.6
	75	1.058	1.6824	3.03	9.7	0.2403	-0.18	-15.1	1.6178	2.89	9.0	0.2694	-0.48	-21.1
	76	1.072	1.7267	3.16	9.3	0.2361	-0.39	-13.8	1.6601	3.03	9.4	0.2604	-0.76	-18.8
	77	1.087	1.7729	3.29	8.8	0.2291	-0.58	-11.9	1.7044	3.16	9.8	0.2475	-1.02	-16.2
	78	1.101	1.8209	3.42	8.2	0.2197	-0.73	-9.7	1.7506	3.31	10.3	0.2314	-1.23	-13.3
	79	1.115	1.8706	3.53	7.5	0.2082	-0.85	-7.3	1.7990	3.46	10.8	0.2124	-1.40	-9.9
	80	1.130	1.9218	3.63	6.9	0.1953	-0.94	-4.5	1.8495	3.62	11.2	0.1914	-1.51	-5.8
	81	1.144	1.9745	3.72	6.2	0.1814	-0.98	-1.5	1.9024	3.78	11.3	0.1692	-1.56	-1.4
	82	1.158	2.0284	3.81	5.5	0.1672	-0.98	1.7	1.9576	3.94	10.8	0.1467	-1.55	3.0
	83	1.173	2.0834	3.88	4.6	0.1533	-0.93	5.0	2.0150	4.09	9.5	0.1249	-1.48	6.9
	84	1.187	2.1394	3.94	3.3	0.1405	-0.84	8.1	2.0745	4.21	7.4	0.1044	-1.35	10.2
	85	1.201	2.1961	3.98	1.3	0.1293	-0.70	10.6	2.1354	4.30	4.5	0.0862	-1.19	12.8
	86	1.215	2.2531	3.98	-1.7	0.1204	-0.54	11.8	2.1974	4.34	0.6	0.0705	-0.99	14.5
	87	1.230	2.3098	3.93	-5.8	0.1140	-0.36	11.6	2.2595	4.32	-4.2	0.0579	-0.77	15.0
	88	1.244	2.3654	3.81	-10.9	0.1100	-0.21	10.2	2.3208	4.22	-10.1	0.0485	-0.56	14.3
	89	1.258	2.4188	3.61	-16.6	0.1081	-0.07	7.9	2.3801	4.03	-16.9	0.0420	-0.36	12.4
	90	1.273	2.4688	3.33	-22.3	0.1079	0.02	4.8	2.4360	3.73	-24.2	0.0381	-0.20	9.5
	91	1.287	2.5142	2.98	-26.8	0.1087	0.07	1.1	2.4869	3.34	-30.9	0.0362	-0.09	6.0
	92	1.301	2.5539	2.57	-29.4	0.1098	0.05	-2.6	2.5314	2.85	-35.8	0.0356	-0.03	2.6
	93	1.316	2.5876	2.14	-29.4	0.1103	-0.01	-5.3	2.5684	2.31	-38.0	0.0354	-0.01	0.2
	94	1.330	2.6151	1.73	-27.1	0.1095	-0.10	-6.1	2.5975	1.76	-37.1	0.0352	-0.02	-0.5
	95	1.344	2.6369	1.36	-22.7	0.1074	-0.18	-4.5	2.6188	1.25	-33.1	0.0347	-0.03	0.3
	96	1.358	2.6541	1.08	-17.4	0.1043	-0.23	-1.6	2.6333	0.82	-26.6	0.0343	-0.01	1.7
HCR	97	1.373	2.6677	0.87	-12.5	0.1009	-0.23	1.2	2.6422	0.49	-19.1	0.0343	0.02	2.4
	98	1.387	2.6789	0.72	-9.3	0.0978	-0.19	2.9	2.6473	0.27	-12.1	0.0349	0.05	2.0
	99	1.401	2.6882	0.60	-8.0	0.0954	-0.15	3.1	2.6500	0.14	-6.9	0.0359	0.08	0.8
	100	1.416	2.6960	0.49	-7.8	0.0936	-0.10	2.5	2.6514	0.07	-3.7	0.0371	0.08	-0.6
	101	1.430	2.7023	0.38	-7.8	0.0924	-0.08	1.5	2.6521	0.04	-2.2	0.0381	0.06	-1.7
	102	1.444	2.7068	0.27	-7.7	0.0915	-0.06	0.7	2.6524	0.01	-1.5	0.0388	0.03	-2.3
	103	1.459	2.7098	0.16	-7.0	0.0907	-0.05	0.4	2.6525	0.00	-1.1	0.0390	0.00	-2.2
	104	1.473	2.7114	0.07	-5.5	0.0899	-0.05	0.5	2.6523	-0.02	-0.8	0.0387	-0.03	-1.5
	105	1.487	2.7117	0.00	-3.4	0.0893	-0.04	0.7	2.6519	-0.03	-0.4	0.0381	-0.05	-0.6
	106	1.501	2.7114	-0.03	-1.2	0.0888	-0.03	0.8	2.6515	-0.03	-0.1	0.0373	-0.05	0.2

TABLE A.2(*d*) Filtered Marker Kinematics—Fifth Metatarsal and Toe

TABLE A.2(*d*) Filtered Marker Kinematics—Fifth Metatarsal and Toe

			RIGHT METATARSAL						RIGHT TOE					
	FRAME	TIME S	X M	VX M/S	AX M/S/S	Y M	VY M/S	AY M/S/S	X M	VX M/S	AX M/S/S	Y M	VY M/S	AY M/S/S
TOR	1	0.000	0.0845	1.63	33.7	0.0926	0.83	-3.5	0.1283	1.34	38.1	0.0464	0.35	5.6
	2	0.014	0.1114	2.12	33.5	0.1042	0.74	-9.6	0.1515	1.92	40.8	0.0521	0.40	1.3
	3	0.029	0.1452	2.59	30.8	0.1137	0.56	-14.4	0.1833	2.51	39.0	0.0580	0.38	-4.1
	4	0.043	0.1854	3.00	26.7	0.1201	0.33	-17.1	0.2232	3.04	34.1	0.0630	0.29	-8.4
	5	0.057	0.2311	3.35	22.4	0.1230	0.07	-17.4	0.2702	3.48	28.1	0.0662	0.14	-10.3
	6	0.072	0.2813	3.64	18.4	0.1222	-0.17	-15.9	0.3229	3.84	22.3	0.0671	-0.01	-9.8
	7	0.086	0.3353	3.88	15.0	0.1180	-0.38	-13.0	0.3801	4.12	17.1	0.0659	-0.14	-7.7
	8	0.100	0.3923	4.07	12.0	0.1112	-0.55	-9.2	0.4407	4.33	12.7	0.0631	-0.23	-4.6
	9	0.114	0.4517	4.22	9.4	0.1024	-0.65	-4.8	0.5039	4.48	8.9	0.0594	-0.27	-1.2
	10	0.129	0.5130	4.34	7.1	0.0927	-0.68	-0.5	0.5689	4.59	5.6	0.0554	-0.26	2.1
	11	0.143	0.5758	4.43	5.0	0.0828	-0.66	3.2	0.6351	4.64	2.8	0.0518	-0.21	5.2
	12	0.157	0.6396	4.48	3.0	0.0737	-0.59	6.5	0.7018	4.67	0.3	0.0493	-0.12	7.8
	13	0.172	0.7040	4.51	0.9	0.0659	-0.48	9.2	0.7685	4.65	-2.1	0.0485	0.01	10.1
	14	0.186	0.7686	4.51	-1.3	0.0601	-0.33	11.1	0.8349	4.61	-4.5	0.0497	0.17	11.7
	15	0.200	0.8330	4.47	-3.6	0.0565	-0.16	12.3	0.9003	4.52	-7.1	0.0534	0.35	12.6
	16	0.215	0.8966	4.41	-5.8	0.0555	0.02	12.7	0.9642	4.40	-9.9	0.0596	0.53	12.8
	17	0.229	0.9590	4.31	-8.5	0.0572	0.20	12.5	1.0262	4.24	-13.2	0.0686	0.71	11.9
	18	0.243	1.0197	4.16	-12.1	0.0614	0.38	11.2	1.0855	4.03	-16.8	0.0801	0.87	9.5
	19	0.257	1.0780	3.96	-16.6	0.0680	0.53	8.7	1.1414	3.76	-20.2	0.0936	0.99	5.7
	20	0.272	1.1330	3.69	-21.4	0.0765	0.63	4.5	1.1930	3.45	-23.1	0.1083	1.04	0.6
	21	0.286	1.1835	3.35	-25.7	0.0859	0.66	-0.9	1.2400	3.10	-25.2	0.1232	1.00	-5.5
	22	0.300	1.2288	2.95	-28.9	0.0952	0.60	-7.2	1.2817	2.73	-26.4	0.1370	0.88	-12.1
	23	0.315	1.2680	2.52	-30.8	0.1031	0.45	-13.2	1.3180	2.35	-26.8	0.1483	0.66	-18.5
	24	0.329	1.3009	2.07	-31.0	0.1081	0.22	-17.6	1.3488	1.96	-26.1	0.1558	0.35	-23.5
	25	0.343	1.3274	1.63	-29.5	0.1094	-0.05	-19.2	1.3741	1.60	-23.9	0.1583	-0.01	-25.5
	26	0.357	1.3477	1.23	-25.9	0.1066	-0.33	-17.7	1.3945	1.28	-20.3	0.1554	-0.38	-24.1
	27	0.372	1.3626	0.89	-21.0	0.1001	-0.56	-13.5	1.4107	1.02	-15.8	0.1474	-0.70	-19.4
HCR	28	0.386	1.3732	0.63	-15.6	0.0907	-0.71	-7.6	1.4237	0.83	-11.9	0.1353	-0.94	-12.6
	29	0.400	1.3807	0.45	-10.9	0.0797	-0.78	-1.4	1.4343	0.68	-9.2	0.1206	-1.06	-4.9
	30	0.415	1.3860	0.32	-7.3	0.0685	-0.75	4.0	1.4432	0.56	-7.7	0.1050	-1.08	2.4
	31	0.429	1.3899	0.24	-5.1	0.0582	-0.66	7.9	1.4504	0.46	-6.9	0.0898	-0.99	8.2
	32	0.443	1.3928	0.18	-3.7	0.0496	-0.53	9.9	1.4563	0.37	-6.2	0.0765	-0.84	11.9

33	0.458	1.3949	0.13	-2.8	0.0431	-0.38	10.2	1.4609	0.28	-5.6	0.0658	-0.65	13.2
34	0.472	1.3965	0.10	-1.9	0.0387	-0.24	9.0	1.4644	0.21	-4.8	0.0578	-0.46	12.3
35	0.486	1.3977	0.08	-1.0	0.0363	-0.12	6.8	1.4668	0.15	-3.7	0.0525	-0.30	10.3
36	0.500	1.3987	0.07	-0.4	0.0353	-0.04	4.2	1.4685	0.10	-2.5	0.0492	-0.17	7.7
37	0.515	1.3997	0.07	-0.5	0.0351	0.00	2.0	1.4697	0.07	-1.8	0.0476	-0.08	5.0
38	0.529	1.4006	0.06	-1.2	0.0353	0.01	0.5	1.4706	0.05	-1.7	0.0469	-0.03	2.4
39	0.543	1.4013	0.03	-1.9	0.0355	0.01	-0.1	1.4712	0.02	-1.7	0.0467	-0.01	0.5
40	0.558	1.4016	0.00	-1.8	0.0357	0.01	-0.2	1.4713	0.00	-1.4	0.0466	-0.01	-0.5
41	0.572	1.4014	-0.02	-1.2	0.0359	0.01	-0.1	1.4712	-0.02	-0.9	0.0463	-0.02	-0.7
42	0.586	1.4010	-0.03	-0.3	0.0360	0.01	-0.2	1.4709	-0.02	-0.4	0.0459	-0.03	-0.5
43	0.601	1.4005	-0.03	0.3	0.0361	0.00	-0.5	1.4705	-0.03	0.1	0.0453	-0.04	-0.1
44	0.615	1.4001	-0.02	0.6	0.0361	-0.01	-0.9	1.4701	-0.02	0.5	0.0448	-0.04	0.0
45	0.629	1.3998	-0.01	0.6	0.0359	-0.02	-0.8	1.4699	-0.01	0.7	0.0443	-0.04	-0.1
46	0.643	1.3997	0.00	0.5	0.0355	-0.03	-0.3	1.4698	0.00	0.7	0.0437	-0.04	-0.3
47	0.658	1.3997	0.00	0.3	0.0351	-0.03	0.2	1.4698	0.01	0.7	0.0431	-0.05	-0.3
48	0.672	1.3997	0.00	0.5	0.0346	-0.02	0.6	1.4700	0.02	0.7	0.0424	-0.05	-0.2
49	0.686	1.3998	0.01	0.9	0.0344	-0.01	0.6	1.4703	0.03	0.6	0.0417	-0.05	0.0
50	0.701	1.4001	0.03	1.2	0.0343	-0.01	0.2	1.4708	0.04	0.4	0.0410	-0.05	0.3
51	0.715	1.4006	0.05	1.0	0.0342	-0.01	-0.2	1.4714	0.04	0.0	0.0403	-0.04	0.5
52	0.729	1.4014	0.06	0.5	0.0340	-0.01	-0.1	1.4719	0.03	-0.5	0.0397	-0.03	0.6
53	0.744	1.4023	0.06	-0.1	0.0338	-0.01	0.6	1.4724	0.02	-0.8	0.0393	-0.03	0.6
54	0.758	1.4031	0.05	-0.6	0.0337	0.00	1.3	1.4726	0.01	-0.9	0.0390	-0.02	0.8
55	0.772	1.4038	0.04	-1.1	0.0339	0.03	1.6	1.4727	0.00	-1.0	0.0388	0.00	0.8
56	0.786	1.4043	0.02	-1.6	0.0345	0.05	1.1	1.4725	-0.02	-1.3	0.0389	0.01	0.7
57	0.801	1.4045	0.00	-1.9	0.0353	0.06	0.3	1.4721	-0.04	-1.4	0.0390	0.01	0.2
58	0.815	1.4042	-0.03	-1.5	0.0361	0.06	-0.4	1.4714	-0.06	-1.0	0.0393	0.01	-0.3
59	0.829	1.4035	-0.05	-0.4	0.0369	0.05	-0.5	1.4705	-0.07	-0.1	0.0394	0.00	-1.0
60	0.844	1.4028	-0.04	1.2	0.0374	0.04	0.2	1.4695	-0.06	1.0	0.0394	-0.02	-1.7
61	0.858	1.4023	-0.01	3.0	0.0380	0.05	1.8	1.4687	-0.04	2.1	0.0390	-0.04	-2.1
62	0.872	1.4024	0.04	4.6	0.0389	0.09	4.0	1.4684	0.00	2.8	0.0381	-0.08	-1.8
63	0.887	1.4035	0.12	5.8	0.0406	0.17	6.4	1.4687	0.04	3.2	0.0368	-0.10	-0.8
64	0.901	1.4058	0.21	7.2	0.0437	0.27	8.3	1.4696	0.09	3.5	0.0353	-0.10	1.0
65	0.915	1.4094	0.32	9.2	0.0485	0.40	9.3	1.4712	0.14	4.6	0.0340	-0.07	3.0
66	0.929	1.4149	0.47	12.4	0.0552	0.54	9.0	1.4737	0.22	7.3	0.0333	-0.01	5.0
67	0.944	1.4229	0.68	16.9	0.0639	0.66	7.3	1.4775	0.35	12.7	0.0337	0.07	6.3
68	0.958	1.4343	0.95	22.1	0.0741	0.75	4.1	1.4837	0.58	20.7	0.0354	0.17	6.4
69	0.972	1.4502	1.31	26.8	0.0852	0.78	-0.2	1.4941	0.94	29.5	0.0385	0.26	5.4

TABLE A.2(*d*) (*Continued*)

	FRAME	TIME	RIGHT METATARSAL						RIGHT TOE					
			X	VX	AX	Y	VY	AY	X	VX	AX	Y	VY	AY
		S	M	M/S	M/S/S	M	M/S	M/S/S	M	M/S	M/S/S	M	M/S	M/S/S
TOR	70	0.987	1.4717	1.72	29.7	0.0963	0.74	-5.4	1.5107	1.43	36.3	0.0428	0.32	3.4
	71	1.001	1.4994	2.16	30.2	0.1063	0.62	-10.6	1.5349	1.98	39.0	0.0478	0.35	0.3
	72	1.015	1.5335	2.59	28.5	0.1142	0.44	-14.7	1.5673	2.54	37.5	0.0530	0.33	-3.4
	73	1.030	1.5734	2.97	25.3	0.1188	0.20	-16.9	1.6075	3.05	33.2	0.0574	0.26	-7.0
	74	1.044	1.6185	3.31	21.7	0.1200	-0.05	-16.6	1.6546	3.49	27.9	0.0603	0.13	-9.3
	75	1.058	1.6681	3.59	18.2	0.1175	-0.27	-14.3	1.7074	3.85	22.5	0.0612	-0.01	-9.9
	76	1.072	1.7213	3.83	15.2	0.1121	-0.45	-10.8	1.7647	4.13	17.5	0.0601	-0.15	-8.5
	77	1.087	1.7776	4.03	12.7	0.1046	-0.58	-7.3	1.8256	4.35	13.2	0.0570	-0.25	-5.7
	78	1.101	1.8365	4.19	10.4	0.0955	-0.66	-4.3	1.8892	4.51	9.7	0.0528	-0.31	-2.1
	79	1.115	1.8975	4.33	8.3	0.0856	-0.70	-1.4	1.9547	4.63	6.7	0.0482	-0.31	1.5
	80	1.130	1.9603	4.43	6.4	0.0754	-0.70	1.7	2.0215	4.70	4.0	0.0439	-0.27	4.7
	81	1.144	2.0243	4.51	4.7	0.0655	-0.66	4.9	2.0892	4.74	1.0	0.0406	-0.18	7.5
	82	1.158	2.0892	4.57	3.1	0.0566	-0.56	8.1	2.1571	4.73	-1.9	0.0388	-0.05	9.9
	83	1.173	2.1548	4.60	1.5	0.0493	-0.43	10.9	2.2246	4.69	-4.5	0.0391	0.11	12.0
	84	1.187	2.2208	4.61	-0.6	0.0444	-0.25	13.2	2.2912	4.61	-6.5	0.0418	0.29	13.4
	85	1.201	2.2866	4.58	-3.4	0.0421	-0.05	14.7	2.3564	4.50	-8.4	0.0474	0.49	14.0
	86	1.215	2.3518	4.51	-7.0	0.0430	0.17	15.1	2.4199	4.37	-10.7	0.0558	0.69	13.7
	87	1.230	2.4155	4.38	-11.3	0.0469	0.38	14.2	2.4813	4.20	-13.9	0.0671	0.88	12.1
	88	1.244	2.4771	4.19	-16.3	0.0539	0.57	11.6	2.5399	3.97	-18.0	0.0811	1.04	8.4
	89	1.258	2.5353	3.91	-21.8	0.0634	0.71	6.8	2.5948	3.68	-22.4	0.0968	1.12	2.6
	90	1.273	2.5890	3.56	-27.2	0.0743	0.77	0.1	2.6451	3.33	-26.3	0.1132	1.11	-4.7
	91	1.287	2.6371	3.14	-31.4	0.0853	0.72	-7.3	2.6900	2.93	-28.8	0.1286	0.99	-12.6
	92	1.301	2.6787	2.66	-33.6	0.0948	0.56	-14.1	2.7289	2.51	-29.5	0.1415	0.75	-20.0
	93	1.316	2.7133	2.18	-33.4	0.1013	0.31	-18.8	2.7617	2.09	-28.2	0.1501	0.42	-25.4
	94	1.330	2.7410	1.71	-30.8	0.1038	0.02	-20.6	2.7886	1.70	-25.4	0.1534	0.02	-27.6
	95	1.344	2.7622	1.30	-26.1	0.1019	-0.28	-19.0	2.8103	1.36	-21.4	0.1508	-0.37	-25.8
	96	1.358	2.7781	0.96	-20.4	0.0959	-0.52	-14.5	2.8275	1.09	-16.8	0.1427	-0.72	-20.6
HCR	97	1.373	2.7898	0.71	-15.0	0.0869	-0.69	-8.3	2.8413	0.88	-12.6	0.1304	-0.96	-12.9
	98	1.387	2.7985	0.53	-11.0	0.0762	-0.76	-1.9	2.8527	0.73	-10.0	0.1153	-1.08	-4.2
	99	1.401	2.8050	0.40	-8.5	0.0651	-0.75	3.5	2.8621	0.60	-9.0	0.0994	-1.08	3.6
	100	1.416	2.8099	0.29	-7.2	0.0548	-0.66	7.1	2.8697	0.47	-9.0	0.0843	-0.98	9.3
	101	1.430	2.8134	0.20	-6.2	0.0461	-0.54	8.6	2.8755	0.34	-8.9	0.0714	-0.81	12.4
	102	1.444	2.8155	0.11	-5.4	0.0393	-0.42	8.5	2.8794	0.21	-8.0	0.0610	-0.63	12.9
	103	1.459	2.8166	0.04	-4.4	0.0342	-0.30	7.4	2.8816	0.11	-6.3	0.0535	-0.44	11.6
	104	1.473	2.8167	-0.01	-2.9	0.0306	-0.21	6.1	2.8825	0.03	-4.1	0.0483	-0.29	9.2
	105	1.487	2.8162	-0.04	-1.1	0.0283	-0.13	4.9	2.8825	-0.01	-1.8	0.0451	-0.18	6.5
	106	1.501	2.8155	-0.04	0.3	0.0270	-0.07	3.5	2.8822	-0.02	-0.1	0.0431	-0.11	4.1

TABLE A.3(*a*) Linear and Angular Kinematics—Foot

TABLE A.3(*a*) Linear and Angular Kinematics—Foot

	FRAME	TIME S	THETA DEG	OMEGA R/S	ALPHA R/S/S	CofM-X M	VEL-X M/S	ACC-X M/S/S	CofM-Y M	VEL-Y M/S	ACC-Y M/S/S
TOR	1	0.000	85.6	-5.01	118.27	0.089	1.937	26.39	0.153	0.814	-4.92
	2	0.014	82.2	-3.11	138.72	0.120	2.310	24.92	0.165	0.707	-9.81
	3	0.029	80.5	-1.04	142.01	0.155	2.650	22.11	0.174	0.533	-13.77
	4	0.043	80.5	0.96	132.71	0.195	2.942	18.72	0.180	0.313	-16.13
	5	0.057	82.1	2.75	115.36	0.239	3.185	15.54	0.183	0.072	-16.70
	6	0.072	85.0	4.25	93.06	0.287	3.386	12.91	0.182	-0.164	-15.72
	7	0.086	89.1	5.41	68.58	0.336	3.555	10.83	0.178	-0.378	-13.58
	8	0.100	93.9	6.22	44.92	0.388	3.696	9.20	0.171	-0.553	-10.64
	9	0.114	99.2	6.70	25.05	0.442	3.818	7.94	0.162	-0.682	-7.27
	10	0.129	104.9	6.93	10.88	0.497	3.923	6.90	0.152	-0.761	-3.80
	11	0.143	110.6	7.01	2.33	0.554	4.015	5.88	0.140	-0.791	-0.37
	12	0.157	116.3	7.00	-2.26	0.612	4.092	4.69	0.129	-0.771	2.99
	13	0.172	122.1	6.94	-4.67	0.671	4.149	3.15	0.118	-0.705	6.11
	14	0.186	127.7	6.87	-5.67	0.731	4.182	1.18	0.109	-0.597	8.65
	15	0.200	133.3	6.78	-5.01	0.791	4.183	-1.12	0.101	-0.457	10.42
	16	0.215	138.8	6.72	-3.24	0.851	4.150	-3.73	0.096	-0.299	11.39
	17	0.229	144.3	6.69	-3.16	0.909	4.076	-6.80	0.093	-0.132	11.57
	18	0.243	149.8	6.63	-8.11	0.967	3.955	-10.45	0.092	0.032	10.78
	19	0.257	155.2	6.46	-19.17	1.023	3.778	-14.54	0.094	0.177	8.80
	20	0.272	160.4	6.08	-34.96	1.075	3.539	-18.79	0.097	0.284	5.61
	21	0.286	165.2	5.46	-53.64	1.124	3.240	-22.83	0.102	0.337	1.47
	22	0.300	169.3	4.55	-73.48	1.168	2.886	-26.24	0.107	0.326	-3.24
	23	0.315	172.6	3.36	-91.19	1.206	2.490	-28.53	0.111	0.245	-7.83
	24	0.329	174.8	1.94	-102.29	1.239	2.070	-29.23	0.114	0.102	-11.35
	25	0.343	175.8	0.43	-104.08	1.266	1.654	-27.90	0.114	-0.080	-12.86
	26	0.357	175.5	-1.03	-97.50	1.286	1.272	-24.51	0.111	-0.266	-11.90
	27	0.372	174.1	-2.36	-85.55	1.302	0.953	-19.64	0.106	-0.420	-8.61
HCR	28	0.386	171.7	-3.48	-68.80	1.314	0.710	-14.43	0.099	-0.512	-3.92
	29	0.400	168.4	-4.32	-43.96	1.322	0.540	-10.02	0.092	-0.532	0.78
	30	0.415	164.6	-4.74	-10.50	1.329	0.424	-7.03	0.084	-0.490	4.32
	31	0.429	160.6	-4.63	24.51	1.334	0.339	-5.31	0.078	-0.409	6.26
	32	0.443	157.0	-4.04	52.05	1.339	0.272	-4.39	0.073	-0.311	6.77
	33	0.458	154.0	-3.14	67.24	1.342	0.214	-3.77	0.069	-0.215	6.26

34	0.472	151.9	-2.11	68.18	1.345	0.164	-3.06	0.066	-0.132	5.12
35	0.486	150.6	-1.19	55.74	1.347	0.126	-2.15	0.065	-0.069	3.71
36	0.500	149.9	-0.52	35.86	1.348	0.103	-1.38	0.064	-0.026	2.36
37	0.515	149.7	-0.16	17.08	1.350	0.087	-1.21	0.064	-0.001	1.25
38	0.529	149.7	-0.03	4.56	1.351	0.068	-1.64	0.064	0.010	0.46
39	0.543	149.7	-0.03	-1.79	1.352	0.040	-2.06	0.065	0.012	0.10
40	0.558	149.6	-0.08	-4.87	1.352	0.009	-1.91	0.065	0.013	0.16
41	0.572	149.5	-0.17	-7.39	1.352	-0.015	-1.21	0.065	0.017	0.43
42	0.586	149.3	-0.29	-10.28	1.352	-0.026	-0.40	0.065	0.025	0.54
43	0.601	149.0	-0.46	-12.53	1.351	-0.026	0.26	0.066	0.033	0.31
44	0.615	148.6	-0.65	-12.14	1.351	-0.018	0.76	0.066	0.034	-0.11
45	0.629	148.0	-0.81	-8.07	1.351	-0.004	1.07	0.067	0.029	-0.47
46	0.643	147.2	-0.88	-1.88	1.351	0.012	1.10	0.067	0.021	-0.60
47	0.658	146.5	-0.86	3.55	1.351	0.027	0.96	0.067	0.012	-0.45
48	0.672	145.8	-0.78	6.31	1.352	0.039	0.86	0.067	0.008	-0.13
49	0.686	145.2	-0.68	5.50	1.352	0.052	0.88	0.067	0.009	0.13
50	0.701	144.7	-0.62	1.74	1.353	0.065	0.85	0.068	0.012	0.22
51	0.715	144.2	-0.63	-2.68	1.354	0.076	0.55	0.068	0.015	0.32
52	0.729	143.7	-0.70	-5.69	1.355	0.080	0.05	0.068	0.021	0.66
53	0.744	143.1	-0.80	-7.38	1.356	0.077	-0.32	0.068	0.034	1.25
54	0.758	142.4	-0.91	-10.45	1.357	0.071	-0.42	0.069	0.056	1.91
55	0.772	141.6	-1.10	-17.81	1.358	0.065	-0.41	0.070	0.088	2.38
56	0.786	140.6	-1.42	-28.50	1.359	0.059	-0.42	0.072	0.125	2.38
57	0.801	139.2	-1.91	-38.25	1.360	0.053	-0.24	0.074	0.156	1.90
58	0.815	137.4	-2.52	-45.01	1.361	0.053	0.57	0.076	0.179	1.39
59	0.829	135.1	-3.20	-49.94	1.362	0.069	2.14	0.079	0.196	1.33
60	0.844	132.2	-3.94	-52.91	1.363	0.114	4.22	0.082	0.217	1.88
61	0.858	128.6	-4.71	-52.39	1.365	0.190	6.40	0.085	0.250	2.88
62	0.872	124.5	-5.44	-49.61	1.368	0.297	8.37	0.089	0.300	4.16
63	0.887	119.7	-6.13	-47.15	1.373	0.429	9.90	0.093	0.369	5.46
64	0.901	114.4	-6.79	-43.01	1.380	0.580	11.08	0.099	0.456	6.34
65	0.915	108.6	-7.36	-31.69	1.390	0.746	12.24	0.106	0.550	6.44
66	0.929	102.4	-7.70	-10.02	1.402	0.930	13.82	0.115	0.640	5.61
67	0.944	96.0	-7.65	21.54	1.417	1.141	16.00	0.125	0.711	3.93
68	0.958	89.8	-7.08	60.42	1.434	1.388	18.51	0.135	0.752	1.50
69	0.972	84.4	-5.92	100.83	1.456	1.671	20.60	0.146	0.754	-1.62

TABLE A.3(*a*) (*Continued*)

	FRAME	TIME S	THETA DEG	OMEGA R/S	ALPHA R/S/S	CofM-X M	VEL-X M/S	ACC-X M/S/S	CofM-Y M	VEL-Y M/S	ACC-Y M/S/S
TOR	70	0.987	80.1	-4.20	133.15	1.482	1.977	21.55	0.157	0.706	-5.36
	71	1.001	77.5	-2.11	148.09	1.513	2.287	21.20	0.166	0.600	-9.39
	72	1.015	76.7	0.04	142.55	1.548	2.583	19.88	0.174	0.437	-13.01
	73	1.030	77.6	1.97	121.35	1.587	2.856	18.00	0.179	0.229	-15.37
	74	1.044	79.9	3.51	93.05	1.629	3.098	15.92	0.181	-0.002	-15.93
	75	1.058	83.3	4.63	65.70	1.675	3.311	13.95	0.179	-0.227	-14.68
	76	1.072	87.5	5.39	44.39	1.724	3.497	12.25	0.174	-0.422	-12.28
	77	1.087	92.1	5.90	29.92	1.775	3.661	10.74	0.167	-0.578	-9.59
	78	1.101	97.1	6.24	20.34	1.829	3.804	9.31	0.158	-0.697	-7.02
	79	1.115	102.4	6.48	13.82	1.884	3.927	7.94	0.147	-0.779	-4.37
	80	1.130	107.8	6.64	9.60	1.941	4.031	6.65	0.135	-0.822	-1.42
	81	1.144	113.3	6.75	7.77	1.999	4.118	5.44	0.123	-0.820	1.72
	82	1.158	118.8	6.86	7.86	2.059	4.187	4.29	0.112	-0.772	4.87
	83	1.173	124.5	6.98	7.28	2.119	4.240	3.03	0.101	-0.680	7.92
	84	1.187	130.3	7.07	3.78	2.180	4.274	1.33	0.092	-0.546	10.66
	85	1.201	136.1	7.09	-0.95	2.241	4.278	-1.08	0.086	-0.375	12.64
	86	1.215	141.9	7.04	-3.31	2.302	4.243	-4.36	0.082	-0.184	13.44
	87	1.230	147.6	6.99	-3.75	2.363	4.154	-8.57	0.080	0.009	12.89
	88	1.244	153.3	6.93	-8.63	2.421	3.998	-13.62	0.082	0.184	10.90
	89	1.258	159.0	6.74	-23.94	2.477	3.764	-19.21	0.086	0.321	7.37
	90	1.273	164.4	6.25	-48.50	2.529	3.448	-24.71	0.091	0.395	2.47
	91	1.287	169.2	5.36	-75.19	2.576	3.057	-29.09	0.097	0.391	-3.11
	92	1.301	173.2	4.10	-96.87	2.616	2.616	-31.47	0.102	0.306	-8.34
	93	1.316	175.9	2.59	-110.55	2.650	2.157	-31.41	0.106	0.153	-12.08
	94	1.330	177.4	0.94	-117.61	2.678	1.718	-28.90	0.107	-0.039	-13.34
	95	1.344	177.5	-0.78	-118.51	2.700	1.331	-24.40	0.105	-0.229	-11.78
	96	1.358	176.1	-2.45	-109.12	2.716	1.020	-18.90	0.100	-0.376	-8.07
HCR	97	1.373	173.5	-3.90	-84.77	2.729	0.790	-13.77	0.094	-0.460	-3.57
	98	1.387	169.7	-4.88	-46.31	2.739	0.626	-10.14	0.087	-0.478	0.46
	99	1.401	165.5	-5.22	-1.86	2.747	0.500	-8.25	0.080	-0.446	3.31
	100	1.416	161.2	-4.93	36.59	2.753	0.390	-7.47	0.074	-0.384	4.77
	101	1.430	157.4	-4.18	59.94	2.758	0.286	-7.03	0.069	-0.310	5.07
	102	1.444	154.3	-3.22	66.68	2.761	0.189	-6.52	0.065	-0.239	4.60
	103	1.459	152.1	-2.27	61.71	2.763	0.100	-5.67	0.062	-0.178	3.88
	104	1.473	150.6	-1.45	51.41	2.764	0.027	-4.22	0.060	-0.128	3.28
	105	1.487	149.7	-0.80	39.26	2.764	-0.021	-2.28	0.059	-0.084	2.79
	106	1.501	149.3	-0.33	25.63	2.763	-0.038	-0.45	0.058	-0.048	2.13

TABLE A.3(*b*) Linear and Angular Kinematics—Leg

TABLE A.3(*b*) Linear and Angular Kinematics—Leg

	FRAME	TIME S	THETA DEG	OMEGA R/S	ALPHA R/S/S	CofM-X M	VEL-X M/S	ACC-X M/S/S	CofM-Y M	VEL-Y M/S	ACC-Y M/S/S
TOR	1	0.000	39.8	-2.41	40.67	0.272	2.479	11.66	0.362	0.268	0.58
	2	0.014	38.0	-1.70	56.27	0.308	2.618	7.99	0.366	0.277	0.37
	3	0.029	37.0	-0.80	66.77	0.347	2.708	4.97	0.370	0.279	-0.29
	4	0.043	36.7	0.21	71.22	0.386	2.760	2.66	0.374	0.268	-1.25
	5	0.057	37.3	1.24	70.05	0.425	2.784	1.03	0.378	0.243	-2.27
	6	0.072	38.8	2.21	64.35	0.465	2.789	-0.07	0.381	0.203	-3.27
	7	0.086	41.0	3.08	55.75	0.505	2.782	-0.78	0.384	0.150	-4.09
	8	0.100	43.8	3.81	46.45	0.545	2.767	-1.20	0.385	0.086	-4.55
	9	0.114	47.2	4.41	38.02	0.584	2.748	-1.41	0.386	0.020	-4.61
	10	0.129	51.0	4.90	30.50	0.624	2.727	-1.52	0.386	-0.045	-4.42
	11	0.143	55.2	5.28	23.32	0.662	2.704	-1.61	0.385	-0.107	-4.07
	12	0.157	59.7	5.56	16.47	0.701	2.681	-1.73	0.383	-0.162	-3.47
	13	0.172	64.3	5.75	10.44	0.739	2.655	-1.99	0.380	-0.206	-2.63
	14	0.186	69.1	5.86	5.57	0.777	2.624	-2.49	0.377	-0.237	-1.74
	15	0.200	74.0	5.91	1.65	0.814	2.584	-3.23	0.374	-0.256	-0.98
	16	0.215	78.8	5.91	-2.00	0.851	2.531	-4.22	0.370	-0.265	-0.37
	17	0.229	83.6	5.85	-6.16	0.886	2.463	-5.38	0.366	-0.266	0.11
	18	0.243	88.4	5.73	-11.89	0.921	2.378	-6.44	0.362	-0.262	0.45
	19	0.257	93.0	5.51	-20.20	0.954	2.279	-7.20	0.358	-0.253	0.66
	20	0.272	97.4	5.16	-31.54	0.986	2.171	-7.72	0.355	-0.243	0.87
	21	0.286	101.5	4.61	-45.55	1.017	2.058	-8.18	0.352	-0.228	1.17
	22	0.300	105.0	3.85	-60.60	1.045	1.938	-8.64	0.348	-0.209	1.35
	23	0.315	107.8	2.88	-73.46	1.072	1.811	-8.98	0.346	-0.190	1.08
	24	0.329	109.7	1.75	-80.46	1.097	1.681	-9.03	0.343	-0.178	0.43
	25	0.343	110.7	0.58	-79.07	1.120	1.553	-8.67	0.340	-0.178	-0.16
	26	0.357	110.7	-0.51	-68.79	1.141	1.433	-7.92	0.338	-0.183	-0.14
	27	0.372	109.8	-1.39	-51.73	1.161	1.326	-6.93	0.335	-0.182	0.76
HCR	28	0.386	108.4	-1.99	-32.87	1.179	1.235	-5.87	0.333	-0.161	2.26
	29	0.400	106.6	-2.33	-18.01	1.196	1.158	-4.90	0.331	-0.117	3.67
	30	0.415	104.6	-2.50	-9.85	1.212	1.095	-4.10	0.329	-0.056	4.33
	31	0.429	102.5	-2.61	-6.21	1.228	1.041	-3.70	0.329	0.007	3.99
	32	0.443	100.3	-2.68	-2.78	1.242	0.989	-4.01	0.330	0.058	2.75
	33	0.458	98.1	-2.69	2.90	1.256	0.926	-5.05	0.331	0.086	0.98

34	0.472	95.9	-2.60	10.00	1.269	0.844	-6.23	0.332	0.086	-0.67
35	0.486	93.8	-2.41	15.87	1.280	0.748	-6.81	0.333	0.066	-1.61
36	0.500	91.9	-2.15	18.47	1.290	0.650	-6.62	0.334	0.040	-1.74
37	0.515	90.3	-1.88	17.56	1.299	0.559	-6.09	0.334	0.016	-1.43
38	0.529	88.9	-1.64	14.51	1.306	0.475	-5.57	0.334	-0.001	-0.98
39	0.543	87.6	-1.46	10.97	1.312	0.399	-4.93	0.334	-0.012	-0.39
40	0.558	86.5	-1.33	7.87	1.317	0.334	-3.89	0.334	-0.012	0.40
41	0.572	85.4	-1.24	5.13	1.322	0.288	-2.49	0.334	0.000	1.14
42	0.586	84.4	-1.18	2.56	1.326	0.263	-1.08	0.334	0.020	1.31
43	0.601	83.5	-1.16	0.58	1.329	0.257	0.06	0.334	0.037	0.64
44	0.615	82.5	-1.17	-0.22	1.333	0.265	0.82	0.335	0.038	-0.53
45	0.629	81.6	-1.17	0.10	1.337	0.281	1.24	0.336	0.022	-1.50
46	0.643	80.6	-1.16	0.69	1.341	0.300	1.34	0.336	-0.005	-1.86
47	0.658	79.7	-1.15	0.65	1.345	0.319	1.27	0.335	-0.031	-1.62
48	0.672	78.7	-1.14	-0.22	1.350	0.336	1.21	0.335	-0.051	-0.95
49	0.686	77.8	-1.16	-1.19	1.355	0.353	1.17	0.334	-0.059	-0.09
50	0.701	76.8	-1.18	-1.64	1.360	0.370	1.00	0.333	-0.053	0.72
51	0.715	75.9	-1.20	-1.93	1.366	0.382	0.66	0.332	-0.038	1.30
52	0.729	74.9	-1.23	-2.99	1.371	0.389	0.45	0.332	-0.016	1.62
53	0.744	73.8	-1.29	-5.30	1.377	0.395	0.80	0.332	0.009	1.74
54	0.758	72.7	-1.39	-8.47	1.383	0.411	1.72	0.332	0.034	1.74
55	0.772	71.6	-1.53	-11.61	1.389	0.444	2.77	0.333	0.058	1.66
56	0.786	70.2	-1.72	-13.88	1.395	0.491	3.58	0.334	0.081	1.43
57	0.801	68.7	-1.93	-14.88	1.403	0.547	4.29	0.335	0.099	1.00
58	0.815	67.1	-2.14	-14.71	1.411	0.613	5.39	0.337	0.110	0.53
59	0.829	65.2	-2.35	-14.08	1.420	0.701	7.07	0.338	0.114	0.23
60	0.844	63.2	-2.55	-13.69	1.431	0.816	9.13	0.340	0.116	0.09
61	0.858	61.1	-2.74	-13.68	1.443	0.962	11.25	0.342	0.117	0.00
62	0.872	58.7	-2.94	-13.46	1.458	1.137	13.11	0.343	0.116	0.06
63	0.887	56.2	-3.13	-11.96	1.476	1.337	14.27	0.345	0.118	0.37
64	0.901	53.6	-3.28	-8.16	1.497	1.545	14.31	0.347	0.127	0.82
65	0.915	50.9	-3.36	-1.88	1.520	1.746	13.29	0.349	0.142	1.13
66	0.929	48.1	-3.33	6.33	1.547	1.926	11.68	0.351	0.159	1.26
67	0.944	45.4	-3.18	15.67	1.575	2.080	9.99	0.353	0.178	1.39
68	0.958	42.9	-2.88	25.25	1.606	2.211	8.47	0.356	0.199	1.66
69	0.972	40.7	-2.46	34.20	1.639	2.322	7.11	0.359	0.225	1.88

TABLE A.3(*b*) (*Continued*)

	FRAME	TIME S	THETA DEG	OMEGA R/S	ALPHA R/S/S	CofM-X M	VEL-X M/S	ACC-X M/S/S	CofM-Y M	VEL-Y M/S	ACC-Y M/S/S
TOR	70	0.987	38.9	-1.91	42.20	1.672	2.415	5.93	0.362	0.253	1.75
	71	1.001	37.6	-1.25	49.54	1.708	2.492	5.10	0.366	0.275	1.03
	72	1.015	36.8	-0.49	56.03	1.744	2.560	4.60	0.370	0.282	-0.23
	73	1.030	36.8	0.35	60.38	1.781	2.624	4.17	0.374	0.269	-1.71
	74	1.044	37.4	1.24	61.40	1.819	2.680	3.53	0.378	0.234	-2.97
	75	1.058	38.8	2.11	58.96	1.857	2.725	2.64	0.381	0.184	-3.79
	76	1.072	40.9	2.92	53.72	1.897	2.755	1.59	0.383	0.125	-4.25
	77	1.087	43.6	3.65	46.69	1.936	2.770	0.57	0.385	0.062	-4.57
	78	1.101	46.8	4.26	38.96	1.976	2.771	-0.31	0.385	-0.006	-4.80
	79	1.115	50.6	4.76	31.29	2.016	2.761	-0.96	0.384	-0.075	-4.73
	80	1.130	54.6	5.15	24.07	2.055	2.744	-1.39	0.383	-0.141	-4.26
	81	1.144	59.0	5.45	17.67	2.094	2.722	-1.68	0.380	-0.197	-3.51
	82	1.158	63.6	5.66	12.52	2.133	2.696	-1.93	0.377	-0.241	-2.66
	83	1.173	68.3	5.81	8.73	2.171	2.667	-2.22	0.373	-0.273	-1.78
	84	1.187	73.1	5.91	6.21	2.209	2.632	-2.66	0.369	-0.292	-0.96
	85	1.201	78.0	5.98	4.43	2.246	2.590	-3.35	0.365	-0.301	-0.35
	86	1.215	82.9	6.04	1.90	2.283	2.536	-4.35	0.361	-0.302	0.05
	87	1.230	87.9	6.04	-3.69	2.319	2.466	-5.61	0.356	-0.299	0.43
	88	1.244	92.8	5.93	-14.21	2.354	2.376	-6.98	0.352	-0.290	0.92
	89	1.258	97.6	5.63	-29.76	2.387	2.266	-8.35	0.348	-0.273	1.38
	90	1.273	102.0	5.08	-48.61	2.418	2.137	-9.47	0.344	-0.250	1.67
	91	1.287	105.9	4.24	-67.64	2.448	1.996	-9.96	0.341	-0.225	1.72
	92	1.301	109.0	3.14	-82.66	2.476	1.853	-9.67	0.338	-0.201	1.54
	93	1.316	111.0	1.88	-89.51	2.501	1.719	-8.82	0.335	-0.181	1.24
	94	1.330	112.0	0.58	-85.71	2.525	1.600	-7.71	0.333	-0.166	1.20
	95	1.344	112.0	-0.57	-71.95	2.547	1.499	-6.53	0.331	-0.147	1.66
	96	1.358	111.1	-1.47	-52.52	2.568	1.414	-5.46	0.329	-0.118	2.38
HCR	97	1.373	109.6	-2.08	-33.63	2.587	1.343	-4.69	0.327	-0.079	2.96
	98	1.387	107.7	-2.43	-20.13	2.606	1.279	-4.45	0.326	-0.033	3.19
	99	1.401	105.6	-2.65	-12.52	2.624	1.215	-4.84	0.326	0.012	2.88
	100	1.416	103.4	-2.79	-7.44	2.641	1.141	-5.82	0.327	0.049	1.96
	101	1.430	101.0	-2.86	-1.54	2.656	1.049	-7.22	0.328	0.068	0.66
	102	1.444	98.7	-2.84	5.68	2.671	0.934	-8.71	0.329	0.068	-0.54
	103	1.459	96.4	-2.70	12.64	2.683	0.800	-9.67	0.330	0.053	-1.23
	104	1.473	94.2	-2.48	18.05	2.694	0.658	-9.59	0.330	0.033	-1.29
	105	1.487	92.3	-2.19	21.86	2.702	0.526	-8.48	0.330	0.016	-0.96
	106	1.501	90.7	-1.85	24.55	2.709	0.415	-6.99	0.331	0.005	-0.61

TABLE A.3(c) Linear and Angular Kinematics—Thigh

TABLE A.3(c) Linear and Angular Kinematics—Thigh

	FRAME	TIME S	THETA DEG	OMEGA R/S	ALPHA R/S/S	CofM-X M	VEL-X M/S	ACC-X M/S/S	CofM-Y M	VEL-Y M/S	ACC-Y M/S/S
TOR	1	0.000	82.7	3.29	24.42	0.430	2.081	1.36	0.652	-0.052	4.40
	2	0.014	85.5	3.59	18.34	0.460	2.073	-1.89	0.652	0.026	6.06
	3	0.029	88.6	3.81	12.48	0.489	2.027	-3.83	0.653	0.122	6.86
	4	0.043	91.8	3.95	5.97	0.518	1.963	-4.53	0.655	0.222	6.72
	5	0.057	95.1	3.98	-1.93	0.546	1.898	-4.32	0.659	0.314	5.74
	6	0.072	98.3	3.89	-10.64	0.572	1.840	-3.64	0.664	0.386	4.11
	7	0.086	101.4	3.68	-18.57	0.598	1.793	-2.92	0.670	0.432	2.20
	8	0.100	104.3	3.36	-24.25	0.624	1.756	-2.45	0.677	0.449	0.42
	9	0.114	106.9	2.99	-27.38	0.648	1.724	-2.28	0.683	0.444	-1.15
	10	0.129	109.2	2.58	-28.82	0.673	1.691	-2.31	0.689	0.417	-2.60
	11	0.143	111.2	2.16	-29.42	0.697	1.658	-2.41	0.695	0.369	-3.94
	12	0.157	112.8	1.74	-29.22	0.720	1.622	-2.53	0.700	0.304	-4.97
	13	0.172	114.0	1.33	-27.99	0.743	1.585	-2.69	0.704	0.227	-5.63
	14	0.186	114.9	0.94	-26.18	0.766	1.545	-2.83	0.706	0.143	-6.02
	15	0.200	115.6	0.58	-24.68	0.787	1.504	-2.79	0.708	0.055	-6.22
	16	0.215	115.9	0.23	-23.85	0.809	1.466	-2.51	0.708	-0.035	-6.23
	17	0.229	115.9	-0.11	-23.01	0.829	1.433	-2.04	0.707	-0.123	-5.97
	18	0.243	115.7	-0.43	-20.96	0.850	1.407	-1.40	0.704	-0.206	-5.34
	19	0.257	115.2	-0.70	-16.94	0.870	1.393	-0.55	0.701	-0.276	-4.22
	20	0.272	114.6	-0.91	-11.04	0.889	1.392	0.42	0.696	-0.326	-2.51
	21	0.286	113.7	-1.02	-4.04	0.909	1.405	1.38	0.692	-0.348	-0.34
	22	0.300	112.9	-1.03	2.83	0.930	1.431	2.28	0.687	-0.336	1.68
	23	0.315	112.1	-0.94	8.12	0.950	1.470	3.03	0.682	-0.300	2.98
	24	0.329	111.3	-0.79	10.77	0.972	1.518	3.42	0.678	-0.251	3.50
	25	0.343	110.8	-0.63	10.49	0.994	1.568	3.26	0.675	-0.200	3.62
	26	0.357	110.3	-0.49	7.65	1.016	1.611	2.48	0.672	-0.148	3.69
	27	0.372	110.0	-0.41	3.48	1.040	1.639	1.20	0.671	-0.094	3.92
HCR	28	0.386	109.6	-0.39	0.13	1.063	1.645	-0.31	0.670	-0.036	4.29
	29	0.400	109.3	-0.41	-0.49	1.087	1.630	-1.67	0.669	0.029	4.62
	30	0.415	109.0	-0.41	1.69	1.110	1.597	-2.61	0.670	0.096	4.64
	31	0.429	108.6	-0.36	4.27	1.133	1.555	-3.28	0.672	0.161	4.04
	32	0.443	108.4	-0.29	3.58	1.154	1.503	-4.05	0.675	0.212	2.63
	33	0.458	108.2	-0.26	-2.78	1.175	1.439	-4.94	0.678	0.236	0.62

34	0.472	107.9	-0.37	-13.38	1.196	1.362	-5.49	0.682	0.230	-1.26
35	0.486	107.6	-0.64	-23.43	1.214	1.282	-5.38	0.685	0.200	-2.26
36	0.500	106.9	-1.04	-28.36	1.232	1.208	-4.85	0.687	0.165	-2.32
37	0.515	105.9	-1.45	-26.92	1.249	1.143	-4.43	0.690	0.134	-2.00
38	0.529	104.5	-1.81	-20.92	1.265	1.082	-4.34	0.691	0.108	-1.77
39	0.543	102.9	-2.05	-12.96	1.280	1.019	-4.36	0.693	0.083	-1.52
40	0.558	101.2	-2.18	-4.84	1.294	0.957	-4.13	0.694	0.064	-1.01
41	0.572	99.3	-2.19	2.16	1.307	0.901	-3.50	0.695	0.054	-0.44
42	0.586	97.6	-2.11	6.84	1.320	0.857	-2.56	0.695	0.052	-0.31
43	0.601	95.9	-1.99	8.52	1.332	0.828	-1.53	0.696	0.045	-0.97
44	0.615	94.3	-1.87	7.81	1.344	0.813	-0.65	0.697	0.024	-2.12
45	0.629	92.8	-1.77	6.10	1.355	0.809	-0.07	0.697	-0.015	-2.99
46	0.643	91.4	-1.70	4.22	1.367	0.811	0.26	0.696	-0.062	-3.02
47	0.658	90.0	-1.65	2.22	1.378	0.817	0.61	0.695	-0.102	-2.35
48	0.672	88.7	-1.63	0.01	1.390	0.828	1.15	0.693	-0.129	-1.46
49	0.686	87.4	-1.65	-2.10	1.402	0.850	1.70	0.691	-0.143	-0.63
50	0.701	86.0	-1.69	-3.38	1.414	0.877	1.85	0.689	-0.147	0.01
51	0.715	84.6	-1.75	-3.24	1.427	0.902	1.51	0.687	-0.143	0.45
52	0.729	83.1	-1.79	-1.71	1.440	0.920	1.16	0.685	-0.134	0.67
53	0.744	81.7	-1.79	0.98	1.453	0.936	1.38	0.683	-0.124	0.79
54	0.758	80.2	-1.76	4.63	1.467	0.960	2.18	0.681	-0.112	0.78
55	0.772	78.8	-1.66	8.63	1.481	0.998	3.02	0.680	-0.101	0.58
56	0.786	77.5	-1.51	11.95	1.495	1.046	3.58	0.679	-0.095	0.20
57	0.801	76.3	-1.32	13.92	1.511	1.100	4.10	0.677	-0.096	-0.19
58	0.815	75.3	-1.11	14.82	1.527	1.163	5.02	0.676	-0.101	-0.46
59	0.829	74.5	-0.90	15.41	1.544	1.244	6.42	0.674	-0.109	-0.73
60	0.844	73.8	-0.67	16.49	1.562	1.347	7.98	0.673	-0.121	-1.12
61	0.858	73.4	-0.43	18.57	1.583	1.472	9.40	0.671	-0.141	-1.49
62	0.872	73.1	-0.14	21.54	1.605	1.616	10.36	0.669	-0.164	-1.58
63	0.887	73.1	0.19	24.59	1.629	1.768	10.40	0.666	-0.186	-1.28
64	0.901	73.4	0.56	27.13	1.655	1.913	9.09	0.663	-0.201	-0.68
65	0.915	74.1	0.97	29.35	1.683	2.028	6.50	0.660	-0.205	0.06
66	0.929	75.0	1.40	31.19	1.713	2.099	3.27	0.658	-0.199	0.94
67	0.944	76.4	1.86	31.71	1.743	2.122	0.24	0.655	-0.179	2.00
68	0.958	78.1	2.31	30.36	1.774	2.106	-2.10	0.652	-0.142	3.27
69	0.972	80.2	2.73	27.73	1.804	2.062	-3.60	0.651	-0.085	4.59

TABLE A.3(*c*) (*Continued*)

	FRAME	TIME S	THETA DEG	OMEGA R/S	ALPHA R/S/S	CofM-X M	VEL-X M/S	ACC-X M/S/S	CofM-Y M	VEL-Y M/S	ACC-Y M/S/S
TOR	70	0.987	82.5	3.10	24.46	1.833	2.003	-4.23	0.650	-0.010	5.65
	71	1.001	85.2	3.43	20.30	1.861	1.941	-3.96	0.650	0.076	6.21
	72	1.015	88.2	3.68	14.67	1.888	1.889	-3.05	0.652	0.167	6.16
	73	1.030	91.3	3.85	7.52	1.915	1.854	-2.02	0.655	0.253	5.61
	74	1.044	94.5	3.90	-0.87	1.941	1.832	-1.28	0.659	0.328	4.76
	75	1.058	97.7	3.82	-10.14	1.967	1.817	-0.93	0.665	0.389	3.66
	76	1.072	100.7	3.61	-19.00	1.993	1.805	-0.88	0.671	0.432	2.15
	77	1.087	103.6	3.28	-25.45	2.019	1.792	-1.09	0.677	0.451	0.23
	78	1.101	106.1	2.88	-28.34	2.045	1.774	-1.51	0.683	0.439	-1.77
	79	1.115	108.3	2.47	-27.92	2.070	1.749	-2.07	0.689	0.400	-3.40
	80	1.130	110.1	2.08	-25.47	2.095	1.715	-2.71	0.695	0.342	-4.45
	81	1.144	111.7	1.74	-22.64	2.119	1.671	-3.30	0.699	0.273	-5.04
	82	1.158	113.0	1.43	-20.83	2.142	1.620	-3.72	0.703	0.198	-5.39
	83	1.173	114.0	1.14	-20.75	2.165	1.565	-3.87	0.705	0.119	-5.64
	84	1.187	114.9	0.84	-22.11	2.187	1.510	-3.74	0.706	0.036	-5.90
	85	1.201	115.4	0.51	-23.89	2.208	1.458	-3.37	0.706	-0.050	-6.16
	86	1.215	115.7	0.16	-24.95	2.229	1.413	-2.72	0.705	-0.140	-6.20
	87	1.230	115.7	-0.20	-24.36	2.249	1.380	-1.76	0.702	-0.227	-5.66
	88	1.244	115.4	-0.54	-21.36	2.268	1.363	-0.58	0.698	-0.302	-4.43
	89	1.258	114.8	-0.81	-15.42	2.288	1.364	0.58	0.693	-0.354	-2.69
	90	1.273	114.0	-0.98	-6.72	2.307	1.380	1.61	0.688	-0.379	-0.70
	91	1.287	113.2	-1.01	3.05	2.327	1.410	2.71	0.682	-0.374	1.30
	92	1.301	112.4	-0.89	10.92	2.348	1.457	3.92	0.677	-0.342	3.07
	93	1.316	111.7	-0.69	14.55	2.369	1.522	4.86	0.673	-0.287	4.38
	94	1.330	111.3	-0.48	13.64	2.391	1.596	5.04	0.669	-0.216	5.19
	95	1.344	110.9	-0.30	9.66	2.415	1.666	4.28	0.667	-0.138	5.50
	96	1.358	110.8	-0.20	5.05	2.439	1.718	2.72	0.665	-0.059	5.33
HCR	97	1.373	110.6	-0.16	2.00	2.464	1.744	0.71	0.665	0.014	4.88
	98	1.387	110.5	-0.14	1.10	2.489	1.739	-1.30	0.666	0.080	4.42
	99	1.401	110.4	-0.13	0.73	2.513	1.706	-3.03	0.667	0.141	3.89
	100	1.416	110.3	-0.12	-1.58	2.537	1.652	-4.52	0.670	0.192	2.92
	101	1.430	110.2	-0.17	-7.15	2.561	1.577	-5.90	0.673	0.224	1.39
	102	1.444	110.0	-0.33	-14.74	2.583	1.483	-7.13	0.676	0.231	-0.34
	103	1.459	109.6	-0.59	-21.06	2.603	1.373	-8.01	0.679	0.215	-1.65
	104	1.473	109.0	-0.93	-22.96	2.622	1.254	-8.45	0.682	0.184	-2.18
	105	1.487	108.1	-1.25	-19.15	2.639	1.132	-8.72	0.684	0.152	-2.11
	106	1.501	107.0	-1.48	-10.16	2.654	1.005	-9.40	0.686	0.124	-1.89

TABLE A.3(*d*) Linear and Angular Kinematics—$\frac{1}{2}$ HAT

TABLE A.3(*d*) Linear and Angular Kinematics—$\frac{1}{2}$ HAT

	FRAME	TIME S	THETA DEG	OMEGA R/S	ALPHA R/S/S	CofM-X M	VEL-X M/S	ACC-X M/S/S	CofM-Y M	VEL-Y M/S	ACC-Y M/S/S
TOR	1	0.000	85.1	0.89	-11.61	0.473	1.377	1.23	1.080	0.035	2.76
	2	0.014	85.7	0.71	-13.17	0.492	1.379	-0.70	1.081	0.084	3.89
	3	0.029	86.2	0.52	-11.89	0.512	1.357	-2.11	1.083	0.146	4.49
	4	0.043	86.6	0.37	-8.26	0.531	1.319	-2.87	1.085	0.212	4.46
	5	0.057	86.8	0.28	-3.23	0.550	1.275	-2.98	1.089	0.274	3.80
	6	0.072	87.0	0.27	1.64	0.568	1.234	-2.50	1.093	0.321	2.60
	7	0.086	87.3	0.33	4.85	0.585	1.203	-1.64	1.098	0.348	1.19
	8	0.100	87.6	0.41	6.05	0.602	1.187	-0.74	1.103	0.355	-0.18
	9	0.114	88.0	0.50	6.17	0.619	1.182	-0.15	1.108	0.343	-1.50
	10	0.129	88.4	0.59	6.19	0.636	1.183	0.06	1.113	0.312	-2.81
	11	0.143	88.9	0.68	6.10	0.653	1.184	0.06	1.117	0.262	-3.89
	12	0.157	89.5	0.76	5.21	0.670	1.184	0.06	1.120	0.201	-4.54
	13	0.172	90.2	0.83	3.01	0.687	1.186	0.21	1.123	0.133	-4.81
	14	0.186	90.9	0.85	-0.12	0.704	1.190	0.56	1.124	0.063	-4.87
	15	0.200	91.6	0.82	-3.16	0.721	1.201	1.15	1.124	-0.007	-4.83
	16	0.215	92.2	0.76	-5.47	0.738	1.223	1.93	1.124	-0.075	-4.72
	17	0.229	92.8	0.67	-6.91	0.756	1.257	2.71	1.122	-0.142	-4.48
	18	0.243	93.3	0.56	-7.30	0.774	1.300	3.24	1.120	-0.203	-3.98
	19	0.257	93.7	0.46	-6.57	0.793	1.349	3.38	1.116	-0.255	-3.07
	20	0.272	94.1	0.37	-4.89	0.813	1.397	3.11	1.112	-0.291	-1.63
	21	0.286	94.3	0.32	-2.38	0.833	1.438	2.48	1.108	-0.302	0.10
	22	0.300	94.6	0.31	0.85	0.854	1.468	1.66	1.104	-0.288	1.53
	23	0.315	94.8	0.34	3.98	0.875	1.486	0.95	1.100	-0.259	2.21
	24	0.329	95.1	0.42	5.39	0.896	1.495	0.64	1.096	-0.225	2.40
	25	0.343	95.5	0.50	3.95	0.918	1.504	0.91	1.093	-0.190	2.60
	26	0.357	96.0	0.53	0.20	0.939	1.521	1.52	1.091	-0.151	3.07
	27	0.372	96.4	0.50	-4.14	0.961	1.548	1.98	1.089	-0.102	3.79
HCR	28	0.386	96.8	0.41	-7.80	0.984	1.578	1.91	1.088	-0.042	4.62
	29	0.400	97.1	0.28	-10.40	1.006	1.602	1.32	1.088	0.030	5.28
	30	0.415	97.2	0.12	-11.97	1.029	1.616	0.50	1.089	0.109	5.48
	31	0.429	97.3	-0.06	-12.71	1.053	1.617	-0.32	1.091	0.186	4.93
	32	0.443	97.1	-0.25	-12.51	1.076	1.606	-1.04	1.094	0.250	3.50
	33	0.458	96.9	-0.42	-10.48	1.098	1.587	-1.63	1.098	0.287	1.45

34	0.472	96.4	-0.55	-6.07	1.121	1.560	-1.99	1.102	0.291	-0.50
35	0.486	96.0	-0.59	-0.55	1.143	1.530	-2.10	1.106	0.272	-1.54
36	0.500	95.5	-0.56	3.64	1.165	1.500	-2.10	1.110	0.247	-1.63
37	0.515	95.0	-0.49	4.75	1.186	1.470	-2.14	1.114	0.226	-1.49
38	0.529	94.7	-0.43	2.67	1.207	1.439	-2.14	1.117	0.205	-1.69
39	0.543	94.3	-0.41	-1.41	1.227	1.409	-1.93	1.119	0.178	-1.99
40	0.558	94.0	-0.47	-5.74	1.247	1.384	-1.49	1.122	0.148	-1.98
41	0.572	93.6	-0.58	-8.82	1.267	1.366	-0.95	1.124	0.121	-1.69
42	0.586	93.0	-0.72	-9.90	1.286	1.356	-0.43	1.125	0.099	-1.57
43	0.601	92.4	-0.86	-9.06	1.305	1.354	0.04	1.126	0.076	-1.99
44	0.615	91.6	-0.98	-7.22	1.325	1.357	0.42	1.127	0.042	-2.88
45	0.629	90.8	-1.07	-5.50	1.344	1.366	0.71	1.128	-0.006	-3.64
46	0.643	89.9	-1.14	-4.17	1.364	1.378	0.90	1.127	-0.062	-3.69
47	0.658	88.9	-1.19	-2.51	1.384	1.392	1.02	1.126	-0.112	-3.04
48	0.672	87.9	-1.21	0.15	1.404	1.407	1.07	1.124	-0.149	-2.20
49	0.686	87.0	-1.18	3.16	1.424	1.422	1.00	1.122	-0.175	-1.53
50	0.701	86.0	-1.12	4.79	1.444	1.435	0.80	1.119	-0.192	-1.05
51	0.715	85.1	-1.05	4.15	1.465	1.445	0.57	1.116	-0.205	-0.71
52	0.729	84.3	-1.00	2.30	1.486	1.452	0.52	1.113	-0.213	-0.43
53	0.744	83.5	-0.98	1.15	1.506	1.460	0.71	1.110	-0.217	-0.07
54	0.758	82.7	-0.97	1.51	1.527	1.472	0.94	1.107	-0.215	0.31
55	0.772	81.9	-0.94	2.94	1.549	1.487	0.88	1.104	-0.208	0.52
56	0.786	81.2	-0.88	5.03	1.570	1.497	0.44	1.101	-0.200	0.55
57	0.801	80.5	-0.79	7.77	1.591	1.499	-0.02	1.098	-0.192	0.60
58	0.815	79.9	-0.66	10.87	1.613	1.497	-0.02	1.096	-0.183	0.79
59	0.829	79.4	-0.48	13.59	1.634	1.499	0.59	1.093	-0.169	0.91
60	0.844	79.1	-0.27	15.25	1.656	1.513	1.59	1.091	-0.156	0.79
61	0.858	78.9	-0.05	15.83	1.677	1.544	2.58	1.089	-0.147	0.58
62	0.872	79.0	0.18	15.92	1.700	1.587	3.10	1.086	-0.140	0.55
63	0.887	79.2	0.41	15.99	1.723	1.633	2.64	1.085	-0.131	0.71
64	0.901	79.7	0.64	15.90	1.747	1.663	0.88	1.083	-0.119	0.95
65	0.915	80.3	0.86	15.03	1.770	1.658	-1.90	1.081	-0.104	1.21
66	0.929	81.1	1.07	12.34	1.794	1.608	-4.71	1.080	-0.085	1.50
67	0.944	82.0	1.22	6.89	1.816	1.523	-6.30	1.079	-0.061	1.88
68	0.958	83.1	1.27	-0.81	1.838	1.428	-6.22	1.078	-0.031	2.43
69	0.972	84.1	1.19	-8.51	1.857	1.345	-5.08	1.078	0.008	3.16

TABLE A.3(*d*) (*Continued*)

	FRAME	TIME S	THETA DEG	OMEGA R/S	ALPHA R/S/S	CofM-X M	VEL-X M/S	ACC-X M/S/S	CofM-Y M	VEL-Y M/S	ACC-Y M/S/S
TOR	70	0.987	85.0	1.02	-13.55	1.876	1.283	-3.73	1.078	0.059	3.86
	71	1.001	85.8	0.81	-14.47	1.894	1.239	-2.53	1.079	0.119	4.26
	72	1.015	86.3	0.61	-11.76	1.911	1.210	-1.56	1.082	0.181	4.25
	73	1.030	86.8	0.47	-7.13	1.929	1.194	-0.86	1.085	0.240	3.91
	74	1.044	87.1	0.40	-2.15	1.946	1.185	-0.45	1.089	0.293	3.44
	75	1.058	87.4	0.41	2.49	1.962	1.181	-0.25	1.093	0.339	2.73
	76	1.072	87.8	0.48	6.25	1.979	1.178	-0.13	1.098	0.371	1.47
	77	1.087	88.2	0.59	8.26	1.996	1.177	0.01	1.104	0.381	-0.38
	78	1.101	88.8	0.71	7.82	2.013	1.178	0.23	1.109	0.360	-2.42
	79	1.115	89.4	0.81	4.89	2.030	1.184	0.59	1.114	0.312	-4.04
	80	1.130	90.1	0.85	0.35	2.047	1.195	1.00	1.118	0.245	-4.95
	81	1.144	90.8	0.82	-4.02	2.064	1.212	1.25	1.121	0.170	-5.20
	82	1.158	91.4	0.74	-6.62	2.082	1.231	1.20	1.123	0.096	-5.05
	83	1.173	92.0	0.63	-7.20	2.099	1.247	0.97	1.124	0.026	-4.71
	84	1.187	92.5	0.53	-6.49	2.117	1.259	0.81	1.124	-0.039	-4.37
	85	1.201	92.9	0.45	-5.14	2.135	1.270	0.83	1.123	-0.099	-4.19
	86	1.215	93.2	0.38	-3.23	2.154	1.283	1.00	1.121	-0.159	-4.11
	87	1.230	93.5	0.35	-0.73	2.172	1.299	1.26	1.118	-0.217	-3.86
	88	1.244	93.8	0.36	1.86	2.191	1.319	1.47	1.115	-0.269	-3.18
	89	1.258	94.1	0.41	3.58	2.210	1.341	1.52	1.110	-0.308	-2.09
	90	1.273	94.4	0.47	3.89	2.229	1.362	1.48	1.106	-0.329	-0.83
	91	1.287	94.9	0.52	3.56	2.249	1.383	1.53	1.101	-0.332	0.43
	92	1.301	95.3	0.57	3.94	2.269	1.406	1.67	1.096	-0.317	1.62
	93	1.316	95.8	0.63	5.06	2.289	1.431	1.80	1.092	-0.285	2.69
	94	1.330	96.3	0.71	5.42	2.309	1.457	1.92	1.088	-0.240	3.68
	95	1.344	96.9	0.79	3.53	2.330	1.486	2.11	1.085	-0.180	4.53
	96	1.358	97.6	0.81	-0.84	2.352	1.517	2.27	1.083	-0.110	5.12
HCR	97	1.373	98.3	0.76	-6.35	2.374	1.551	2.12	1.082	-0.034	5.43
	98	1.387	98.9	0.63	-11.08	2.396	1.578	1.49	1.082	0.045	5.56
	99	1.401	99.3	0.45	-13.91	2.419	1.593	0.57	1.083	0.125	5.41
	100	1.416	99.6	0.23	-14.78	2.442	1.594	-0.38	1.086	0.200	4.65
	101	1.430	99.7	0.02	-13.99	2.465	1.582	-1.19	1.089	0.258	3.12
	102	1.444	99.6	-0.17	-11.75	2.487	1.560	-1.90	1.093	0.289	1.13
	103	1.459	99.4	-0.31	-8.21	2.509	1.528	-2.78	1.097	0.291	-0.64
	104	1.473	99.1	-0.40	-3.79	2.531	1.481	-4.11	1.101	0.271	-1.69
	105	1.487	98.8	-0.42	0.37	2.552	1.410	-6.06	1.105	0.242	-2.07
	106	1.501	98.4	-0.39	2.87	2.571	1.308	-8.68	1.108	0.212	-2.24

TABLE A.4 Relative Joint Angular Kinematics—Ankle, Knee, and Hip

TABLE A.4 Relative Joint Angular Kinematics—Ankle, Knee, and Hip

	FRAME	TIME S	ANKLE THETA DEG	ANKLE OMEGA R/S	ANKLE ALPHA R/S/S	KNEE THETA DEG	KNEE OMEGA R/S	KNEE ALPHA R/S/S	HIP THETA DEG	HIP OMEGA R/S	HIP ALPHA R/S/S
TOR	1	0.000	-15.2	-2.29	94.89	46.7	6.74	-21.91	-2.4	2.39	36.03
	2	0.014	-16.4	-0.82	98.72	52.1	6.23	-46.59	-0.2	2.89	31.50
	3	0.029	-16.5	0.54	84.63	56.9	5.41	-65.29	2.3	3.30	24.37
	4	0.043	-15.6	1.60	63.13	61.0	4.37	-77.66	5.2	3.58	14.23
	5	0.057	-13.9	2.34	41.31	64.1	3.19	-84.77	8.2	3.70	1.30
	6	0.072	-11.7	2.78	20.10	66.2	1.94	-87.62	11.3	3.62	-12.28
	7	0.086	-9.3	2.92	-0.74	67.3	0.68	-86.54	14.2	3.35	-23.42
	8	0.100	-6.9	2.76	-18.70	67.3	-0.53	-81.99	16.8	2.95	-30.30
	9	0.114	-4.8	2.38	-30.23	66.4	-1.66	-75.17	19.0	2.48	-33.55
	10	0.129	-3.0	1.90	-34.13	64.6	-2.68	-67.64	20.8	1.99	-35.01
	11	0.143	-1.7	1.41	-32.12	62.0	-3.60	-60.31	22.2	1.48	-35.53
	12	0.157	-0.7	0.98	-26.84	58.7	-4.41	-53.09	23.3	0.98	-34.42
	13	0.172	-0.1	0.64	-20.23	54.8	-5.11	-45.78	23.8	0.50	-30.99
	14	0.186	0.3	0.40	-13.59	50.3	-5.72	-38.95	24.1	0.09	-26.06
	15	0.200	0.6	0.25	-8.03	45.4	-6.23	-32.91	24.0	-0.25	-21.52
	16	0.215	0.7	0.17	-3.87	40.1	-6.66	-26.57	23.7	-0.53	-18.38
	17	0.229	0.9	0.14	-0.56	34.5	-6.99	-18.05	23.1	-0.77	-16.10
	18	0.243	1.0	0.15	2.33	28.7	-7.17	-5.87	22.4	-0.99	-13.65
	19	0.257	1.1	0.21	4.60	22.8	-7.16	10.65	21.5	-1.16	-10.37
	20	0.272	1.3	0.29	5.47	16.9	-6.87	31.09	20.5	-1.28	-6.16
	21	0.286	1.6	0.36	3.83	11.5	-6.27	53.59	19.4	-1.34	-1.66
	22	0.300	1.9	0.40	-0.84	6.7	-5.34	74.76	18.3	-1.33	1.98
	23	0.315	2.2	0.34	-8.36	2.7	-4.13	90.62	17.2	-1.28	4.14
	24	0.329	2.5	0.16	-19.22	-0.1	-2.75	97.96	16.2	-1.21	5.38
	25	0.343	2.5	-0.21	-33.28	-1.8	-1.33	95.28	15.2	-1.13	6.53
	26	0.357	2.1	-0.79	-46.39	-2.3	-0.02	82.64	14.3	-1.03	7.45
	27	0.372	1.2	-1.54	-50.65	-1.8	1.04	62.40	13.6	-0.92	7.62
HCR	28	0.386	-0.4	-2.24	-38.30	-0.6	1.76	40.55	12.8	-0.81	7.93
	29	0.400	-2.5	-2.63	-7.99	1.1	2.20	24.37	12.2	-0.69	9.91
	30	0.415	-4.7	-2.47	30.12	3.0	2.46	16.62	11.7	-0.52	13.66
	31	0.429	-6.5	-1.77	59.60	5.1	2.67	12.93	11.4	-0.30	16.99
	32	0.443	-7.6	-0.77	70.70	7.4	2.83	5.97	11.2	-0.04	16.09

33	0.458	-7.8	0.25	64.54	9.8	2.84	-8.17	11.3	0.16	7.69
34	0.472	-7.2	1.08	46.87	12.1	2.60	-27.06	11.5	0.18	-7.32
35	0.486	-6.0	1.59	24.33	14.0	2.07	-44.04	11.6	-0.05	-22.88
36	0.500	-4.6	1.77	4.14	15.4	1.34	-53.13	11.4	-0.47	-32.00
37	0.515	-3.1	1.71	-8.73	16.2	0.55	-52.57	10.8	-0.96	-31.67
38	0.529	-1.8	1.52	-13.85	16.3	-0.17	-44.30	9.8	-1.38	-23.59
39	0.543	-0.6	1.31	-13.32	15.9	-0.72	-31.51	8.6	-1.64	-11.54
40	0.558	0.4	1.14	-9.71	15.2	-1.07	-17.12	7.2	-1.71	0.90
41	0.572	1.2	1.04	-6.42	14.2	-1.21	-3.98	5.8	-1.61	10.98
42	0.586	2.1	0.96	-6.79	13.2	-1.18	5.07	4.5	-1.40	16.74
43	0.601	2.8	0.84	-10.34	12.3	-1.06	8.50	3.5	-1.13	17.58
44	0.615	3.4	0.66	-12.36	11.4	-0.94	7.49	2.7	-0.89	15.03
45	0.629	3.9	0.49	-9.52	10.7	-0.85	4.99	2.0	-0.70	11.60
46	0.643	4.2	0.39	-3.26	10.1	-0.80	3.03	1.5	-0.56	8.39
47	0.658	4.6	0.40	3.33	9.4	-0.76	1.90	1.1	-0.46	4.74
48	0.672	4.9	0.49	7.14	8.8	-0.74	1.14	0.8	-0.43	-0.13
49	0.686	5.3	0.60	4.92	8.2	-0.73	0.45	0.4	-0.47	-5.26
50	0.701	5.9	0.63	-3.15	7.6	-0.73	0.18	0.0	-0.58	-8.17
51	0.715	6.4	0.51	-11.50	7.0	-0.72	1.21	-0.5	-0.70	-7.39
52	0.729	6.7	0.30	-14.34	6.4	-0.69	4.06	-1.2	-0.79	-4.00
53	0.744	6.9	0.10	-10.68	5.9	-0.61	8.77	-1.8	-0.81	-0.17
54	0.758	6.9	-0.01	-4.17	5.4	-0.44	15.06	-2.5	-0.79	3.12
55	0.772	6.9	-0.02	0.00	5.2	-0.18	22.07	-3.1	-0.73	5.70
56	0.786	6.8	-0.01	-1.14	5.1	0.19	28.41	-3.7	-0.63	6.92
57	0.801	6.8	-0.05	-7.42	5.5	0.63	32.98	-4.2	-0.53	6.16
58	0.815	6.8	-0.22	-17.46	6.2	1.13	35.54	-4.6	-0.45	3.95
59	0.829	6.5	-0.55	-28.75	7.3	1.65	36.64	-4.9	-0.41	1.81
60	0.844	5.9	-1.04	-37.44	8.9	2.18	37.19	-5.2	-0.40	1.24
61	0.858	4.8	-1.62	-40.93	10.9	2.71	37.90	-5.6	-0.38	2.74
62	0.872	3.2	-2.21	-40.94	13.3	3.26	39.00	-5.9	-0.32	5.62
63	0.887	1.2	-2.79	-41.67	16.2	3.83	40.04	-6.1	-0.22	8.61
64	0.901	-1.4	-3.40	-43.21	19.6	4.41	39.95	-6.2	-0.08	11.23
65	0.915	-4.4	-4.03	-38.18	23.5	4.97	37.59	-6.2	0.10	14.31
66	0.929	-8.0	-4.50	-17.72	27.8	5.48	32.25	-6.0	0.33	18.85
67	0.944	-11.8	-4.54	18.41	32.4	5.90	23.44	-5.7	0.64	24.81
68	0.958	-15.4	-3.97	59.42	37.4	6.15	10.87	-5.0	1.04	31.17
69	0.972	-18.3	-2.84	91.15	42.5	6.21	-4.80	-4.0	1.53	36.24

TABLE A.4 (*Continued*)

	FRAME	TIME	ANKLE			KNEE			HIP		
			THETA	OMEGA	ALPHA	THETA	OMEGA	ALPHA	THETA	OMEGA	ALPHA
		S	DEG	R/S	R/S/S	DEG	R/S	R/S/S	DEG	R/S	R/S/S
TOR	70	0.987	-20.1	-1.36	104.21	47.6	6.02	-21.92	-2.5	2.08	38.01
	71	1.001	-20.5	0.14	96.45	52.4	5.58	-38.75	-0.5	2.62	34.78
	72	1.015	-19.8	1.40	72.74	56.7	4.91	-53.72	1.8	3.07	26.42
	73	1.030	-18.2	2.22	42.50	60.4	4.04	-65.55	4.5	3.38	14.66
	74	1.044	-16.2	2.61	15.01	63.4	3.03	-74.32	7.3	3.49	1.29
	75	1.058	-14.0	2.65	-3.93	65.4	1.92	-80.71	10.2	3.41	-12.63
	76	1.072	-11.8	2.50	-13.24	66.5	0.73	-83.96	12.9	3.13	-25.25
	77	1.087	-9.9	2.28	-16.29	66.6	-0.48	-82.70	15.3	2.69	-33.71
	78	1.101	-8.1	2.03	-17.51	65.7	-1.64	-76.99	17.3	2.17	-36.16
	79	1.115	-6.5	1.77	-18.73	63.9	-2.69	-68.11	18.9	1.66	-32.81
	80	1.130	-5.2	1.50	-19.65	61.3	-3.59	-57.74	20.1	1.23	-25.82
	81	1.144	-4.1	1.21	-19.37	58.0	-4.34	-48.02	20.9	0.92	-18.62
	82	1.158	-3.2	0.94	-17.21	54.2	-4.96	-40.91	21.6	0.70	-14.21
	83	1.173	-2.5	0.72	-13.87	49.9	-5.51	-37.09	22.1	0.51	-13.55
	84	1.187	-2.0	0.55	-10.79	45.2	-6.02	-35.35	22.4	0.31	-15.62
	85	1.201	-1.6	0.41	-7.88	40.0	-6.52	-32.90	22.6	0.06	-18.75
	86	1.215	-1.3	0.32	-3.38	34.5	-6.96	-26.57	22.5	-0.23	-21.73
	87	1.230	-1.1	0.31	3.27	28.6	-7.28	-14.15	22.2	-0.56	-23.63
	88	1.244	-0.8	0.42	9.14	22.6	-7.37	4.76	21.6	-0.90	-23.22
	89	1.258	-0.4	0.58	10.06	16.5	-7.14	28.98	20.7	-1.22	-19.01
	90	1.273	0.1	0.70	4.80	10.9	-6.54	56.01	19.6	-1.45	-10.61
	91	1.287	0.7	0.71	-4.67	5.8	-5.54	81.46	18.3	-1.52	-0.51
	92	1.301	1.3	0.57	-16.27	1.8	-4.21	99.89	17.1	-1.46	6.98
	93	1.316	1.7	0.25	-29.67	-1.1	-2.68	107.07	15.9	-1.32	9.49
	94	1.330	1.7	-0.28	-44.20	-2.6	-1.15	101.58	14.9	-1.19	8.22
	95	1.344	1.2	-1.02	-54.99	-3.0	0.22	85.39	14.0	-1.09	6.14
	96	1.358	0.0	-1.85	-53.10	-2.3	1.30	63.88	13.1	-1.01	5.89
HCR	97	1.373	-1.8	-2.53	-31.93	-0.8	2.05	43.96	12.3	-0.92	8.35
	98	1.387	-4.1	-2.77	4.96	1.1	2.55	29.90	11.6	-0.78	12.18
	99	1.401	-6.4	-2.39	43.33	3.4	2.90	20.18	11.1	-0.57	14.65
	100	1.416	-8.0	-1.53	67.62	5.9	3.13	9.21	10.7	-0.36	13.20
	101	1.430	-8.9	-0.46	71.62	8.5	3.17	-6.47	10.5	-0.20	6.83
	102	1.444	-8.8	0.52	59.88	11.0	2.94	-24.49	10.4	-0.16	-2.99
	103	1.459	-8.0	1.25	41.13	13.3	2.47	-39.25	10.2	-0.28	-12.86
	104	1.473	-6.7	1.70	22.10	15.1	1.82	-46.90	9.9	-0.53	-19.17
	105	1.487	-5.2	1.89	5.23	16.3	1.13	-47.03	9.4	-0.83	-19.53
	106	1.501	-3.7	1.85	-10.10	16.9	0.48	-40.67	8.5	-1.09	-13.02

TABLE A.5(*a*) Reaction Forces and Moments of Force—Ankle and Knee

TABLE A.5(*a*) Reaction Forces and Moments of Force—Ankle and Knee

	FRAME	TIME	GROUND		FOOT SEGMENT ANKLE		GROUND	ANKLE	LEG SEGMENT ANKLE		LEG SEGMENT KNEE		ANKLE	KNEE
			RX	RY	RX	RY	CofP X	MOMENT	RX	RY	RX	RY	MOMENT	MOMENT
		S	N	N	N	N	M	N.M	N	N	N	N	N.M	N.M
TOR	1	0.000	0.0	0.0	20.9	3.9	0.000	1.6	-20.9	-3.9	52.0	31.6	-1.6	7.0
	2	0.014	0.0	0.0	19.8	0.0	0.000	1.6	-19.8	0.0	41.1	27.1	-1.6	7.0
	3	0.029	0.0	0.0	17.6	-3.1	0.000	1.5	-17.6	3.1	30.8	22.2	-1.5	6.9
	4	0.043	0.0	0.0	14.9	-5.0	0.000	1.3	-14.9	5.0	22.0	17.8	-1.3	6.5
	5	0.057	0.0	0.0	12.3	-5.5	0.000	1.1	-12.3	5.5	15.1	14.6	-1.1	5.8
	6	0.072	0.0	0.0	10.3	-4.7	0.000	0.9	-10.3	4.7	10.1	12.7	-0.9	4.8
	7	0.086	0.0	0.0	8.6	-3.0	0.000	0.7	-8.6	3.0	6.5	12.2	-0.7	3.6
	8	0.100	0.0	0.0	7.3	-0.7	0.000	0.6	-7.3	0.7	4.1	13.4	-0.6	2.3
	9	0.114	0.0	0.0	6.3	2.0	0.000	0.5	-6.3	-2.0	2.5	15.9	-0.5	1.0
	10	0.129	0.0	0.0	5.5	4.8	0.000	0.4	-5.5	-4.8	1.4	19.1	-0.4	-0.1
	11	0.143	0.0	0.0	4.7	7.5	0.000	0.4	-4.7	-7.5	0.4	22.8	-0.4	-1.0
	12	0.157	0.0	0.0	3.7	10.2	0.000	0.5	-3.7	-10.2	-0.9	27.1	-0.5	-1.9
	13	0.172	0.0	0.0	2.5	12.6	0.000	0.5	-2.5	-12.6	-2.8	31.8	-0.5	-2.7
	14	0.186	0.0	0.0	0.9	14.7	0.000	0.6	-0.9	-14.7	-5.7	36.2	-0.6	-3.5
	15	0.200	0.0	0.0	-0.9	16.1	0.000	0.6	0.9	-16.1	-9.5	39.6	-0.6	-4.2
	16	0.215	0.0	0.0	-3.0	16.8	0.000	0.6	3.0	-16.8	-14.2	42.0	-0.6	-4.9
	17	0.229	0.0	0.0	-5.4	17.0	0.000	0.6	5.4	-17.0	-19.7	43.4	-0.6	-5.8
	18	0.243	0.0	0.0	-8.3	16.3	0.000	0.6	8.3	-16.3	-25.5	43.7	-0.6	-6.8
	19	0.257	0.0	0.0	-11.5	14.8	0.000	0.5	11.5	-14.8	-30.7	42.7	-0.5	-8.0
	20	0.272	0.0	0.0	-14.9	12.2	0.000	0.3	14.9	-12.2	-35.5	40.7	-0.3	-9.4
	21	0.286	0.0	0.0	-18.1	9.0	0.000	0.1	18.1	-9.0	-39.9	38.2	-0.1	-11.1
	22	0.300	0.0	0.0	-20.8	5.2	0.000	-0.1	20.8	-5.2	-43.9	34.9	0.1	-12.8
	23	0.315	0.0	0.0	-22.7	1.6	0.000	-0.3	22.7	-1.6	-46.6	30.6	0.3	-14.3
	24	0.329	0.0	0.0	-23.2	-1.2	0.000	-0.5	23.2	1.2	-47.3	26.1	0.5	-15.1
	25	0.343	0.0	0.0	-22.2	-2.4	0.000	-0.5	22.2	2.4	-45.3	23.3	0.5	-14.6
	26	0.357	0.0	0.0	-19.5	-1.7	0.000	-0.5	19.5	1.7	-40.6	24.1	0.5	-12.7
	27	0.372	0.0	0.0	-15.6	1.0	0.000	-0.3	15.6	-1.0	-34.1	29.1	0.3	-9.5
HCR	28	0.386	37.3	87.1	-48.7	-82.4	1.227	-1.7	48.7	82.4	-64.3	-50.2	1.7	-33.8
	29	0.400	-4.2	192.6	-3.8	-184.1	1.244	4.6	3.8	184.1	-16.8	-148.2	-4.6	-19.9
	30	0.415	-43.7	304.1	38.2	-292.9	1.261	8.3	-38.2	292.9	27.2	-255.2	-8.3	-7.6
	31	0.429	-74.0	404.2	69.8	-391.4	1.291	2.8	-69.8	391.4	59.9	-354.6	-2.8	-4.5

32	0.443	-91.9	476.6	88.4	-463.4	1.290	6.9	-88.4	463.4	77.7	-429.9	-6.9	7.8
33	0.458	-102.7	521.7	99.7	-508.9	1.301	4.6	-99.7	508.9	86.3	-480.2	-4.6	14.5
34	0.472	-110.5	552.9	108.1	-541.1	1.304	5.0	-108.1	541.1	91.5	-516.7	-5.0	24.8
35	0.486	-114.2	579.8	112.5	-569.0	1.311	2.6	-112.5	569.0	94.3	-547.2	-2.6	31.6
36	0.500	-110.5	599.6	109.4	-589.9	1.317	-0.5	-109.4	589.9	91.7	-568.4	0.5	35.0
37	0.515	-98.1	604.5	97.1	-595.7	1.316	0.2	-97.1	595.7	80.9	-573.4	-0.2	37.8
38	0.529	-79.8	589.9	78.5	-581.7	1.321	-3.5	-78.5	581.7	63.7	-558.2	3.5	32.5
39	0.543	-62.0	558.1	60.3	-550.2	1.322	-5.1	-60.3	550.2	47.2	-525.1	5.1	28.1
40	0.558	-48.7	516.7	47.2	-508.7	1.325	-6.6	-47.2	508.7	36.8	-481.5	6.6	24.8
41	0.572	-39.8	473.4	38.8	-465.3	1.333	-10.5	-38.8	465.3	32.2	-436.1	10.5	20.2
42	0.586	-33.0	433.4	32.7	-425.2	1.335	-10.6	-32.7	425.2	29.8	-395.6	10.6	19.8
43	0.601	-27.0	400.3	27.2	-392.2	1.345	-14.1	-27.2	392.2	27.4	-364.4	14.1	15.8
44	0.615	-21.8	377.1	22.4	-369.4	1.358	-18.7	-22.4	369.4	24.6	-344.7	18.7	10.9
45	0.629	-18.1	365.0	19.0	-357.6	1.369	-22.5	-19.0	357.6	22.3	-335.4	22.5	7.8
46	0.643	-16.6	362.3	17.4	-355.0	1.377	-25.3	-17.4	355.0	21.0	-333.8	25.3	6.6
47	0.658	-16.8	366.5	17.6	-359.1	1.384	-27.9	-17.6	359.1	21.0	-337.2	27.9	6.6
48	0.672	-16.8	375.0	17.5	-367.3	1.390	-30.3	-17.5	367.3	20.7	-343.7	30.3	7.0
49	0.686	-14.5	386.1	15.2	-378.2	1.396	-33.5	-15.2	378.2	18.3	-352.3	33.5	5.9
50	0.701	-9.6	400.3	10.3	-392.4	1.406	-38.8	-10.3	392.4	13.0	-364.3	38.8	2.3
51	0.715	-3.8	418.8	4.2	-410.8	1.417	-45.1	-4.2	410.8	6.0	-381.2	45.1	-2.3
52	0.729	2.2	441.0	-2.1	-432.7	1.412	-45.4	2.1	432.7	-0.9	-402.2	45.4	-0.1
53	0.744	8.7	465.8	-9.0	-457.0	1.424	-53.5	9.0	457.0	-6.8	-426.3	53.5	-5.0
54	0.758	16.5	493.7	-16.8	-484.4	1.428	-58.8	16.8	484.4	-12.3	-453.7	58.8	-6.2
55	0.772	26.8	523.9	-27.1	-514.2	1.432	-64.8	27.1	514.2	-19.7	-483.6	64.8	-7.9
56	0.786	40.1	552.6	-40.4	-542.9	1.436	-71.2	40.4	542.9	-30.9	-513.0	71.2	-10.4
57	0.801	54.8	576.8	-55.0	-567.5	1.440	-77.3	55.0	567.5	-43.5	-538.7	77.3	-12.5
58	0.815	68.1	595.8	-67.6	-586.9	1.444	-82.7	67.6	586.9	-53.3	-559.3	82.7	-12.4
59	0.829	79.6	608.6	-77.9	-599.7	1.447	-87.0	77.9	599.7	-59.1	-573.0	87.0	-10.0
60	0.844	90.8	612.1	-87.4	-602.8	1.451	-89.7	87.4	602.8	-63.1	-576.4	89.7	-6.4
61	0.858	101.5	602.3	-96.4	-592.2	1.455	-89.8	96.4	592.2	-66.5	-566.1	89.8	-2.2
62	0.872	110.4	576.1	-103.8	-565.0	1.459	-86.7	103.8	565.0	-68.9	-538.7	86.7	2.3
63	0.887	115.6	530.3	-107.7	-518.2	1.463	-79.7	107.7	518.2	-69.7	-491.1	79.7	6.5
64	0.901	114.5	463.0	-105.7	-450.2	1.467	-68.6	105.7	450.2	-67.5	-421.9	68.6	10.1
65	0.915	105.2	377.3	-95.5	-364.4	1.470	-54.3	95.5	364.4	-60.1	-335.2	54.3	12.5
66	0.929	88.2	282.1	-77.2	-269.9	1.474	-38.8	77.2	269.9	-46.1	-240.4	38.8	13.3
67	0.944	65.7	190.1	-53.0	-179.1	1.478	-24.2	53.0	179.1	-26.4	-149.3	24.2	12.5
68	0.958	41.4	110.8	-26.7	-101.8	1.482	-12.3	26.7	101.8	-4.1	-71.2	12.3	10.6
69	0.972	18.4	44.4	-2.1	-37.9	1.486	-3.6	2.1	37.9	16.9	-6.8	3.6	6.8

TABLE A.5(*a*) (*Continued*)

	FRAME	TIME	GROUND		FOOT SEGMENT ANKLE		GROUND	ANKLE	LEG SEGMENT ANKLE		LEG SEGMENT KNEE		ANKLE	KNEE
			RX	RY	RX	RY	CofP X	MOMENT	RX	RY	RX	RY	MOMENT	MOMENT
		S	N	N	N	N	M	N.M	N	N	N	N	N.M	N.M
TOR	70	0.987	0.0	0.0	17.1	3.5	0.000	1.4	-17.1	-3.5	32.9	34.3	-1.4	3.6
	71	1.001	0.0	0.0	16.8	0.3	0.000	1.4	-16.8	-0.3	30.4	29.2	-1.4	4.6
	72	1.015	0.0	0.0	15.8	-2.5	0.000	1.4	-15.8	2.5	28.0	23.0	-1.4	5.7
	73	1.030	0.0	0.0	14.3	-4.4	0.000	1.3	-14.3	4.4	25.4	17.2	-1.3	6.4
	74	1.044	0.0	0.0	12.6	-4.9	0.000	1.1	-12.6	4.9	22.1	13.4	-1.1	6.3
	75	1.058	0.0	0.0	11.1	-3.9	0.000	0.9	-11.1	3.9	18.1	12.2	-0.9	5.5
	76	1.072	0.0	0.0	9.7	-2.0	0.000	0.7	-9.7	2.0	14.0	12.8	-0.7	4.3
	77	1.087	0.0	0.0	8.5	0.2	0.000	0.6	-8.5	-0.2	10.0	14.1	-0.6	3.0
	78	1.101	0.0	0.0	7.4	2.2	0.000	0.5	-7.4	-2.2	6.6	15.6	-0.5	1.8
	79	1.115	0.0	0.0	6.3	4.3	0.000	0.5	-6.3	-4.3	3.7	17.8	-0.5	0.6
	80	1.130	0.0	0.0	5.3	6.7	0.000	0.5	-5.3	-6.7	1.6	21.5	-0.5	-0.5
	81	1.144	0.0	0.0	4.3	9.2	0.000	0.5	-4.3	-9.2	-0.2	25.9	-0.5	-1.5
	82	1.158	0.0	0.0	3.4	11.7	0.000	0.6	-3.4	-11.7	-1.7	30.7	-0.6	-2.2
	83	1.173	0.0	0.0	2.4	14.1	0.000	0.6	-2.4	-14.1	-3.5	35.5	-0.6	-2.7
	84	1.187	0.0	0.0	1.1	16.3	0.000	0.7	-1.1	-16.3	-6.0	39.8	-0.7	-3.1
	85	1.201	0.0	0.0	-0.9	17.8	0.000	0.8	0.9	-17.8	-9.8	43.0	-0.8	-3.6
	86	1.215	0.0	0.0	-3.5	18.5	0.000	0.8	3.5	-18.5	-15.1	44.7	-0.8	-4.3
	87	1.230	0.0	0.0	-6.8	18.0	0.000	0.7	6.8	-18.0	-21.7	45.3	-0.7	-5.6
	88	1.244	0.0	0.0	-10.8	16.4	0.000	0.6	10.8	-16.4	-29.4	45.0	-0.6	-7.4
	89	1.258	0.0	0.0	-15.3	13.6	0.000	0.4	15.3	-13.6	-37.5	43.5	-0.4	-9.7
	90	1.273	0.0	0.0	-19.6	9.7	0.000	0.1	19.6	-9.7	-44.8	40.3	-0.1	-12.3
	91	1.287	0.0	0.0	-23.1	5.3	0.000	-0.2	23.1	-5.3	-49.6	36.0	0.2	-14.5
	92	1.301	0.0	0.0	-25.0	1.2	0.000	-0.4	25.0	-1.2	-50.8	31.4	0.4	-15.9
	93	1.316	0.0	0.0	-24.9	-1.8	0.000	-0.5	24.9	1.8	-48.4	27.6	0.5	-15.9
	94	1.330	0.0	0.0	-22.9	-2.8	0.000	-0.6	22.9	2.8	-43.5	26.5	0.6	-14.5
	95	1.344	0.0	0.0	-19.4	-1.6	0.000	-0.5	19.4	1.6	-36.8	29.0	0.5	-11.7
	96	1.358	0.0	0.0	-15.0	1.4	0.000	-0.3	15.0	-1.4	-29.5	33.9	0.3	-8.2
HCR	97	1.373	0.0	0.0	-10.9	4.9	0.000	0.0	10.9	-4.9	-23.4	39.0	0.0	-4.9
	98	1.387	0.0	0.0	-8.1	8.2	0.000	0.3	8.1	-8.2	-19.9	42.8	-0.3	-2.8
	99	1.401	0.0	0.0	-6.5	10.4	0.000	0.5	6.5	-10.4	-19.4	44.2	-0.5	-2.0
	100	1.416	0.0	0.0	-5.9	11.6	0.000	0.6	5.9	-11.6	-21.4	42.9	-0.6	-2.4
	101	1.430	0.0	0.0	-5.6	11.8	0.000	0.7	5.6	-11.8	-24.8	39.7	-0.7	-3.2
	102	1.444	0.0	0.0	-5.2	11.4	0.000	0.7	5.2	-11.4	-28.4	36.1	-0.7	-4.1
	103	1.459	0.0	0.0	-4.5	10.9	0.000	0.6	4.5	-10.9	-30.3	33.7	-0.6	-4.6
	104	1.473	0.0	0.0	-3.4	10.4	0.000	0.6	3.4	-10.4	-28.9	33.1	-0.6	-4.2
	105	1.487	0.0	0.0	-1.8	10.0	0.000	0.6	1.8	-10.0	-24.4	33.6	-0.6	-3.2
	106	1.501	0.0	0.0	-0.4	9.5	0.000	0.6	0.4	-9.5	-19.0	34.0	-0.6	-2.1

TABLE A.5(*b*) Reaction Forces and Moments of Force—Hip

TABLE A.5(*b*) Reaction Forces and Moments of Force—Hip

			THIGH SEGMENT					
			KNEE		HIP		KNEE	HIP
	FRAME	TIME S	RX N	RY N	RX N	RY N	MOMENT N.M	MOMENT N.M
TOR	1	0.000	-52.0	-31.6	59.7	112.1	-7.0	22.9
	2	0.014	-41.1	-27.1	30.3	117.1	-7.0	17.8
	3	0.029	-30.8	-22.2	9.1	116.7	-6.9	13.8
	4	0.043	-22.0	-17.8	-3.8	111.5	-6.5	10.8
	5	0.057	-15.1	-14.6	-9.4	102.8	-5.8	8.5
	6	0.072	-10.1	-12.7	-10.6	91.7	-4.8	6.7
	7	0.086	-6.5	-12.2	-10.0	80.4	-3.6	5.0
	8	0.100	-4.1	-13.4	-9.8	71.3	-2.3	3.4
	9	0.114	-2.5	-15.9	-10.4	65.0	-1.0	2.0
	10	0.129	-1.4	-19.1	-11.7	60.0	0.1	0.9
	11	0.143	-0.4	-22.8	-13.3	56.1	1.0	0.0
	12	0.157	0.9	-27.1	-15.2	54.5	1.9	-0.8
	13	0.172	2.8	-31.8	-18.1	55.5	2.7	-1.5
	14	0.186	5.7	-36.2	-21.7	57.6	3.5	-2.5
	15	0.200	9.5	-39.6	-25.3	59.9	4.2	-3.6
	16	0.215	14.2	-42.0	-28.4	62.3	4.9	-5.0
	17	0.229	19.7	-43.4	-31.3	65.1	5.8	-6.7
	18	0.243	25.5	-43.7	-33.4	69.0	6.8	-8.7
	19	0.257	30.7	-42.7	-33.9	74.4	8.0	-10.5
	20	0.272	35.5	-40.7	-33.1	82.1	9.4	-12.2
	21	0.286	39.9	-38.2	-32.1	91.9	11.1	-14.0
	22	0.300	43.9	-34.9	-30.9	100.1	12.8	-16.0
	23	0.315	46.6	-30.6	-29.4	103.1	14.3	-17.9
	24	0.329	47.3	-26.1	-27.9	101.5	15.1	-19.1
	25	0.343	45.3	-23.3	-26.8	99.4	14.6	-18.7
	26	0.357	40.6	-24.1	-26.5	100.7	12.7	-16.1
	27	0.372	34.1	-29.1	-27.3	106.9	9.5	-11.7
HCR	28	0.386	64.3	50.2	-66.1	29.7	33.8	-54.4
	29	0.400	16.8	148.2	-26.3	-66.4	19.9	-37.6
	30	0.415	-27.2	255.2	12.4	-173.3	7.6	-23.5
	31	0.429	-59.9	354.6	41.3	-276.1	4.5	-20.8
	32	0.443	-77.7	429.9	54.7	-359.4	-7.8	-11.1

33	0.458	-86.3	480.2	58.3	-421.0	-14.5	-7.8
34	0.472	-91.5	516.7	60.4	-468.3	-24.8	-0.4
35	0.486	-94.3	547.2	63.8	-504.4	-31.6	4.7
36	0.500	-91.7	568.4	64.2	-526.0	-35.0	7.3
37	0.515	-80.9	573.4	55.7	-529.1	-37.8	9.9
38	0.529	-63.7	558.2	39.0	-512.7	-32.5	5.0
39	0.543	-47.2	525.1	22.5	-478.1	-28.1	3.0
40	0.558	-36.8	481.5	13.4	-431.6	-24.8	4.7
41	0.572	-32.2	436.1	12.3	-383.0	-20.2	6.5
42	0.586	-29.8	395.6	15.2	-341.7	-19.8	12.0
43	0.601	-27.4	364.4	18.7	-314.3	-15.8	12.5
44	0.615	-24.6	344.7	21.0	-301.1	-10.9	10.9
45	0.629	-22.3	335.4	21.9	-296.7	-7.8	10.1
46	0.643	-21.0	333.8	22.5	-295.3	-6.6	11.2
47	0.658	-21.0	337.2	24.4	-295.0	-6.6	13.8
48	0.672	-20.7	343.7	27.3	-296.3	-7.0	16.7
49	0.686	-18.3	352.3	27.9	-300.2	-5.9	17.7
50	0.701	-13.0	364.3	23.4	-308.6	-2.3	15.2
51	0.715	-6.0	381.2	14.5	-323.0	2.3	11.2
52	0.729	0.9	402.2	5.7	-342.8	0.1	14.6
53	0.744	6.8	426.3	1.0	-366.2	5.0	12.2
54	0.758	12.3	453.7	0.1	-393.6	6.2	14.7
55	0.772	19.7	483.6	-2.5	-424.7	7.9	16.6
56	0.786	30.9	513.0	-10.6	-456.2	10.4	16.5
57	0.801	43.5	538.7	-20.3	-484.2	12.5	16.1
58	0.815	53.3	559.3	-24.8	-506.3	12.4	18.3
59	0.829	59.1	573.0	-22.7	-521.5	10.0	23.6
60	0.844	63.1	576.4	-17.8	-527.1	6.4	29.5
61	0.858	66.5	566.1	-13.2	-519.0	2.2	34.4
62	0.872	68.9	538.7	-10.1	-492.1	-2.3	37.3
63	0.887	69.7	491.1	-10.7	-442.7	-6.5	37.2
64	0.901	67.5	421.9	-16.0	-370.1	-10.1	33.6
65	0.915	60.1	335.2	-23.2	-279.2	-12.5	27.5
66	0.929	46.1	240.4	-27.5	-179.5	-13.3	20.7
67	0.944	26.4	149.3	-25.0	-82.3	-12.5	15.2
68	0.958	4.1	71.2	-16.0	2.9	-10.6	11.9
69	0.972	-16.9	6.8	-3.5	74.9	-6.8	9.3

TABLE A.5(*b*) (*Continued*)

			THIGH SEGMENT					
			KNEE		HIP		KNEE	HIP
	FRAME	TIME	RX	RY	RX	RY	MOMENT	MOMENT
		S	N	N	N	N	N.M	N.M
TOR	70	0.987	-32.9	-34.3	8.9	122.0	-3.6	9.0
	71	1.001	-30.4	-29.2	7.9	120.0	-4.6	10.4
	72	1.015	-28.0	-23.0	10.7	113.5	-5.7	12.3
	73	1.030	-25.4	-17.2	14.0	104.6	-6.4	13.6
	74	1.044	-22.1	-13.4	14.8	96.0	-6.3	13.4
	75	1.058	-18.1	-12.2	12.8	88.5	-5.5	11.8
	76	1.072	-14.0	-12.8	9.0	80.7	-4.3	9.4
	77	1.087	-10.0	-14.1	3.9	71.1	-3.0	6.8
	78	1.101	-6.6	-15.6	-2.0	61.1	-1.8	4.2
	79	1.115	-3.7	-17.8	-8.0	54.2	-0.6	2.0
	80	1.130	-1.6	-21.5	-13.8	51.8	0.5	0.4
	81	1.144	0.2	-25.9	-18.9	53.0	1.5	-0.7
	82	1.158	1.7	-30.7	-22.8	55.8	2.2	-1.4
	83	1.173	3.5	-35.5	-25.5	59.1	2.7	-1.7
	84	1.187	6.0	-39.8	-27.3	62.0	3.1	-2.1
	85	1.201	9.8	-43.0	-28.9	63.7	3.6	-3.0
	86	1.215	15.1	-44.7	-30.5	65.2	4.3	-4.5
	87	1.230	21.7	-45.3	-31.7	68.8	5.6	-6.7
	88	1.244	29.4	-45.0	-32.7	75.5	7.4	-9.4
	89	1.258	37.5	-43.5	-34.2	83.8	9.7	-12.8
	90	1.273	44.8	-40.3	-35.7	92.0	12.3	-16.4
	91	1.287	49.6	-36.0	-34.2	99.0	14.5	-19.0
	92	1.301	50.8	-31.4	-28.5	104.4	15.9	-19.7
	93	1.316	48.4	-27.6	-20.9	108.1	15.9	-18.6
	94	1.330	43.5	-26.5	-14.9	111.6	14.5	-15.7
	95	1.344	36.8	-29.0	-12.5	115.8	11.7	-11.5
	96	1.358	29.5	-33.9	-14.1	119.7	8.2	-6.7
HCR	97	1.373	23.4	-39.0	-19.4	122.3	4.9	-2.9
	98	1.387	19.9	-42.8	-27.3	123.5	2.8	-1.0
	99	1.401	19.4	-44.2	-36.7	121.9	2.0	-1.4
	100	1.416	21.4	-42.9	-47.1	115.1	2.4	-3.9
	101	1.430	24.8	-39.7	-58.3	103.2	3.2	-7.8
	102	1.444	28.4	-36.1	-68.8	89.8	4.1	-12.0
	103	1.459	30.3	-33.7	-75.7	80.0	4.6	-14.7
	104	1.473	28.9	-33.1	-76.9	76.4	4.2	-14.9
	105	1.487	24.4	-33.6	-73.9	77.2	3.2	-12.8
	106	1.501	19.0	-34.0	-72.3	78.9	2.1	-10.4

TABLE A.6 Segment Potential, Kinetic, and Total Energies—Foot, Leg, Thigh, and $\frac{1}{2}$ HAT

TABLE A.6 Segment Potential, Kinetic, and Total Energies—Foot, Leg, Thigh, and $\frac{1}{2}$ HAT

		FOOT SEGMENT				LEG SEGMENT				THIGH SEGMENT				H.A.T. SEGMENT			
FRAME	TIME	PE	TKE	RKE	TOTAL	PE	TKE	RKE	TOTAL	PE	TKE	RKE	TOTAL	PE	TKE	RKE	TOTAL
	S	J	J	J	J	J	J	J	J	J	J	J	J	J	J	J	J
1	0.000	1.2	1.8	0.0	3.0	9.5	8.3	0.1	17.9	36.3	12.3	0.3	48.8	203.6	18.2	0.4	222.3
2	0.014	1.3	2.3	0.0	3.6	9.6	9.2	0.1	18.9	36.2	12.2	0.3	48.8	203.8	18.3	0.3	222.4
3	0.029	1.4	2.9	0.0	4.3	9.7	9.9	0.0	19.6	36.3	11.7	0.4	48.4	204.1	17.9	0.1	222.1
4	0.043	1.4	3.5	0.0	4.9	9.8	10.2	0.0	20.0	36.4	11.1	0.4	47.9	204.6	17.2	0.1	221.8
5	0.057	1.4	4.0	0.0	5.5	9.9	10.4	0.0	20.3	36.7	10.5	0.4	47.6	205.2	16.3	0.0	221.6
6	0.072	1.4	4.6	0.0	6.0	10.0	10.4	0.1	20.5	36.9	10.0	0.4	47.4	206.0	15.6	0.0	221.7
7	0.086	1.4	5.1	0.0	6.5	10.0	10.3	0.2	20.5	37.3	9.6	0.4	47.3	207.0	15.1	0.1	222.1
8	0.100	1.3	5.5	0.1	6.9	10.1	10.2	0.3	20.5	37.6	9.3	0.3	47.2	207.9	14.8	0.1	222.8
9	0.114	1.3	6.0	0.1	7.3	10.1	10.1	0.3	20.5	38.0	9.0	0.2	47.2	208.9	14.6	0.1	223.6
10	0.129	1.2	6.3	0.1	7.6	10.1	9.9	0.4	20.4	38.3	8.6	0.2	47.1	209.8	14.4	0.2	224.3
11	0.143	1.1	6.6	0.1	7.8	10.1	9.8	0.5	20.3	38.7	8.2	0.1	47.0	210.6	14.1	0.2	224.9
12	0.157	1.0	6.9	0.1	8.0	10.0	9.6	0.5	20.2	38.9	7.7	0.1	46.7	211.2	13.9	0.3	225.4
13	0.172	0.9	7.0	0.1	8.0	9.9	9.4	0.6	20.0	39.1	7.3	0.0	46.5	211.6	13.7	0.4	225.7
14	0.186	0.8	7.1	0.1	8.0	9.9	9.2	0.6	19.7	39.3	6.8	0.0	46.1	211.9	13.7	0.4	225.9
15	0.200	0.8	7.0	0.1	7.9	9.8	9.0	0.6	19.4	39.4	6.4	0.0	45.8	212.0	13.9	0.4	226.2
16	0.215	0.7	6.9	0.1	7.7	9.7	8.6	0.6	18.9	39.4	6.1	0.0	45.5	211.9	14.4	0.3	226.6
17	0.229	0.7	6.6	0.1	7.4	9.6	8.2	0.6	18.3	39.3	5.9	0.0	45.2	211.6	15.4	0.2	227.2
18	0.243	0.7	6.2	0.1	7.0	9.5	7.6	0.6	17.7	39.2	5.7	0.0	44.9	211.1	16.7	0.2	227.9
19	0.257	0.7	5.7	0.1	6.5	9.4	7.0	0.5	16.9	39.0	5.7	0.0	44.7	210.5	18.1	0.1	228.7
20	0.272	0.8	5.0	0.1	5.8	9.3	6.4	0.5	16.1	38.7	5.8	0.0	44.6	209.7	19.6	0.1	229.4
21	0.286	0.8	4.2	0.0	5.0	9.2	5.7	0.4	15.3	38.5	5.9	0.0	44.4	208.9	20.8	0.1	229.7
22	0.300	0.8	3.3	0.0	4.2	9.1	5.1	0.3	14.4	38.2	6.1	0.0	44.3	208.1	21.5	0.0	229.7
23	0.315	0.9	2.5	0.0	3.4	9.0	4.4	0.1	13.6	37.9	6.4	0.0	44.3	207.4	21.9	0.1	229.3
24	0.329	0.9	1.7	0.0	2.6	9.0	3.8	0.1	12.8	37.7	6.7	0.0	44.4	206.7	22.0	0.1	228.8
25	0.343	0.9	1.1	0.0	2.0	8.9	3.3	0.0	12.2	37.5	7.1	0.0	44.6	206.1	22.1	0.1	228.4
26	0.357	0.9	0.7	0.0	1.5	8.8	2.8	0.0	11.6	37.4	7.4	0.0	44.8	205.7	22.5	0.1	228.3
27	0.372	0.8	0.4	0.0	1.3	8.8	2.4	0.0	11.2	37.3	7.6	0.0	44.9	205.3	23.1	0.1	228.6
28	0.386	0.8	0.3	0.0	1.1	8.7	2.1	0.1	10.8	37.2	7.7	0.0	44.9	205.1	23.9	0.1	229.2
29	0.400	0.7	0.2	0.0	1.0	8.6	1.8	0.1	10.5	37.2	7.5	0.0	44.8	205.1	24.7	0.0	229.8
30	0.415	0.7	0.2	0.0	0.9	8.6	1.6	0.1	10.3	37.3	7.3	0.0	44.5	205.3	25.2	0.0	230.5
31	0.429	0.6	0.1	0.0	0.7	8.6	1.4	0.1	10.2	37.4	6.9	0.0	44.3	205.7	25.5	0.0	231.1
32	0.443	0.6	0.1	0.0	0.7	8.6	1.3	0.1	10.0	37.5	6.5	0.0	44.1	206.3	25.4	0.0	231.7

33	0.458	0.5	0.0	0.0	0.6	8.6	1.2	0.1	9.9	37.7	6.0	0.0	43.8	207.0	25.0	0.1	232.1
34	0.472	0.5	0.0	0.0	0.5	8.7	1.0	0.1	9.8	37.9	5.4	0.0	43.3	207.8	24.2	0.2	232.2
35	0.486	0.5	0.0	0.0	0.5	8.7	0.8	0.1	9.6	38.1	4.8	0.0	42.9	208.6	23.2	0.2	232.0
36	0.500	0.5	0.0	0.0	0.5	8.7	0.6	0.1	9.4	38.2	4.2	0.0	42.5	209.3	22.2	0.2	231.7
37	0.515	0.5	0.0	0.0	0.5	8.7	0.4	0.1	9.2	38.4	3.8	0.1	42.2	209.9	21.3	0.1	231.3
38	0.529	0.5	0.0	0.0	0.5	8.7	0.3	0.0	9.1	38.4	3.3	0.1	41.9	210.5	20.3	0.1	230.9
39	0.543	0.5	0.0	0.0	0.5	8.7	0.2	0.0	9.0	38.5	3.0	0.1	41.6	211.0	19.4	0.1	230.5
40	0.558	0.5	0.0	0.0	0.5	8.7	0.1	0.0	8.9	38.6	2.6	0.1	41.3	211.5	18.6	0.1	230.2
41	0.572	0.5	0.0	0.0	0.5	8.7	0.1	0.0	8.9	38.6	2.3	0.1	41.1	211.8	18.1	0.2	230.1
42	0.586	0.5	0.0	0.0	0.5	8.7	0.1	0.0	8.8	38.7	2.1	0.1	40.9	212.1	17.8	0.3	230.2
43	0.601	0.5	0.0	0.0	0.5	8.7	0.1	0.0	8.9	38.7	1.9	0.1	40.8	212.4	17.7	0.4	230.4
44	0.615	0.5	0.0	0.0	0.5	8.8	0.1	0.0	8.9	38.7	1.9	0.1	40.7	212.6	17.7	0.5	230.8
45	0.629	0.5	0.0	0.0	0.5	8.8	0.1	0.0	8.9	38.7	1.9	0.1	40.7	212.6	17.9	0.6	231.1
46	0.643	0.5	0.0	0.0	0.5	8.8	0.1	0.0	8.9	38.7	1.9	0.1	40.7	212.5	18.3	0.7	231.5
47	0.658	0.5	0.0	0.0	0.5	8.8	0.1	0.0	8.9	38.6	1.9	0.1	40.6	212.3	18.7	0.7	231.7
48	0.672	0.5	0.0	0.0	0.5	8.8	0.2	0.0	8.9	38.6	2.0	0.1	40.6	211.9	19.2	0.8	231.9
49	0.686	0.5	0.0	0.0	0.5	8.7	0.2	0.0	8.9	38.4	2.1	0.1	40.6	211.5	19.7	0.7	231.9
50	0.701	0.5	0.0	0.0	0.5	8.7	0.2	0.0	8.9	38.3	2.2	0.1	40.6	211.0	20.2	0.6	231.8
51	0.715	0.5	0.0	0.0	0.5	8.7	0.2	0.0	8.9	38.2	2.4	0.1	40.7	210.4	20.5	0.6	231.5
52	0.729	0.5	0.0	0.0	0.5	8.7	0.2	0.0	8.9	38.1	2.5	0.1	40.6	209.9	20.7	0.5	231.1
53	0.744	0.5	0.0	0.0	0.5	8.7	0.2	0.0	8.9	38.0	2.5	0.1	40.6	209.3	20.9	0.5	230.7
54	0.758	0.5	0.0	0.0	0.5	8.7	0.2	0.0	8.9	37.9	2.6	0.1	40.6	208.7	21.3	0.5	230.5
55	0.772	0.5	0.0	0.0	0.6	8.7	0.3	0.0	9.0	37.8	2.9	0.1	40.7	208.1	21.7	0.5	230.2
56	0.786	0.6	0.0	0.0	0.6	8.7	0.3	0.1	9.1	37.7	3.1	0.1	40.9	207.6	21.9	0.4	229.9
57	0.801	0.6	0.0	0.0	0.6	8.8	0.4	0.1	9.2	37.7	3.5	0.0	41.2	207.1	22.0	0.3	229.3
58	0.815	0.6	0.0	0.0	0.6	8.8	0.5	0.1	9.4	37.6	3.9	0.0	41.5	206.5	21.9	0.2	228.6
59	0.829	0.6	0.0	0.0	0.6	8.8	0.7	0.1	9.6	37.5	4.4	0.0	41.9	206.1	21.9	0.1	228.1
60	0.844	0.6	0.0	0.0	0.7	8.9	0.9	0.1	9.9	37.4	5.2	0.0	42.6	205.6	22.3	0.0	227.9
61	0.858	0.7	0.0	0.0	0.7	8.9	1.3	0.1	10.3	37.3	6.2	0.0	43.5	205.2	23.1	0.0	228.3
62	0.872	0.7	0.1	0.0	0.8	9.0	1.7	0.2	10.9	37.2	7.5	0.0	44.7	204.8	24.4	0.0	229.3
63	0.887	0.7	0.1	0.1	0.9	9.0	2.4	0.2	11.6	37.1	9.0	0.0	46.0	204.5	25.8	0.1	230.4
64	0.901	0.8	0.2	0.1	1.1	9.1	3.2	0.2	12.5	36.9	10.5	0.0	47.4	204.1	26.7	0.2	231.1
65	0.915	0.8	0.3	0.1	1.2	9.1	4.1	0.2	13.4	36.7	11.8	0.0	48.5	203.8	26.5	0.4	230.7
66	0.929	0.9	0.5	0.1	1.5	9.2	5.0	0.2	14.3	36.6	12.6	0.1	49.2	203.6	24.9	0.6	229.1
67	0.944	1.0	0.7	0.1	1.8	9.2	5.8	0.2	15.2	36.4	12.9	0.1	49.4	203.4	22.3	0.8	226.5
68	0.958	1.1	1.0	0.1	2.1	9.3	6.6	0.1	16.0	36.3	12.6	0.1	49.1	203.2	19.6	0.8	223.7
69	0.972	1.1	1.3	0.0	2.5	9.4	7.3	0.1	16.7	36.2	12.1	0.2	48.5	203.2	17.4	0.7	221.3

TABLE A.6 (*Continued*)

FRAME	TIME	FOOT SEGMENT				LEG SEGMENT				THIGH SEGMENT				H.A.T. SEGMENT			
		PE	TKE	RKE	TOTAL	PE	TKE	RKE	TOTAL	PE	TKE	RKE	TOTAL	PE	TKE	RKE	TOTAL
	S	J	J	J	J	J	J	J	J	J	J	J	J	J	J	J	J
70	0.987	1.2	1.7	0.0	3.0	9.5	7.9	0.1	17.4	36.1	11.4	0.3	47.8	203.3	15.8	0.5	219.7
71	1.001	1.3	2.2	0.0	3.5	9.6	8.4	0.0	18.0	36.2	10.7	0.3	47.2	203.5	14.9	0.3	218.7
72	1.015	1.4	2.7	0.0	4.1	9.7	8.8	0.0	18.5	36.3	10.2	0.4	46.8	203.9	14.4	0.2	218.5
73	1.030	1.4	3.3	0.0	4.7	9.8	9.3	0.0	19.1	36.4	9.9	0.4	46.8	204.5	14.3	0.1	218.9
74	1.044	1.4	3.8	0.0	5.2	9.9	9.6	0.0	19.5	36.7	9.8	0.4	46.9	205.2	14.3	0.1	219.6
75	1.058	1.4	4.4	0.0	5.8	10.0	9.9	0.1	20.0	37.0	9.8	0.4	47.1	206.1	14.5	0.1	220.7
76	1.072	1.4	4.9	0.0	6.3	10.0	10.1	0.2	20.3	37.3	9.8	0.3	47.4	207.0	14.7	0.1	221.8
77	1.087	1.3	5.5	0.0	6.8	10.1	10.2	0.2	20.5	37.6	9.7	0.3	47.6	208.1	14.7	0.2	223.0
78	1.101	1.2	5.9	0.1	7.2	10.1	10.2	0.3	20.6	38.0	9.5	0.2	47.7	209.1	14.6	0.3	224.0
79	1.115	1.1	6.4	0.1	7.6	10.0	10.2	0.4	20.6	38.3	9.1	0.2	47.6	210.0	14.4	0.3	224.8
80	1.130	1.1	6.7	0.1	7.8	10.0	10.1	0.5	20.5	38.6	8.7	0.1	47.4	210.8	14.3	0.4	225.5
81	1.144	1.0	7.0	0.1	8.0	9.9	9.9	0.5	20.4	38.9	8.1	0.1	47.1	211.3	14.4	0.4	226.1
82	1.158	0.9	7.2	0.1	8.1	9.9	9.8	0.6	20.2	39.1	7.6	0.1	46.7	211.7	14.7	0.3	226.6
83	1.173	0.8	7.3	0.1	8.2	9.8	9.6	0.6	19.9	39.2	7.0	0.0	46.2	211.9	14.9	0.2	227.0
84	1.187	0.7	7.4	0.1	8.2	9.7	9.3	0.6	19.6	39.3	6.5	0.0	45.8	211.8	15.2	0.1	227.2
85	1.201	0.7	7.3	0.1	8.1	9.5	9.1	0.6	19.2	39.3	6.0	0.0	45.3	211.6	15.6	0.1	227.4
86	1.215	0.6	7.2	0.1	7.9	9.4	8.7	0.6	18.8	39.2	5.7	0.0	44.9	211.3	16.1	0.1	227.4
87	1.230	0.6	6.8	0.1	7.5	9.3	8.2	0.6	18.2	39.0	5.5	0.0	44.6	210.8	16.7	0.1	227.5
88	1.244	0.6	6.4	0.1	7.1	9.2	7.6	0.6	17.5	38.8	5.5	0.0	44.4	210.1	17.4	0.1	227.6
89	1.258	0.7	5.7	0.1	6.4	9.1	6.9	0.6	16.6	38.6	5.6	0.0	44.2	209.3	18.2	0.1	227.6
90	1.273	0.7	4.8	0.1	5.5	9.0	6.2	0.5	15.6	38.3	5.8	0.0	44.1	208.5	18.9	0.1	227.5
91	1.287	0.8	3.8	0.0	4.6	8.9	5.4	0.3	14.6	38.0	6.0	0.0	44.0	207.6	19.4	0.1	227.2
92	1.301	0.8	2.8	0.0	3.6	8.8	4.6	0.2	13.6	37.7	6.4	0.0	44.0	206.7	20.0	0.2	226.8
93	1.316	0.8	1.9	0.0	2.7	8.8	4.0	0.1	12.8	37.4	6.8	0.0	44.2	205.9	20.5	0.2	226.5
94	1.330	0.8	1.2	0.0	2.0	8.7	3.5	0.0	12.2	37.2	7.4	0.0	44.6	205.1	21.0	0.3	226.4
95	1.344	0.8	0.7	0.0	1.5	8.6	3.0	0.0	11.7	37.1	7.9	0.0	45.0	204.6	21.5	0.3	226.4
96	1.358	0.8	0.5	0.0	1.3	8.6	2.7	0.0	11.3	37.0	8.4	0.0	45.4	204.2	22.3	0.3	226.8
97	1.373	0.7	0.3	0.0	1.1	8.6	2.4	0.1	11.0	37.0	8.6	0.0	45.6	204.0	23.1	0.3	227.4
98	1.387	0.7	0.2	0.0	1.0	8.5	2.2	0.1	10.8	37.0	8.6	0.0	45.6	204.0	24.0	0.2	228.2
99	1.401	0.6	0.2	0.0	0.8	8.5	2.0	0.1	10.6	37.1	8.3	0.0	45.4	204.2	24.6	0.1	228.9
100	1.416	0.6	0.1	0.0	0.7	8.5	1.7	0.1	10.4	37.2	7.8	0.0	45.1	204.7	24.8	0.0	229.5
101	1.430	0.5	0.1	0.0	0.6	8.6	1.5	0.1	10.2	37.4	7.2	0.0	44.6	205.3	24.7	0.0	230.0
102	1.444	0.5	0.0	0.0	0.6	8.6	1.2	0.1	9.9	37.6	6.4	0.0	44.0	206.1	24.2	0.0	230.3
103	1.459	0.5	0.0	0.0	0.5	8.6	0.9	0.1	9.6	37.8	5.5	0.0	43.3	206.9	23.3	0.1	230.2
104	1.473	0.5	0.0	0.0	0.5	8.6	0.6	0.1	9.3	37.9	4.6	0.0	42.5	207.6	21.8	0.1	229.5
105	1.487	0.5	0.0	0.0	0.5	8.6	0.4	0.1	9.1	38.1	3.7	0.0	41.8	208.3	19.7	0.1	228.1
106	1.501	0.5	0.0	0.0	0.5	8.6	0.2	0.1	8.9	38.2	2.9	0.1	41.1	208.9	16.9	0.1	225.9

TABLE A.7 Power Generation/Absorption and Transfer—Ankle, Knee, and Hip

TABLE A.7 Power Generation/Absorption and Transfer—Ankle, Knee, and Hip

	FRAME	TIME	MUSCLE POWER GEN(+)/ABS(-)			RATE OF TRANSFER ACROSS JOINTS AND MUSCLE						SEGMENT ANGULAR VELOCITY			
						LEG TO FOOT		THIGH TO LEG		PELVIS TO THIGH					
			ANKLE	KNEE	HIP	JOINT	MUSCLE	JOINT	MUSCLE	JOINT	MUSCLE	FOOT	LEG	THIGH	HAT
		S	W	W	W	W	W	W	W	W	W	R/S	R/S	R/S	R/S
	2	0.014	-2.8	-37.3	51.4	49.4	-2.7	110.5	0.0	56.1	12.6	-3.46	-1.70	3.59	0.71
	3	0.029	-0.4	-32.0	45.6	46.0	-1.2	85.6	0.0	29.6	7.2	-1.06	-0.80	3.81	0.52
	4	0.043	1.3	-24.4	38.7	41.3	0.3	62.9	1.4	17.6	4.0	1.18	0.21	3.95	0.37
	5	0.057	2.1	-15.9	31.6	36.8	1.4	44.7	7.2	14.8	2.4	3.14	1.24	3.98	0.28
	6	0.072	2.3	-8.0	24.3	32.8	2.0	31.6	10.6	15.2	1.8	4.74	2.21	3.89	0.27
	7	0.086	2.0	-2.1	16.8	28.9	2.2	22.6	11.0	14.6	1.6	5.91	3.08	3.68	0.33
	8	0.100	1.6	1.0	10.0	24.6	2.1	17.4	7.6	12.2	1.4	6.66	3.81	3.36	0.41
	9	0.114	1.2	1.4	5.1	20.1	2.1	14.9	3.0	8.2	1.0	7.03	4.41	2.99	0.50
	10	0.129	1.0	-0.2	1.9	15.2	2.1	13.7	-0.2	2.6	0.6	7.16	4.90	2.58	0.59
	11	0.143	0.8	-3.3	0.0	10.0	2.3	12.5	-2.3	-3.8	0.0	7.17	5.28	2.16	0.68
	12	0.157	0.7	-7.3	-0.8	4.1	2.7	10.3	-3.3	-10.6	-0.6	7.13	5.56	1.74	0.76
	13	0.172	0.7	-12.1	-0.8	-2.3	3.1	6.1	-3.6	-18.4	-1.3	7.08	5.75	1.33	0.83
	14	0.186	0.7	-17.3	-0.2	-9.1	3.4	-0.8	-3.3	-27.4	-2.1	7.06	5.86	0.94	0.85
	15	0.200	0.7	-22.4	0.9	-15.6	3.7	-10.0	-2.4	-36.6	-2.1	7.06	5.91	0.58	0.82
	16	0.215	0.7	-27.9	2.6	-22.0	3.8	-20.9	-1.1	-45.3	-1.2	7.06	5.91	0.23	0.76
	17	0.229	0.7	-34.4	5.2	-28.7	3.7	-32.6	0.0	-54.1	0.0	7.02	5.85	-0.11	0.67
	18	0.243	0.7	-42.0	8.5	-36.2	3.3	-43.6	0.0	-62.3	0.0	6.90	5.73	-0.43	0.56
	19	0.257	0.5	-49.9	12.2	-44.1	2.6	-52.6	0.0	-68.6	0.0	6.66	5.51	-0.70	0.46
	20	0.272	0.3	-57.2	15.7	-51.3	1.6	-59.6	0.0	-73.1	0.0	6.25	5.16	-0.91	0.37
	21	0.286	0.1	-62.3	18.7	-56.6	0.4	-65.4	0.0	-76.3	0.0	5.61	4.61	-1.02	0.32
	22	0.300	-0.1	-62.4	21.3	-58.4	-0.5	-69.8	0.0	-76.3	0.0	4.71	3.85	-1.03	0.31
	23	0.315	-0.2	-54.6	23.0	-55.6	-1.0	-72.5	0.0	-72.4	0.0	3.53	2.88	-0.94	0.34
	24	0.329	-0.2	-38.4	23.2	-47.9	-0.9	-73.4	0.0	-66.8	0.0	2.11	1.75	-0.79	0.42
	25	0.343	0.0	-17.7	21.1	-36.8	-0.3	-71.5	0.0	-61.7	0.0	0.49	0.58	-0.63	0.50
	26	0.357	0.4	0.2	16.6	-25.2	0.2	-65.9	6.3	-58.0	0.0	-1.27	-0.51	-0.49	0.53
	27	0.372	0.4	9.3	10.8	-16.1	0.4	-56.4	3.9	-55.4	0.0	-2.99	-1.39	-0.41	0.50
HCR	28	0.386	4.0	53.9	44.0	-12.7	3.3	-99.0	13.3	-113.1	0.0	-4.40	-1.99	-0.39	0.41
	29	0.400	-13.3	38.2	25.9	50.6	-10.7	-28.3	8.1	-47.0	0.0	-5.24	-2.33	-0.41	0.28
	30	0.415	-23.5	16.0	12.3	86.6	-20.7	22.6	3.1	0.8	0.0	-5.34	-2.50	-0.41	0.12
	31	0.429	-6.2	10.2	6.2	91.8	-7.4	43.2	1.6	15.1	1.3	-4.80	-2.61	-0.36	-0.06
	32	0.443	-8.3	-18.6	0.4	76.0	-18.5	39.0	-2.2	-2.7	2.7	-3.88	-2.68	-0.29	-0.25
	33	0.458	-0.7	-35.3	-1.3	55.5	-12.4	30.3	-3.8	-29.2	2.0	-2.85	-2.69	-0.26	-0.42

	34	0.472	3.5	-55.3	-0.1	39.6	-9.3	31.1	-9.1	-43.5	0.1	-1.88	-2.60	-0.37	-0.55
	35	0.486	3.3	-55.8	-0.2	29.2	-2.9	40.9	-20.3	-41.8	-2.8	-1.12	-2.41	-0.64	-0.59
	36	0.500	-0.8	-38.8	-3.5	20.5	0.3	51.5	-36.2	-36.0	-4.1	-0.60	-2.15	-1.04	-0.56
	37	0.515	0.3	-16.0	-9.5	11.5	-0.1	55.7	-54.9	-38.9	-4.9	-0.29	-1.88	-1.45	-0.49
	38	0.529	-5.4	5.3	-6.9	3.2	0.4	52.9	-53.3	-48.4	-2.1	-0.12	-1.64	-1.81	-0.43
	39	0.543	-7.3	16.6	-5.0	-2.5	0.1	46.3	-41.1	-51.6	-1.3	-0.02	-1.46	-2.05	-0.41
	40	0.558	-9.0	21.0	-8.0	-6.7	0.0	37.1	-33.0	-42.9	-2.2	0.03	-1.33	-2.18	-0.47
	41	0.572	-13.6	19.3	-10.4	-11.9	0.0	25.0	-25.0	-27.6	-3.7	0.06	-1.24	-2.19	-0.58
	42	0.586	-12.8	18.4	-16.7	-18.5	0.0	12.9	-23.4	-12.7	-8.6	0.03	-1.18	-2.11	-0.72
	43	0.601	-15.2	13.1	-14.2	-24.7	1.2	6.0	-18.3	-0.1	-10.8	-0.09	-1.16	-1.99	-0.86
	44	0.615	-16.8	7.7	-9.7	-28.2	5.0	8.3	-12.8	12.1	-10.7	-0.27	-1.17	-1.87	-0.98
	45	0.629	-16.6	4.7	-7.1	-28.3	9.7	18.3	-9.1	26.1	-10.8	-0.43	-1.17	-1.77	-1.07
	46	0.643	-16.5	3.5	-6.3	-24.7	12.9	31.5	-7.7	41.5	-12.7	-0.51	-1.16	-1.70	-1.14
	47	0.658	-18.4	3.3	-6.4	-18.7	13.7	43.7	-7.6	56.4	-16.4	-0.49	-1.15	-1.65	-1.19
	48	0.672	-22.5	3.4	-7.1	-13.0	12.3	52.2	-8.0	69.0	-20.2	-0.41	-1.14	-1.63	-1.21
	49	0.686	-28.1	2.9	-8.3	-10.0	10.7	54.6	-6.9	76.9	-20.9	-0.32	-1.16	-1.65	-1.18
	50	0.701	-32.6	1.2	-8.7	-10.9	13.1	50.2	-2.7	78.3	-16.9	-0.34	-1.18	-1.69	-1.12
	51	0.715	-31.2	-1.2	-7.8	-15.4	23.1	40.2	2.7	74.2	-11.7	-0.51	-1.20	-1.75	-1.05
	52	0.729	-20.0	0.0	-11.5	-23.7	36.0	27.5	0.1	69.5	-14.6	-0.79	-1.23	-1.79	-1.00
	53	0.744	-10.8	-2.5	-10.0	-36.4	58.2	14.7	6.4	68.6	-12.0	-1.09	-1.29	-1.79	-0.98
	54	0.758	-3.8	-2.3	-11.7	-54.6	77.7	3.0	8.6	70.4	-14.2	-1.32	-1.39	-1.76	-0.97
	55	0.772	-1.8	-1.0	-12.1	-80.0	97.4	-8.1	12.1	68.7	-15.6	-1.50	-1.53	-1.66	-0.94
	56	0.786	-0.8	2.2	-10.4	-113.2	121.4	-19.1	15.7	59.9	-14.6	-1.71	-1.72	-1.51	-0.88
	57	0.801	6.1	7.6	-8.5	-150.5	149.1	-27.8	16.4	48.6	-12.8	-2.01	-1.93	-1.32	-0.79
	58	0.815	26.5	12.8	-8.3	-187.0	177.3	-31.0	13.8	42.8	-12.1	-2.46	-2.14	-1.11	-0.66
	59	0.829	65.3	14.5	-9.8	-222.4	204.4	-28.5	9.0	44.0	-11.4	-3.10	-2.35	-0.90	-0.48
	60	0.844	120.8	12.0	-11.8	-260.9	228.2	-22.8	4.3	49.0	-8.0	-3.89	-2.55	-0.67	-0.27
	61	0.858	181.5	5.1	-13.0	-303.1	246.0	-15.9	0.9	54.8	-1.6	-4.76	-2.74	-0.43	-0.05
	62	0.872	232.2	-6.3	-12.0	-344.4	254.6	-10.7	-0.3	57.2	0.0	-5.62	-2.94	-0.14	0.18
	63	0.887	263.5	-21.7	-8.1	-376.4	249.1	-12.8	0.0	49.2	7.1	-6.43	-3.13	0.19	0.41
	64	0.901	272.4	-38.9	-2.6	-387.8	225.0	-24.0	0.0	27.1	18.9	-7.25	-3.28	0.56	0.64
	65	0.915	254.1	-54.0	2.8	-366.0	182.6	-36.6	0.0	-3.4	23.8	-8.03	-3.36	0.97	0.86
	66	0.929	203.4	-62.7	6.9	-307.3	129.2	-39.1	0.0	-28.9	22.2	-8.58	-3.33	1.40	1.07
	67	0.944	130.7	-62.9	9.8	-221.6	77.0	-24.4	0.0	-37.9	18.5	-8.57	-3.18	1.86	1.22
	68	0.958	60.7	-54.9	12.4	-125.8	35.5	5.8	0.0	-29.0	15.0	-7.81	-2.88	2.31	1.27
	69	0.972	14.0	-35.3	14.2	-31.8	8.9	44.1	0.0	-8.1	11.0	-6.31	-2.46	2.73	1.19
TOR	70	0.987	-3.3	-18.2	18.7	40.6	-2.6	81.8	0.0	18.1	9.2	-4.28	-1.91	3.10	1.02

APPENDIX B

UNITS AND DEFINITIONS RELATED TO BIOMECHANICAL AND ELECTROMYOGRAPHICAL MEASUREMENTS

All units used are SI (Système International d'Unités). The system is based on seven well-defined base units and two supplementary units. Only one measurement unit is needed for any physical quantity, whether the quantity is a base unit or a derived unit (which is the product and/or quotient of two or more of the base units).

TABLE B.1 Base SI Units

Physical Quantity	Symbol	Name of SI Unit	Symbol of Unit
Length	l	meter	m
Mass	m	kilogram	kg
Time	t	second	s
Electric current	I	ampere	A
Temperature	T	kelvin	K
Amount of substance	n	mole	mol
Luminous intensity	I	candela	cd
Plane angle	θ, ϕ, etc.	radian	rad
Solid angle	Ω	steradian	sr

TABLE B.2 Derived SI Units

Physical Quantity	Symbol	Name of SI Unit	Definition
Velocity	v	$m \cdot s^{-1}$	Time rate of change of position.
Acceleration	a	$m \cdot s^{-2}$	Time rate of change of velocity.

TABLE B.2 (*Continued*)

Physical Quantity	Symbol	Name of SI Unit	Definition
Acceleration	g	$m \cdot s^{-2}$	Acceleration of a freely falling body in a vacuum due to gravity. At sea level $g = 9.80665\ m \cdot s^{-2}$.
Angular velocity	ω	$rad \cdot s^{-1}$	Time rate of change of orientation of a line segment in a plane.
Angular acceleration	α	$rad \cdot s^{-2}$	Time rate of change of angular velocity.
Angular displacement	θ	radian (rad)	Change in orientation of a line segment, which is given by the plane angle between initial and final orientations.
Period	T	second (s)	Time to complete one cycle of a periodic event or, more generally, time duration of any event or phase of an event.
Frequency	f	hertz (Hz)	Number of repetitions of a periodic event that occurs in a given time interval. 1 Hz equals 1 repetition or cycle per second. ($1\ Hz = 1\ s^{-1}$.)
Density	ρ	$kg \cdot m^{-3}$	Mass per unit volume of an object or substance.
Specific gravity	d		Ratio of the density of a substance to the density of water at 4°C.
Force	F	newton (N)	Effect of one body on another that causes the bodies to accelerate relative to an inertial reference frame. 1 N is that force that, when applied to 1 kg of mass, causes it to accelerate at 1 $m \cdot s^{-2}$ in the direction of the force application relative to the inertial reference frame. ($1\ N = 1\ kg \cdot m \cdot s^{-2}$.)
Weight	G	N	Force exerted on a mass due to gravitational attraction; equal to the product of the mass of the body and the acceleration due to gravity. ($G = m \cdot g$.)

TABLE B.2 (*Continued*)

Physical Quantity	Symbol	Name of SI Unit	Definition
Mass moment of inertia	I	$kg \cdot m^2$	Measure of a body's resistance to accelerated angular motion about a given axis; equal to the sum of the products of the masses of its differential elements and the squares of their individual distances from that axis.
Linear momentum	p	$kg \cdot m \cdot s^{-1}$	Vector quantity possessed by a moving rigid body, quantified by the product of its mass and the velocity of its mass center.
Angular momentum	L	$kg \cdot m^2 \cdot s^{-1}$	Vector of the linear momentum of a rigid body about a point; the product of the linear momentum and the perpendicular distance of the linear momentum from the point. For planar movement, the angular momentum is the product of the moment of inertia in the plane about its centroid and the angular velocity in the plane.
Moment of force	M	$N \cdot m$	Turning effect of a force about a point; the product of the force and the perpendicular distance from its line of action to that point.
Pressure, normal stress, shear stress	p	pascal (Pa)	Intensity of a force applied to, or distributed over, a surface; the force per unit area. 1 Pa is the pressure resulting from 1 N applied uniformly and in a directional perpendicular over an area of 1 m^2. (1 Pa = 1 $N \cdot m^{-2}$.)
Linear strain Shear strain	ϵ γ		Deformation resulting from stress measured by the percentage change in length of a line (linear strain) or the change in angle of an initially perpendicular line (shear strain)

TABLE B.2 (*Continued*)

Physical Quantity	Symbol	Name of SI Unit	Definition
Young's modulus Shear modulus	E G	Pa	Ratio of stress to strain over the initial linear portion of a stress–strain curve.
Work	W	joule (J)	Energy change over a period of time as a result of a force acting through a displacement in the direction of the force. 1 J is the work done when a force of 1 N is displaced a distance of 1 m in the direction of the force. (1 J = 1 N · m.) Work is also the time integral of power. (1 J = 1 W · s.)
Mechanical energy	E	J	Capacity of a rigid body to do work; quantified by the sum of its potential and kinetic energies.
Potential energy	V	J	Energy of a mass or spring associated with its position or configuration relative to a spatial reference. Gravitational potential energy of a mass m raised a distance h above the reference level equals mgh, where g is the acceleration due to gravity. Elastic potential energy of a linear spring with stiffness k stretched or compressed a distance e equals $k \cdot e^2/2$.
Kinetic energy	T	J	Energy of a mass associated with its translational and rotational velocities. The translational kinetic energy T of a mass m with a velocity v is $1/2\ mv^2$. The rotational kinetic energy T of a body rotating in a plane with a moment of inertia I rotating with a angular velocity ω is $1/2\ I\omega^2$.

TABLE B.2 (*Continued*)

Physical Quantity	Symbol	Name of SI Unit	Definition
Power	P	watt (W)	Rate at which work is done or energy is expended. The power generated by a force is the dot product of the force and the velocity at the point of application of the force. ($P = F \cdot V$.) The power generated by a moment is the dot product of the moment and the angular velocity of the rigid body. ($P = M \cdot \omega$.)
Coefficient of friction	μ	—	For two objects in contact over a surface, the ratio of the contact force parallel to the surface to the contact force perpendicular to the surface.
Coefficient of viscosity	η	$N \cdot s \cdot m^{-2}$	Resistance of a substance to change in form; calculated by the ratio of shear stress to its rate of deformation.
Electrical charge	q	coulomb (C)	Quantity of a negative or positive charge on any mass. The charge on an electron or proton is 1.602×10^{-19} C, or 1 C, has the charge of 6.242×10^{18} electrons or protons. [1 A (ampere) = 1 $C \cdot s^{-1}$.]
Voltage, electrical potential	E	volt (V)	Potential for an electrical charge to do work. (1 V = 1 $J \cdot C^{-1}$.)
Electrical resistance	R	ohm (Ω)	Property of a conducting element that opposes the flow of electrical charge in response to an applied voltage. ($1\ \Omega = 1\ V \cdot A^{-1}$.)
Electrical capacitance	C	farad (F)	Property of an electrical element that quantifies its ability to store electrical charge. A capacitance of 1 F means that 1 C of charge is stored with a voltage change of 1 V. ($1\ F = 1\ C \cdot V^{-1}$.)

NOTES TO TABLES B.1 AND B.2

1. Prefixes are used to designate multiples or submultiples of units.

Prefix	Multiplier	Symbol	Examples
mega	10^6	M	megahertz (MHz)
kilo	10^3	k	kilowatt (kW)
centi	10^{-2}	c	centimeter (cm)
milli	10^{-3}	m	millisecond (ms)
micro	10^{-6}	μ	microvolt (μV)

2. (a) When a compound unit is formed by multiplication of two or more units, the symbol for the compound unit can be indicated as follows:

$$\mathrm{N \cdot m} \text{ or } \mathrm{N\ m}, \text{ but not } \mathrm{Nm}$$

   (b) When a compound unit is formed by dividing one unit by another, the symbol for the compound unit can be indicated as follows:

$$\mathrm{kg/m^3} \text{ or as a product of kg and } \mathrm{m^{-3}},\ \mathrm{kg \cdot m^{-3}}$$

3. Because the symbol for second is s, not sec, the pluralization of symbols should not be done; e.g., kgs may be mistaken for kg · s, or cms may be mistaken for cm · s.

INDEX

of her life without ever getting to know
hed heavily.
look so unhappy, little one." Jeremy's long
d her chin. "I'm sorry I denigrated your
ion."
ht." Pippa mustered a smile. "Maybe I'll
ronics so we can agree on something."
gers gently stroked her cheek. "I may be a
t as you secretly believe, but it doesn't go
t want you to pay me lip service—at least
He grinned briefly before becoming seri-
u're bright, and sweet, and fun to be with.
or anyone."
constriction in her throat. "I just wish

dn't finish, Jeremy said, "What do you

. It isn't important. Do you want to go
t of the shops?"
ndered up and down the short street,
gradually lessened. Only once did it re-
caught her high heel in the uneven pav-
hed out immediately, saving her from a
arm around her waist was steadying,
inst his hard, lean body for a breathless

went to his broad shoulders, feeling their
through the soft fabric of his sport coat.
d moment in time she remained in his
face to his like a flower thirsty for the
ht his breath, lowering his head.
were jolted by a little boy on a
y," the child trilled, racing away, com-
of what he had destroyed.
d her immediately. "Are you all right?"
as a little ragged.
his brick is murder on high heels."
oking at him.

She avoided his mocking eyes. "It's interesting that you still regard California as home."

"It always will be," he said simply.

"Don't you like Washington?"

Jeremy shrugged. "It has its compensations, but I never intended to live there permanently. When my job is finished, I'll be back."

"Did you ever consider running for office?" He'd certainly get all the female vote! Pippa had noticed how women passing their sidewalk table had stolen surreptitious glances at the tall, broad-shouldered man with the unusual green eyes and the sensuous mouth. His blatant masculinity ensnared even strangers.

"Good Lord, no!" He answered her question with an emphatic negative. "I can't think of anything I'd hate more."

"You're working for the government *now,*" she pointed out.

"A different thing entirely. I'm conducting a private investigation, or at least trying to," he noted grimly. "The result won't be influenced by a need to win votes, or a desire for publicity."

"You really don't like being in the public eye, do you?" she asked slowly.

"No, I don't." His answer was clipped. "That's for entertainers—or people who need it for their ego."

"Still, by taking on this job, you have no right to avoid it completely," Pippa replied stubbornly. "How about the public's right to know?"

Jeremy's lip curled contemptuously. "A classic media cop-out; you parrot it very well. But how about the rights of the people under investigation? Don't the innocent deserve a chance to be cleared before being tainted forevermore by the mere *suspicion* of wrongdoing?"

"Of course, but—"

"And how about the right of the public to a fair shake? Is it reasonable for an investigation to be jeopardized just to fill up a lot of pulp paper, or to give a blithering newscaster something to read off a monitor?"

"You act as though all reporters are the enemy," Pippa said hotly. "There are ethics in the news industry like anywhere else."

Jeremy's eyes were chips of green glass. "If there are, I've never run across them."

"That's patently unfair! How can you indict a whole profession?"

"Because I have the battle scars to prove my contention." His mouth curved derisively before straightening to a thin line. "I've told you the stratagems the media have employed to harass me."

"Only because you're uncooperative," she protested. "I can't believe *everything* is top secret. If you'd part with just a little information, they wouldn't hound you so much."

"You don't honestly believe that?"

"Yes, I do. You can't blame them for trying to do their job. It's just as honorable in its way as yours is."

Jeremy raised his eyes to the sky. "Why am I arguing with a starry-eyed innocent?" He fixed her with a stern gaze. "Do you have any idea what those noble people you're defending would give to be sitting where you are right now? To be able to catch me in an unguarded moment when I might reveal some nugget of classified information?"

Pippa looked down at the bread crumb she was rolling around and around in her slim fingers. "No decent newsperson would take advantage of you like that."

"The two words aren't synonymous," he said dryly. "No, my dear, I'm afraid reporters and I are natural enemies."

Her downcast eyelashes fluttered. "Maybe if you actually got to know one, you'd change your mind."

Jeremy chuckled. "That's right, you're a journalism major, aren't you? I'd almost forgotten. No wonder your defense was so impassioned. It hasn't changed my mind though."

"But look how well we've gotten along—for the most part," she faltered at his wholly male expression.

"It isn't the same
graph, not a story."
flushed face. "That
was it actually for a
per?"
At least he still b
shiver go up her
treading. Someho
gaze. "I told you I
forced a smile. "A
I'm sure I could
Jeremy relaxed
"Thank God you
anything you wa
Pippa bit her l
nestly. "I would
emy."
He covered he
ing her fingers
honey. You're t
level."
Pippa had ne
Jeremy knew t
ble to contemp
she would neve
it, though, not
meet him. Th
fading in impo
istic career. I
come the pri
In just a
intrigue her.
anything foo
nating man
very many m
ligence, and
because of t
do anythin

him walk ou
him. Pippa si
"Hey, don't
forefinger lifte
chosen profess
"It's all rig
switch to elec
His long fin
male chauvinis
that far. I don
not that way."
ous again. "Yo
Don't change
Pippa felt a
. . ."
When she d
wish, honey?"
"Oh, nothing
through the res
As they wa
Pippa's tension
turn, when she
ing. Jeremy rea
fall. His strong
holding her aga
instant.
Pippa's hands
muscular width
For an enchante
arms, lifting her
sun. Jeremy cau
Suddenly the
skateboard. "Sor
pletely unaware
Jeremy release
His deep voice w
"Yes, I . . . t
Pippa avoided lo

"I'd better hang on to you." He tucked her arm into his, smiling down at her.

She felt golden sparks shooting through her entire body as she clung to his sinewy forearm, storing up the memory against tomorrow, when that was all she would have to remember.

Jeremy bought like a tourist, piling all his little purchases in Pippa's arms. "I'll wrap these for you when I get home," she promised.

"They're for you."

Pippa shook her head. "You have to take them as souvenirs—so you'll remember today."

He stopped in the middle of the street, turning her toward him. His hands framed her face gently, holding her while he examined each feature. "I'm not likely to forget today—or you either."

Pippa swallowed the lump in her throat. She knew what he was telling her. Pinning a bright smile on her face, she said, "I'll think about you once in a while too."

"It sounds like you're saying good-bye."

Her smile faltered. "It's what you want."

"Is it?" His thumbs absently caressed her cheekbones as he stared down at her, a wealth of unreadable expressions confusing Pippa. "We'll talk about it later." Taking her hand, he led her toward a crowded, open-front shop. "Let's see what they have in here."

She followed him in a warm haze of pleasure. Some time since she had picked him up this noon, Jeremy had changed his mind. He was indicating he might want to see her again! It would be terribly unwise. The sensible thing to do was say good-bye now, before he found her out. But Pippa knew she wasn't going to be sensible.

"What are those?" He was peering interestedly at some small, cone-shaped objects in neon-bright colors of red and green.

"Cactus candy!" Pippa exclaimed. "I thought it disappeared along with Mexican jumping beans."

Jeremy pointed to a sign. "They have those too."

"Plastic," she said scornfully. "When I was a little girl,

there used to be the genuine article—little brown beans with a larva inside. When you warmed them in your hands, the little worm woke up and started to jump. They're not allowed to import them anymore though."

"Perhaps it's just as well," he said dryly. "They don't sound wildly appealing."

"You're too sophisticated," she complained. "I'll bet you wouldn't even appreciate cactus candy."

Jeremy accepted the challenge, buying a green one for Pippa and a red one for himself. She tried to stifle her laughter as he manfully licked the sickly sweet confection while they strolled along.

When they reached a lovely, circular mosaic fountain, she took the sucker out of his hand, depositing it in a trash can. "You've proved you're a good sport."

Without warning Jeremy leaned down and kissed her, his lips moving sensuously against hers. An immediate response flooded Pippa as her mouth clung to his. She was oblivious to the people around them, hopelessly enmeshed in the spell of this man who affected her like no other ever had. Jeremy was the one to draw back.

She looked at him in confusion. "Wha-what was that for?"

"I was conducting an experiment." He grinned. "That stuff is so gummy, I wondered if our mouths would stick together."

Pippa ran her tongue over her lips, savoring his special taste. "You're lucky. It—it might have caused problems."

"I could live with them." Jeremy's green eyes glowed with an inner fire.

Pippa was suddenly smitten with doubts. She was the one who was supposed to be the hunter, yet he was turning the tables on her. What had happened to the tough professional she prided herself on being? One casual caress—even a melting look, or a sensuous smile—and she was as vulnerable as the girl he thought her to be.

He drew her down on the tile coping of the fountain at the end of Olvera Street, cupping her chin in his palm so he could look into her troubled face. "Do I frighten you, Pippa?"

"A little," she admitted, although he couldn't know all the complex reasons.

"I don't mean to." His eyes grew cool. "In fact, I'm acting like a prize fool. Maybe we'd better go."

She hesitated. "If you like." Pippa no longer knew what she wanted.

As they walked toward the exit, she tried to break the tension between them. "That's the Avila house over there." She indicated a large white adobe building. "It's the oldest house in Los Angeles. It was a real mansion when it was built back in the 1800's."

Jeremy showed polite interest. "And I believe you said this was the oldest street?"

"Well, close enough. At any rate, it's the only relic of the old days. It was restored in 1929, and all the old traditions are celebrated here. Around Easter there's the blessing of the animals, and at Christmastime they have the ceremony of Los Posados. It's a candelight procession symbolizing Mary's journey to Bethlehem in search of a birthplace for Jesus. They're both very colorful."

Even to her own ears Pippa sounded like a frantic tour guide trying to capture the attention of a bored tourist. Jeremy was courteous and remote. He looked and listened obediently, but it was obvious his thoughts were elsewhere. She had lost him.

When they were back in the car, Pippa's only concern was to prevent an embarrassing silence from falling. She cast wildly about for an innocuous subject.

"Will you have time to visit San Francisco after the investigation is over?" she asked. "Before going back to Washington, I mean."

Jeremy's eyes were on the road. "I expect to, yes."

"I suppose it will seem strange, staying in a hotel in your hometown. Or do you have relatives still living there?"

He turned to look at her then. "I do, but I won't be

staying with them. I told you, I have a home in Woodside."

"Yes, but I thought maybe you'd sold it when you moved away."

"I would never do that." His face relaxed somewhat. "I have twelve acres nestled up against the hills in the most beautiful country you've ever seen. At dusk the deer come down to graze." He grinned suddenly. "They eat the rosebuds, and the raccoons cleaned out the fishpond. It drives the groundskeeper up the wall, but I won't let him put up fences."

"It sounds beautiful," Pippa said softly.

"It is. You've no idea what it's like to ride early in the morning when the grass sparkles with dew and everything is clean and quiet."

"You have horses?"

He nodded. "A few."

"It seems so idyllic; I'm surprised you could bear to leave it."

His smile was self-mocking. "You see before you a latter-day Don Quixote."

"You sound as though you regret it," she said quietly.

Jeremy thought about it for a moment. "Not really. On the whole it's been quite interesting. Although some of the people I've met leave a bad taste in my mouth—the liars and cheats."

Pippa looked down at her tightly interlaced fingers. "People are only human. They make mistakes."

He shrugged. "I suppose."

She stole a glance at his clean-cut profile. It was austere and aloof. This was a man who wouldn't forgive easily. Pippa sighed inwardly.

"Who looks after your house when you're not there?" That seemed like a safe question.

"I have a staff," he answered tersely.

"Do you? Um . . ." Pippa wanted to ask if he lived there alone, but that would be too personal. "It must be a big house."

"It is." He smiled wryly. "My mother never loses an

opportunity to remark how the place cries out for a mistress—in this case meaning a wife—or to point out which rooms would be suitable for a nursery."

"Mothers are like that," Pippa commented lightly, dismissing the unwelcome picture of Jeremy and the serene Calla, knee-deep in babies. Pippa had never taken such an immediate dislike to someone she hadn't even met.

Jeremy turned to her with a mocking grin. "Not your mother surely? She must be praying for just the opposite—at least for the present."

"My mother is dead," Pippa said quietly.

Jeremy's grin faded. "I'm sorry. That was thoughtless of me."

"It's all right. It wasn't recent."

"I know so little about you, Pippa," he said slowly. "Do you have any other family—sisters or brothers?"

"I have two half brothers—they're twins."

"Are they older or younger than you?"

"They're five. My father remarried after Mother died."

Although her terse answers couldn't fail to indicate her reluctance, Jeremy continued his gentle probing. "They're practically babies. Do you see much of them?"

"No, they . . . my father and his second family live in Phoenix." Why was he asking all these questions? He couldn't really be interested. Jeremy was merely being polite, Pippa decided, helping her keep the conversation going. She wished he wouldn't though.

"So, basically, you see them only on holidays," he remarked casually.

"Well, I . . . sometimes."

Holidays were the worst. Like last Christmas. Pippa had volunteered to work because everyone else had someplace special to go. Her friends had been very kind, inviting her to join them, but it was a family time after all.

Not that Pippa couldn't have gone to Phoenix for the holidays. It wasn't that she wouldn't have been welcome, or that she didn't get along with Lita, her stepmother. Lita was only five years older than Pippa, theoretically they should have had a lot in common. It hadn't worked out that way in practice. Lita was completely wrapped up in her husband, her children, and her home. She was totally disinterested in anything outside of her own small world.

Pippa's father thrived in this climate of domesticity. He was proud of having two small sons, and a wife almost twenty years younger than himself. It was only when Pippa was around that he seemed vaguely embarrassed, her presence serving to remind him of his true age. As a consequence, she felt rather constrained also. Her visits became farther and farther apart, until finally their contact was desultory. Pippa had accepted the situation long ago. She just didn't like to talk about it.

"Do you have any other family in Los Angeles?" Jeremy persisted.

"No." Pippa tried to think of some way to change the subject, but he didn't give her a chance.

"It must be very lonely for you," he said gently.

"You don't have to feel sorry for me," she replied stiffly. "I have a lot of friends."

"I don't doubt that for a minute." His voice was soothing. "I just want to be one of them."

"Since when?" she asked bluntly. "You haven't made any secret of the fact that you resent the way I forced myself into your life."

"Strictly a defense mechanism," he assured her. "You scare the daylights out of me."

Her blue eyes were wide with surprise. "How could I possibly do that?"

Jeremy looked in the rearview mirror, edging over into the right-hand lane so he could coast to a stop in front of a small park. Turning off the ignition, he moved sideways in his seat to face Pippa.

"It can't come as any surprise that I'm tremendously attracted to you. Or that it bothers the hell out of me." He ran his hand through his thick, dark hair. "I can't understand it. I know I'm too old for you. What kind of an idiot am I?"

Pippa felt a spreading pleasure replacing her depression. "You're being ridiculous, Jeremy! What difference does age make when it comes to two people being friends?"

"I don't want to be your friend, I want to be your lover," he groaned. "I want to take you to bed and keep you there for a week. I fantasize about your beautiful body, and exactly how I'll caress you. I want to make you part of me, to bring you realms of pleasure no other man has ever shown you."

Pippa's thick lashes made a feathery fan on her flushed cheeks. "I don't mean to sound conceited, but you aren't the first man who has felt this way. Why does it bother you so much, Jeremy?"

"Because you've admitted you're afraid of me—and rightly so." His hand caressed her cheek before gliding down to circle her neck, his thumb idly rotating over the rapidly beating pulse at the hollow of her throat. "To you I'm a glamorous older man. A few romantic kisses would satisfy you and make you the belle of the sorority house." He slowly unfastened the top buttons of her blouse, sliding his hand inside to cup her breast. One finger traced the outline of her bra before slipping inside to touch the taut nipple that betrayed her. "I could so easily convince you to go further."

Pippa forced herself not to react to his enflaming touch, although every inch of her body longed for closer contact. She wanted to put her arms around his neck and pull his head down to hers, to mold her quivering body to the hard length of his.

By a supreme force of will she answered quietly, "I don't think you could make me do anything I didn't want to do."

His other hand was at her waist, urging her pliant body toward his. Pippa lifted her face, her lips unconsciously parting. Jeremy drew a sharp breath before thrusting her away, almost roughly.

"If you believe that, you'll believe anything," he rasped, turning on the ignition. "Just thank your stars that I have a shred of decency left."

Jeremy drove like a madman, weaving in and out of traffic, his eyes grimly fixed on the road. Pippa sat as far away from him as she could get, her flushed face turned toward the window. Once again he had provoked an unfettered reaction from her—and then had been the one to withdraw. When would she learn? The only saving grace was that he thought she was too young to know any better. One thing he was right about, however. They shouldn't see each other anymore.

When they got to the hotel, Jeremy turned to her with a perfunctory smile. "Thank you for a most instructive afternoon."

"It was for me too," she answered stiffly. "I hope your investigation turns out satisfactorily."

He paused with his hand on the door. "Pippa, I—"

"Good-bye, Jeremy," she said firmly.

Various emotions warred in his strong face—indecision, determination, regret—and something else she couldn't fathom. "Pippa," he began again. A horn blared loudly in back of them.

"We're holding up traffic," she pointed out.

"The devil with them!" A muscle worked at the angle of his jaw.

A bellman approached the driver's side. "Would you mind pulling up, sir?" he asked politely.

Pippa leaned over, looking up at the man. "Mr. Hawke is getting out here."

The uniformed man opened the door, leaving Jeremy no choice. Pippa slid swiftly over to the driver's seat, putting the car in gear almost before the door was closed. She drove out to the street without looking in the rearview mirror.

Chaotic thoughts accompanied her all the way home. The day she had looked forward to with such high hopes had turned into a disaster. Jeremy had taken her emotions on a roller-coaster ride, lifting her to the heights, and then plunging her to the depths. No story was worth that.

Pippa's hands gripped the wheel as she faced her own

self-deception. It wasn't the loss of the story that was bothering her, it was Jeremy. Here was a man she could easily fall in love with. It wasn't only his sexual attraction, although that had been immediate and breathtaking. It was so many other things about him—his integrity, his strength, his compassion. As each new facet of this fascinating man was revealed, Pippa felt herself falling deeper under his spell.

A lot of good it would do her! She sighed heavily, pulling into a parking space near her apartment. The only sensible thing she had done all day was drive away without a backward glance. When something was as impossible as the present situation, the best thing was to cut your losses and run.

The phone was ringing as she unlocked the door, but Pippa didn't make a move toward it. The last thing she felt like doing right now was talking to anyone. She wandered disconsolately through the apartment, wondering what she did feel like doing. The answer to that was so unproductive that Pippa took herself in hand. This would be the perfect time to clean out the hall closet, a messy job that had been put off too long.

She was changing into jeans when the phone rang again. Pippa scowled at the inoffensive instrument, ignoring it. When it started a few minutes later she had second thoughts. It was probably only a friend wanting to chat, yet, on the other hand, it might be important.

"Where were you?" Jeremy's deep voice demanded. "I was getting worried."

Her legs suddenly gave way and she sank down on the bed, gripping the phone tightly.

"Pippa? Answer me, damn it!"

She pulled herself together. "Why are you calling?" she asked bluntly.

His voice softened. "I couldn't leave you like that."

"I can't imagine what else there was to say," she commented bitterly. "You made your feelings crystal-clear!"

"I couldn't have, because they aren't clear even to me." His tone deepened.

Pippa knew it would be the height of folly to step on that merry-go-round once more. "I'll make it easy for you, Jeremy. I don't want to see you again."

"I don't blame you."

She didn't want him to be remorseful. Why couldn't he be his most imperious, arrogant self? "You said right from the beginning that it wasn't a good idea," Pippa explained carefully.

"I know."

There was a long pause. She drew a shuddering breath. "Well, good-bye again, Jeremy."

"I have to attend a dinner tonight with the mayor and a crowd of dignitaries," he said abruptly. "Will you go with me?"

"But you said—"

"Stop reminding me of what I said!" he exclaimed impatiently. "Will you come?"

Pippa's immediate inclination was to accept joyfully, until a shred of caution stopped her. The media would be out in force for the event, including her own TV station. There was no way she could avoid being recognized, and, worse, greeted. It destroyed her to have to say, "I can't, Jeremy."

"Do you have another date?" he demanded.

"Well, I—yes. I'm sorry."

"Don't be." His voice was glacial. "It was probably a lousy idea anyway." He was on the point of hanging up.

"Wait!" The exclamation was forced out of her. "I'm free tomorrow during the day. We—we could go to the beach if you'd like."

There was a short silence. Pippa held her breath until Jeremy said softly, "I'd like that very much."

Her heart not only began to beat again, it raced wildly in her breast. "Then I'll pick you up at—"

"No, my dear, this time I'll pick *you* up. Indulge me," he said wryly when she started to protest. "I'm already insecure enough where you're concerned."

Pippa's soft mouth was curved in a smile of pure happiness as she hung up. Jeremy must feel something for her to want to see her again. He must feel a little of the same powerful attraction. Maybe everything wasn't over between them after all. If only nothing happened before they actually got to know each other!

Ignoring the tiny twinge of apprehension that tried to surface, Pippa wrapped her arms around herself. She had a date with Jeremy tomorrow. What could possibly go wrong?

CHAPTER FOUR

Pippa awoke to the sound of rain beating against the windows.

"Oh, no, it can't be!" she groaned, staring at the dripping trees in disbelief. Would Jeremy call off the date because they couldn't go to the beach?

That seemed to be what he had in mind when he phoned a short time later. "It looks like we won't be going swimming today," he announced, much too cheerfully to suit Pippa.

"There ought to be a law against rain on weekends," she muttered wrathfully. "Couldn't you do something about it when you get back to Washington?"

"Afraid not," he laughed. "Man proposes but God disposes, remember?"

"Well, we'll just have to go to plan B," she remarked casually. Pippa could almost see his raised eyebrow.

"Which is?"

"Why don't you come over here for brunch, and afterward we'll listen to some tapes on the stereo." She tried to paint a picture of a relaxed, uneventful afternoon.

"I don't think that would be a very good idea," he answered quietly.

"Are you afraid I can't cook?"

"That's the least of my worries."

She let that go by. "You don't want to stay in a stuffy little hotel room all afternoon," she coaxed.

"It's neither stuffy nor little," he pointed out dryly.

"Come on, Jeremy, you're not going to stand me up, are you?" she wheedled. "That wouldn't be very gentlemanly."

He hesitated. "Tell you what—I'll take you out to brunch."

"Okay, if you'd rather," she agreed promptly "I just thought maybe you were tired of restaurants."

"I am!" His answer was fervent.

"Then aren't you being silly? I make marvelous pecan waffles with real maple syrup. If I promise not to seduce you, will you come?"

He laughed unwillingly. "It wasn't your conduct I was worried about."

"All right, I promise not to let *you* seduce *me.*"

"You're very persuasive." He capitulated with a shade of reluctance. "What can I bring?"

"Just your appetite."

After she had given him directions Pippa flew around getting ready, her spirits very light. She had high hopes for this day. Maybe the rain was a blessing in disguise. In the easy atmosphere of a lazy day at home, they could really get to know each other. The opportunity might even present itself to tell him the truth about everything. Pippa still wanted the story, but in a straightforward manner, with no favors asked or given. She also wanted to set the record straight with Jeremy so she could act like herself, not have to keep playing a part that didn't suit her.

The apartment was redolent with the aroma of crisping bacon when Jeremy arrived with a box of candy and a bouquet of yellow roses.

"Candy and flowers, you *are* an old-fashioned beau," Pippa teased.

"My mother taught me always to bring my hostess a gift, and you're too young for the bottle of wine I would normally have selected," he teased back.

Pippa's original pleasure in his offerings was dimmed. She was more than ever determined to set the record straight as soon as possible.

"The long arm of the law has yet to reach into our homes," she responded tartly. "Why don't you make us each a bloody Mary while I start the waffles."

She had purposely set the round table in the kitchen so Jeremy wouldn't feel threatened, although brunch could scarcely be called a romantic meal—unless it followed a night spent together. Pippa put such thoughts firmly out of her mind.

Jeremy leaned against the counter, sipping his drink and watching her move around the kitchen. There was a puzzled look on his face. "I didn't think of it until just now, but didn't you tell me you lived at the sorority house?"

She continued to pour batter into the waffle iron. "I did for the first two years, until I got too old for all that frantic activity."

"Yes, I can see you're quite ancient," he smiled, touching her forehead lightly. "Isn't that the start of a wrinkle?"

"I wish you'd take me seriously, Jeremy. I'm really quite mature," Pippa insisted. "I've been on my own for a long time."

His amusement vanished. "I know, honey, and I'm sorry."

"I don't want your pity," she cried in exasperation. "I just want to be treated like a woman."

Something flared briefly in his eyes before he turned to pick up his drink. "Why don't I come back in a few years when you really are one?"

Pippa gritted her teeth as she stared at his broad shoulders and the well-shaped back of his head. "Oh, go sit down, Jeremy! Everything's ready."

He did full justice to the meal, consuming an amazing amount of waffles.

Pippa watched in surprise. "Either you didn't have

anything to eat last night, or you really do like my cooking."

"Both," he answered briefly, reaching for the syrup.

"I thought you were at a political dinner."

"If you can call it dinner." His face wore a disgusted look. "It's my private opinion that politics would attract a better class of people if the food were at least edible."

Pippa bubbled with laughter. "Don't tell me—soggy chicken, underdone potatoes, and a spoonful of hard little peas."

"So that's why you wouldn't go with me!" he exclaimed. "You knew ahead of time."

She refused to give him any sympathy. "So did you; it's the reason you asked me. Misery loves company."

"I asked you so that at least I'd have something to feast my eyes on."

"Weren't there any beautiful women there?" she inquired carelessly.

"If I tell you that under their long evening gowns they wore sensible shoes, does that answer your question?"

Pippa pretended to be shocked. "Is that how you spent your time, peering under ladies' dresses?"

His voice dropped a note. "I spent the evening wondering what you were doing."

She looked down at her plate. "Nothing very interesting." Then she remembered the excuse she had given for turning him down. "My date was a washout too."

"It looks like we both wasted our time."

"At least you couldn't help it," she consoled him.

"That's true. It's the part of the job I could do without. All those endless functions are meaningless anyway." He grinned suddenly. "When I move out of the hotel, I think I'll leave a phony forwarding address."

"You're moving?"

He nodded. "I've taken a house at Malibu for greater privacy. I hope to make it harder for all the kooks to bother me."

"Like me?" Pippa looked at him steadily.

"You know better than that." His light eyes darkened as they held hers.

Pippa felt a rush of joy, which she masked with an offhand remark. "Maybe I can help you get settled."

"The place is fully furnished and comes with a housekeeper," he declined.

"It seems you're all set," she commented lightly. "I guess you don't need me for anything."

Jeremy was silent for a long moment, regarding her with an enigmatic expression. "Do you toss out provocative remarks like that on purpose?"

Swift color tinged Pippa's clear skin. "*You're* the one!" she exclaimed angrily. "You read something sexy into everything I say."

He sighed. "Because it's always on my mind when I'm with you."

"That isn't my fault."

His eyes went over the alluring curves her simple, blue silk shirtdress couldn't conceal. Pippa had purposely avoided anything tight or revealing, but nothing could camouflage her enticing figure.

"Maybe not," he answered her criticism. "The fact remains that you disturb me mightily, and I don't know what to do about it."

She grinned impishly. "Would you like to take a cold shower?"

He gave her a slow smile. "Only if you join me."

The idea didn't bear thinking about. "Wouldn't that rather defeat the purpose?" she asked casually.

"You can bet on it, lady!" The little flames in his eyes seemed to scorch her.

This was getting out of hand, Pippa decided. Whenever Jeremy followed his instincts, he had a violent reaction. She didn't want to risk losing him again.

"In that case, why don't you go relax in the living room while I pile these dishes in the sink?"

His rueful smile acknowledged the way she was backing away from danger. He uncoiled his long length, getting to his feet. "I'll help. That's my job, cleaning up messes."

It was the first time Jeremy had ever referred to the investigation without prodding. Pippa was quick to seize the opening. "How are things going?"

"I'm beginning to get a glimmer of light." His firm mouth turned down at the corners. "Without the cooperation of the police, I might add."

She was startled. "I would think they'd be your main source of information."

"That would be logical," he concurred. "But it seems a few of the higher-ups have their noses out of joint. They think bringing in an outsider is a reflection on their ability. That's nonsense of course. They have a whole city to run; this is my only priority."

"Didn't you point that out?"

He shrugged. "I have neither the time nor the inclination to soothe the ruffled feelings of a bunch of prima donnas. If they don't realize how important this thing is, that's their problem. I'll get to the bottom of it if I have to do it singlehandedly."

"I could help," Pippa offered impulsively. "I could dig things out of the files for you and do your legwork."

He raised a questioning eyebrow. "Doesn't college require attendance anymore?"

"We don't have to answer the roll or bring a note from our parents when we're absent," she answered impatiently. "I have plenty of free time."

He cupped her chin in his palm, raising her face to his. "I appreciate your offer, honey, but I'm determined not to contribute to your delinquency."

"That's rather a harsh word for cutting a few classes," she murmured, looking up at him and noticing again how thick his eyelashes were.

"I'm afraid it would turn into more than that." His hand slipped down to circle her throat, stroking the satin skin lingeringly. "If I saw you every day, it would only be a matter of time until I locked the door and said the hell with the investigation."

His touch was feather-light on her skin, but it seemed to inhibit Pippa's breathing. "You wouldn't do that, Jeremy," she managed breathlessly. "You have too much integrity."

"Then why do I always seem to misplace it around you?" he groaned. "Right now all I can think about is making love to you."

Pippa summoned up a faint smile. "No one ever got hurt thinking of things."

He didn't return her smile. "I don't want to hurt you, Pippa, and that's what I'd be doing. You're so very young."

She felt as though her very bones were melting. "Oh, Jeremy, I just wish—"

When she stopped, he removed his hand, smiling bitterly. "You could rearrange the calendar? Make me younger, and you older?"

Her whole body yearned for this man, who was so determined to put obstacles in their path. "No, I wish you weren't such a knucklehead! Go in the other room and let me finish up here," she ordered angrily.

He did as she said, leaving Pippa to slam pots and pans noisily in an attempt to vent her frustration. When the kitchen was neat, she went into the bedroom to freshen her makeup. Jeremy was right of course. Any involvement between them would be disastrous, given his misapprehensions. She could tell him the truth, but Pippa had a feeling this wasn't the time. If he thought she'd tried to make a fool of him, he'd undoubtedly storm out and that would be the end of it.

Jeremy was stretched out on the couch when she finally went into the living room. His eyes were closed and he seemed to be asleep. Pippa envied him. Her own body was still tied up in knots. She stared down at the strong face that looked vulnerable somehow, relaxed now in sleep.

Pippa sighed, reaching for an afghan to cover him. It was folded on the back of the couch. As she leaned over him, Jeremy's eyes opened. He didn't look vulnerable anymore. He looked like a sleepy lion about to have a meal.

His arms closed around her waist, toppling her down next to him.

"Wake up, Jeremy," she cried desperately.

"What makes you think I'm asleep?" One arm circled her shoulders, the other her hips, drawing them close to his.

"You don't know what you're doing!"

He chuckled softly. "I've never had any complaints before."

His mouth moved seductively over her cheek as he clasped her tighter, throwing one leg over both of Pippa's so she was pinned between his taut thighs. Excitement flared inside her at the intimate contact, matching his mounting desire. When his tongue parted her lips, Pippa welcomed his invasion with a sigh of pleasure.

Jeremy's breath mingled with hers as he groaned against her mouth, "I've wanted to do that since I walked in the door."

"What took you so long?" she murmured, trailing her fingers across his face.

He captured her hand, kissing each fingertip. "You're a fine one." His voice was husky. "I thought you said you weren't going to seduce me."

"I lied," she whispered, hooking her arm around his neck.

Apprehension clouded his face, warring with the desire there. "Pippa, I don't—"

"Did anyone ever tell you that you talk too much?"

When she pulled his head down, touching her lips to his, Jeremy was powerless to resist. His arms gathered her in an embrace that made her aware of every hard muscle. His hands wandered restlessly down her back to caress her silken hip before retracing the path. Pippa trembled as his fingers curled around her breast, gently teasing the nipple to a diamond-hard peak.

For just an instant a warning bell sounded in her brain. She had wanted so much for him to kiss her that she had ignored the danger. That was all Jeremy had intended, too, until she had pushed him over the edge. Should she have backed off? The confused thoughts came too late. Pippa was lost in her need for this man, deaf to everything except the sound of thunder in her ears as his mouth stoked the desire that was turning her liquid inside.

Her arms tightened around his neck as Jeremy parted her legs with his knee, moving sensuously against her. She traced the taut muscles of his back, glorying in his hard, masculine feel.

When the doorbell rang, Pippa had a moment of déjà vu. It was like that night in Jeremy's hotel room all over again. She scrambled to her feet, unable to look at him.

"I'll answer it if you like," his low voice offered.

"No, I . . . it's all right." Pippa took a deep breath, straightening her dress and smoothing her rumpled hair.

A little girl stood in the entry. "Would you like to buy some Girl Scout cookies?"

"Oh, well, yes, I guess so." Pippa was having trouble pulling herself together. "I'll get some money."

"Here." Jeremy took a bill out of his pocket.

She took it without glancing at him, turning back to the child. "You ought to have an umbrella," Pippa remarked absently.

"It isn't raining anymore. Gee, thanks!" the little girl exclaimed as she took the bill. She began to pile boxes in Pippa's arms.

After she left, Pippa surveyed her loaded arms, laughing reluctantly. "How much did you give me?"

"I didn't notice." Jeremy put his hands on her shoulders, the cartons a safe barrier between them. "I'm sorry, Pippa. It seems I'm always saying that to you."

"And I'm always telling you it wasn't your fault alone." She managed a shaky laugh. "We'll just have to stop meeting like this."

She was immediately sorry when Jeremy said, "You're right. I'd better go."

"I didn't mean it like that! Look, Jeremy, I think we both need to lighten up. How about a movie?"

"I really think I should—"

"We'll each get a box of popcorn, and that way we can't even hold hands," she coaxed.

Laughter was forced out of him. "How can I trust you? You've already admitted you lied."

Pippa was startled until she realized what he was referring to. "Live dangerously," she advised. "How would you like to walk to the Village? We're only five or six blocks away, and there's no parking there anyway."

"That sounds like a good idea." He gave her a mocking smile. "Maybe the fresh air will be cooling."

The walk and the movie served to lessen their tension. The only problem was that Westwood Village was the college town for U.C.L.A. located on the slope above. They were surrounded by young students. In a way, Pippa hoped that was a plus. Surely Jeremy could see how much more mature she was than the jean-clad youngsters around them. Her frequently stolen glances at him told her nothing.

The movie they chose by tacit consent was a comedy. They came out completely relaxed, to find that it was pouring again.

"No problem," Jeremy assured her. "We'll catch a cab."

"This is California," she informed him. "Cabs don't cruise like they do in the East. Especially not in the Village. Starving students don't take taxis."

"How do you get one then?" he demanded.

"You phone—and then wait interminably."

"Well, at least we can wait in a bar."

She shook her head. "There aren't any. It's too close to campus."

They settled on a coffeeshop, along with all the other stranded people. It was steamy and noisy, they couldn't get waited on, and there was no place to sit down. Jeremy stood it for about five minutes.

"How do you feel about walking in the rain?" he asked.

"I'm game if you are."

He gave her an approving look. "Most women would be afraid of spoiling their hairdo."

"I'm more concerned about your beautiful outfit getting ruined," she answered truthfully, glancing at the elegant tweed jacket and gabardine slacks.

"Don't give it another thought," he said, leading her outside.

At first they ducked their heads and hurried their pace, until it occurred to them that in six blocks they were going to get soaked no matter what they did. Then they slowed down.

"This is kind of fun," Jeremy remarked with surprise. "Sort of a latter-day *Singin' in the Rain.*"

"I'll be Debbie Reynolds and you can be Gene Kelly," Pippa told him.

"I wouldn't have it any other way," he grinned. "Unfortunately, I can't dance."

"I would never have come if I'd known that," she remarked with mock disapproval. "What *can* you do to amuse me on this soggy safari?"

"How about this?" Jeremy suddenly stamped his foot in a puddle, sending water spraying in all directions.

Pippa raised her eyebrows. "It leaves something to be desired."

"That's because you were never a five-year-old boy," he assured her. "What do little girls that age do for amusement?"

"Pray for little boys to grow up," she replied dryly.

Jeremy turned her to face him, eyeing her searchingly. "Were your prayers answered?" he asked softly.

Pippa stared up at him, looking at the thick, dark hair plastered with rain that bounced off his high cheekbones. At that crazy, illogical moment, she realized that she was totally, irrevocably in love with Jeremy Hawke.

"Yes, all my prayers have been answered," she told him quietly, unaware of the sadness in her face. What she told him was true, but her happiness could so easily be swept away in the murky waters of misunderstanding.

He brushed the wet hair off her forehead tenderly, touching the drops that ran down her face with a gentle forefinger. "These look like tears. I'm glad they're not for me. I don't ever want to make you cry, Pippa."

She shook off a horrible premonition, giving him a shaky smile. "I don't hold you responsible for the rain. Come on, Gene Kelly-Hawke, stomp on another puddle for me."

They were both soaked to the skin by the time they got back to Pippa's apartment.

"You have to get out of those wet clothes," she told Jeremy decisively.

His eyes held a gleam of laughter. "What would you suggest I wear instead? I'm afraid nothing of yours is going to fit me."

He was so right. Pippa regarded him with a frown while she attempted to think of a solution. "I guess you'll just have to wrap up in a blanket while I try to do something about your clothes. I can press your shirt and slacks; it's only the jacket that's a problem. If you carry it over your arm, though, at least you'll look presentable when you walk through the lobby."

"You worry too much. I'm not trying to impress anyone."

"I don't want you to catch cold," she insisted, rummaging through the linen closet for a blanket. "Here, go in the bathroom and put this on."

"You're wetter than I am." He was suddenly concerned, noticing how the thin silk dress clung to her.

"I'll change too," she assured him. "And then I'll make us some hot tea."

Pippa put on a woolly yellow bathrobe, toweling her hair dry. It fell in a gold-tipped cloud around her shoulders, a riotous mass of untamed waves and ringlets. Well, it would just have to do, she decided, since her hair dryer was in the bathroom with Jeremy.

She was in the kitchen making tea when he came in

wrapped in a brown blanket. His bare feet were visible below, and dark curling hair showed on the part of his tan chest that wasn't completely covered. Pippa couldn't help wondering what he had on underneath.

To drive the thought out of her mind, she pulled down the built-in ironing board. "If you'll bring me your shirt and slacks, I'll press them."

"No hurry," he said easily.

"I might as well do it now." Pippa felt a compulsion to keep busy.

While she made makeshift repairs to his clothes, Jeremy lounged on a kitchen chair, one ankle resting on his knee. Pippa wished he would cover up, but she didn't know quite how to tell him.

"Aren't we the picture of domesticity?" he laughed. "A casual observer would think we were married."

His mocking words caused a pain between her breasts. "Or living together," she said sharply, to cover the hurt.

His quick frown was instantly replaced by a bland expression. "That's right, I forgot. It's your generation's alternative to marriage."

"We happen to be the same generation, Jeremy," she responded curtly. "And I don't see you rushing to tie the knot."

"True." His enigmatic eyes took in the bright cloud of hair around her heart-shaped face, lingering for a long moment on the soft mouth, thinned now in annoyance. "Since I've waited this long, maybe I'll wait until you grow up."

"By the time that happens, you'll be too old for me," she flared.

"I already am," he said quietly.

Her eyes flew to his face in alarm. When would she learn to curb her sharp tongue. "Jeremy, I didn't mean—"

"It's all right, honey." He stood up, wrapping the blanket sarong fashion around his hips. "Are those ready now?"

"Let them dry out a little longer," she mumbled miserably, hanging the shirt and slacks on a hanger.

Jeremy was in complete control of himself, even if Pippa wasn't. "You've waited on me, now it's my turn. Go curl up in the living room, and I'll bring you a cup of tea."

He brought it on a tray, kneeling down as he put it on the couch next to her. The muscles in his broad shoulders and bare chest rippled powerfully.

A bubble of laughter pierced Pippa's misery. "If you could manage to call me mem sahib, I'd think I was in India, back in the days of the British Raj."

"Your faithful servant, mem sahib," Jeremy grinned, touching his forehead, his throat, and his chest, before taking her hand and kissing it.

"I don't think servants were allowed kissing privileges," she told him.

His arms circled her waist as he knelt before her. "If they had a mistress like you, they'd break the rules," he murmured in a husky voice.

Pippa's hands went to his bare shoulders, meaning to hold him off but unconsciously caressing them. "We're just playing a game, Jeremy. Don't get carried away."

He framed her face in his palms, his warm breath fanning her cheek. "When I'm with you, there isn't any reality."

"I can't go through this again, Jeremy," she pleaded.

"You're right, of course." He widened the lapels of her robe slightly, feasting his eyes on the curve of her breasts. "It's all wrong."

"Only because you think so," she answered helplessly.

He opened her robe suddenly, his eyes glowing as they rested on her rose-tipped ivory skin. "God, you're beautiful!" He drew her down onto the rug, bending his head to kiss each swelling breast.

All of Pippa's reservations were swept away in a rising tide of passion. This was the man she loved. They could bring each other ecstasy if only she could convince him.

"Love me, Jeremy, please," she sighed. "Nothing else matters."

"Nothing," he muttered, capturing her mouth.

His hands moved down her body, tantalizing her with slow caresses that made her flame with desire. She arched her body into his, anchoring her fingers in his hair and murmuring his name over and over.

He raised his head, looking at her with blazing eyes. "You want me, don't you, sweetheart?"

"So much," she whispered. "I never even knew it could be like this."

He stiffened in her arms. "Are you a virgin, Pippa?"

She gazed at him without comprehension for a moment. How could he break the spell that bound them together? Then his question penetrated. Whatever she answered would be wrong. "Don't spoil it, Jeremy," she begged.

But it was too late. He got up, standing over her with a haggard look on his face. "Each encounter brings us a little closer, Pippa. It's time one of us used common sense, and I seem to be elected."

She sat there on the floor, huddled in a little heap with her head resting on her knees while Jeremy got into his clothes.

He paused in front of her on his way to the door. "What can I say, Pippa?"

"I don't think anything more is called for." Her voice was muffled. "Would you please go?"

Not until long after the door closed softly did she raise her head. The leaden lump where her heart used to be told her it was all over, but Pippa found it hard to accept. How could she have been thrown out of Eden by an unguarded phrase? It was useless to tell herself that Jeremy wasn't worth all this agony. When did love ever make sense?

After a long time Pippa got stiffly to her feet. Maybe someday when she was old and gray, this would assume the status of a relatively minor incident. The trouble was, she was very far from being old and gray.